Computational Intelligence Assisted Design

In Industrial Revolution 4.0

Yi Chen

Industry 4.0 Artificial Intelligence Laboratory
Dongguan University of Technology
Songshanhu, Dongguan 523808, China

and

Yun Li

Industry 4.0 Artificial Intelligence Laboratory
Dongguan University of Technology
Songshanhu, Dongguan 523808, China

Professor in Design, Manufacture and Engineering Management
Faculty of Engineering, University of Strathclyde
Glasgow G1 1XJ, UK

CRC Press
Taylor & Francis Group
Boca Raton   London   New York

CRC Press is an imprint of the
Taylor & Francis Group, an **informa** business

A SCIENCE PUBLISHERS BOOK

CRC Press
Taylor & Francis Group
6000 Broken Sound Parkway NW, Suite 300
Boca Raton, FL 33487-2742

© 2018 by Taylor & Francis Group, LLC
CRC Press is an imprint of Taylor & Francis Group, an Informa business

No claim to original U.S. Government works

Printed on acid-free paper
Version Date: 20180314

International Standard Book Number-13: 978-1-4987-6066-9 (Hardback)

Library of Congress Cataloging-in-Publication Data

Names: Chen, Yi (Professor of industrial design), author. | Li, Yun, (Systems engineer), author.
Title: Computational intelligence assisted design in Industrial revolution 4.0 / Yi Chen, School of Engineering and Built Environment, Glasgow Caledonian University, Glasgow, UK, and Yun Li, Professor of Systems Engineering, University of Glasgow, Glasgow, UK.
Description: Boca Raton, FL : CRC Press, Taylor & Francis Group, 2018. | "A science publishers book." | Includes bibliographical references and index.
Identifiers: LCCN 2018006389 | ISBN 9781498760669 (hardback)
Subjects: LCSH: Computer-aided engineering. | Computational intelligence. | Industrial revolution (Electronic resource)
Classification: LCC TA345 .C4834 2018 | DDC 620/.0042028563--dc23
LC record available at https://lccn.loc.gov/2018006389

Visit the Taylor & Francis Web site at
http://www.taylorandfrancis.com

and the CRC Press Web site at
http://www.crcpress.com

Dedication

For Our Families

Foreword

In the era of artificial intelligence and Industry 4.0 (i4), this book is an enthusiastic celebration of many computational intelligence (CI) algorithms and applications, especially those which are of relevance in Science, Technology, Engineering and Mathematics (STEM). It is also a tribute to researchers who are involved in the study and synthesis of Nature-inspired computation. Also, herein are included innumerable historical vignettes that interweave CI algorithms, design approaches and STEM applications in an appealing manner.

Although the emphasis of this work is on selected evolutionary algorithms, it contains much that will be of interest to those outside, indeed to anyone, with a fascination for the world of smart design and CI-assisted creativity. The authors have selected well over 10 multi-disciplinary applications as the key subjects of their essays. Although these represent only a small section of the world of CI-related applications, they amply illustrate what has evolved may help advance i4.

Preface

In recent years, many sectors of industry have experienced a paradigm shift in the way business is conducted, primarily due to the proliferating use of the internet and artificial intelligence. With this shift, Industry 4.0 (i4) is emerging as the first a-priori engineered 'industrial revolution' to bring about the 'Fourth Industrial Revolution', in which the factory floor is set to become a centre of innovation with where customization is possible at economical costs.

A common feature of all trends of i4 is the integration of several features in response to challenges of computerised decision making and big data, which are proliferated via cloud computing, and has led to a race in intelligent design, innovation, and creativity. However, i4 research and development (R&D) so far is mostly spearheaded by manufacturers, such as Siemens, and thus lacks an academic perspective and an 'intelligent design' tool commensurate with such a race. Clearly, 'market informatics' should be used to close the loop of i4 in customised smart manufacturing and smart products. Amongst the most innovative activities, design and engineering are essential to the global economy and manufacturing industries. Due to customization requirements, design efficiency, and responsiveness, demand for a smart design system has increased.

This book presents computational intelligence (CI) for smart design and manufacturing. It starts with the values and principles of CI, providing a brief history of various CI theories and the need for Computational Intelligence Assisted Design (CIAD) and engineering. Furthermore, the book introduces an important conceptual framework of a smart design and manufacturing platform for the future —CIAD and its multidisciplinary applications. Readers will find this book an easy to understand textbook for CIAD courses. It is written for senior students in science, technology, engineering and mathematics (STEM) as well as in the social sciences. Readers are expected to have fulfilled the following basic level of prerequisites: vector-matrix algebra, differential equations, control engineering, circuit analysis, mechanics, dynamics and elementary programming, e.g., MATLAB®, C/C++ or Python.

The book has been written to provide a basic understanding of CI theory and its applications. Highly mathematical arguments have been intentionally avoided. Statement proofs are provided whenever they contribute to the understanding of the subject matter presented. The book has 25 chapters with greater knowledge on multidisciplinary applications. Special efforts have been made to provide examples at strategic points so that the reader will have a clear understanding of the subject matter discussed. The reader is encouraged to study all such worked problems carefully, which will allow him/her to obtain a deeper understanding of the topics discussed. The unsolved problems may be used as homework or quiz problems to further develop skills.

If this book is used as course text, relevant portions of the book may be used, while skipping the rest. Because of the abundance of model problems and worked examples that answer many questions, this book can also serve as a self-study book for practising engineers and industrial designers in the era f i4.

Acknowledgements

The authors would like to acknowledge the support provided by the Dongguan University of Technology High-Level Talent Research Startup Grant (No. G200906-14, KCYXM2017012), the Royal Society of the United Kingdom Newton Fund (NI140008), the Engineering and Physical Sciences Research Council of the United Kingdom (EP/P001246, GR/K24987) and the National Natural Science Foundation of China (51105061, 91123023, 61511130078). We would like to express appreciation to our many hard-working former and present PhD students who provided the necessary material have been adopted in this book and they include, to mention only: Prof. Kay Chen Tan FIEEE, Dr Kim Chwee Ng, Mr Alfredo Alan Flores Saldivar, Dr Jin Yin, Dr Christoph Schöning, Ms Wuqiao Luo, Dr Lin Li, Dr Cindy Goh, and Dr Lipton Chan.

Contents

Part III CIAD for Science and Technology

Part IV CIAD for Social Sciences

Acronyms

ABF   Adaptive Bathtub Failure Rate
AABF   Asymmetric ABF
AI   Artificial Intelligence
ANNs   Artificial Neural Networks
BFR   Bathtub-shaped Failure Rate
CI   Computational Intelligence
CIAD   Computational Intelligence-assisted Design
CIAE   Computational Intelligence-assisted Engineering
CIAM   Computational Intelligence-assisted Manufacture
EC   Evolutionary Computation
FL   Fuzzy Logic
FRF   Failure Rate Function
GAs   Genetic Algorithms
i4   Industry 4.0
RBF   Radial Basis Function
SABF   Symmetric ABF
SI   Swarm Intelligence
STEM   Science, Technology, Engineering and Mathematics
μGA   Micro-genetic Algorithm

1

Introduction

1.1 Introduction

Computational intelligence (CI) is a set of nature-inspired computational approaches that offer a wealth of capabilities in complex problem solving. Compared to traditional optimization methods, the first advantage of CI is that it does not need to reformulate a problem to search a non-linear or non-differentiable space, thereby providing effective or feasible solutions to many multidisciplinary applications with or without analytical representations of the real-world problems. Another advantage of CI is its flexibility in formulating the fitness function, which can be expressed as a function of the system output. This feature is particularly appealing when an explicit objective function is difficult to obtain, where traditional models often fail to address uncertainties in ever-changing conditions or surroundings.

As shown in Figure 1.1, this book considers four paradigms of the CI family: Fuzzy Logic theory (FL), Artificial Neural Networks (ANNs), Evolutionary Computation (EC) and Swarm Intelligence (SI). In recent years, there have been significant developments in various CI-related research works, including the newly developed methods such as Swarm Intelligence, Artificial Immune Systems (AIS), Quantum Computing (QC) and DNA Computing (DNAC).

1.2 History of Computational Intelligence

The terms artificial intelligence (AI), soft computing and natural computing have been widely employed for similar or even the same area in research. John McCarthy, known as the father of AI, has noted that 'CI' is a more suitable name for the subject of AI, which highlights the key role played by computers in AI [Andresen (2002)].

The term 'Computational Intelligence' was formally introduced by Bezdek in 1992 [Bezdek (1992), Bezdek (2013)] in an attempt to attach the term computational intelligence to several activities related to the IEEE Computational Intelligence Society (IEEE CIS); however, Bezdek did not consider himself as the originator of the term CI [Bezdek (2013)]. He also discussed the actual history of CI,

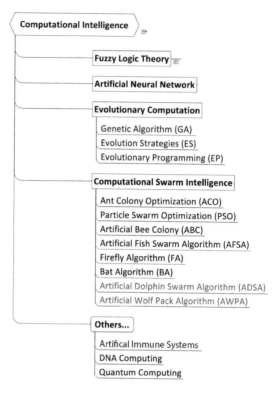

Fig. 1.1 The computational intelligence family tree

which came from various sources and was attributed to various creators; his work simply helped this term become more popular in various research areas.

Since then, CI has been widely studied and utilized in academic and industrial applications under increasing demands, with a few up-to-date keywords being related to it, such as artificial intelligence, Internet of Things (IoT), big data, cloud computing, intelligent robotics and industry 4.0. In this section, we talk about the history of the CI members as listed in Section 1.1 since every member's history contributes to the history of the CI family.

- **Fuzzy Logic Theory**

  The first research on fuzzy logic can be traced back from the 1920s to 1930s, then known as '*infinite-valued logic*', proposed by Łukasiewicz and Tarski. In addition, there was a very strong resurgence of interest in the 1950s and early 1960s [Pelletier (2000)]. Fuzzy logic theory, first introduced by Lotfi A. Zadeh in 1965 [Zadeh (1965a)], was based on the concept of fuzzy sets. Zadeh performed his work on fuzzy mathematics and system theory a few years before he published the concept of fuzzy sets [Zadeh (1962)] in 1962.

In 2015, Zadeh published a novel paper to introduce fuzzy logic theory in a historical perspective [Zadeh (2015)], therein marking the 50th anniversary of the publication of his first paper on fuzzy sets in 1965. Table 1.1 summarizes Zadeh's works (some with other researchers) on fuzzy logic theory from its beginning to the present day.

Table 1.1 Summary of Zadeh's Work on Fuzzy Logic Theory [Zadeh (2015), Zadeh, et al. (1996), Zadeh (2008)]

Year	Work	Reference
1962	Fuzzy mathematics and system theory	[Zadeh (1962)]
1964	Separation theorem	[Zadeh (1964), Zadeh (1973a), Zadeh (1976b)]
1965	Fuzzy sets, membership functions, α-cut	[Zadeh (1965a)]
1965–1982	Fuzzy system	[Zadeh (1965b), Zadeh (1971a), Zadeh (1973b), Zadeh (1973c), Zadeh (1974b), Zadeh (1982)]
1966	Fuzzy relation, shadows of fuzzy sets	[Zadeh (1966), Zadeh (1971b)]
1968	Probability theory with fuzzy sets	[Zadeh (1968b)]
1968	Fuzzy algorithm, fuzzy Turing machine	[Zadeh (1968a), Zadeh (1972a), Zadeh (1975a), Zadeh (1975c), Zadeh (1975d)]
1969	Fuzzy language	[Lee and Zadeh (1969), Zadeh (1971c), Zadeh (1971d), Zadeh (1996)]
1970	Fuzzy decision making, fuzzy dynamical programming	[Bellman and Zadeh (1970), Zadeh (1976a)]
1971	Fuzzy graph, fuzzy finite-state machines and n-dimensional unit hypercubes	[Zadeh (1971a)]
1971	Similarity, fuzzy probabilities	[Zadeh (1971b), Zadeh (1981), Zadeh (1984b), Zadeh (2014)]
1972	Fuzzy control	[Zadeh (1972b), Chang and Zadeh (1972)]
1973	Linguistic variables	[Zadeh (1973a), Zadeh (1975a), Zadeh (1975c), Zadeh (1975d)]
1974	Fuzzy logic	[Zadeh (1975a), Zadeh (1975c), Zadeh (1975d), Zadeh (1974a)]
1975	Approximate reasoning	[Zadeh (1975a), Zadeh (1975c), Zadeh (1975d), Zadeh (1975b), Zadeh (1975e), ?, Zadeh (1979b), Zadeh (1984), Zadeh (1992)]
1975	Extension principle	[Zadeh (1975a), Zadeh (1975c), Zadeh (1975d)]
1975	Information and restriction	[Zadeh (1975e), Zadeh (2013), Zadeh (2015)infoprinciple]
1976	Fuzzy optimization	[Zadeh (1976a)]
1978	Possibility theory and fuzzy propositions	[Zadeh (1978b), Zadeh (1978), Zadeh (1978a)]
1978	Possibilistic relational universal fuzzy (PRUF)	[Zadeh (1978), Zadeh (1978a), Zadeh (1982), Zadeh (1984)]
1979	Granularity	[Zadeh (1979a), Zadeh (1997)]
1983–1988	Approximate reasoning with dispositions	[Zadeh (1983), Zadeh (1985), Zadeh (1988), Zadeh (1987), Zadeh (1988)]
1983	Fuzzy logic in expert systems	[Zadeh (1983), Zadeh (1986), Zadeh (1989)]
1985	Usuality	[Zadeh (1985), Zadeh (1988), Zadeh (1986)]
1993	Soft computing	[Zadeh (1993), Zadeh (1994)]
1996	Computing with words (CWW)	[Zadeh (1996), Zadeh (2002), Zadeh (2012)]
2002–2005	Generalized theory of uncertainty (GTU)	[Zadeh (2002), Zadeh (2005), Zadeh (2006)]
2008	FL-generalization	[Zadeh (2008)]

- **Artificial Neural Networks**

ANNs, as used in artificial intelligence, have traditionally been viewed as simplified models of neural processing in the brain, even though the relation between this model and the biological architecture of the brain remains under debate; it is not clear to what degree ANNs mirror the brain function [Russell (2012)].

The modern era of neural network research is credited to the work done by neuro-physiologists. In 1943, McCulloch and Pitts [Mcllochw and Li (1943)] published the first paper on ANNs, in which they proposed their work on a computational model of Neural Networks (NNs) based on mathematics and algorithms, called 'threshold logic'. Basically, ANNs are biologically-inspired computational models inspired by the morphological and biophysical properties of neurons in the human brain. Such networks can simulate the way in which the human brain processes information and that is in the following two ways: (1) an NN acquires knowledge by learning, and (2) the knowledge of an NN is stored in the connections between neurons, known as synaptic weights. Researchers have been working on Alan Turing's unorganized machines and ANNs since 1948, particularly in the context of their potential to demonstrate intelligent behavior [Webster (2012)]. In the late 1940s, Hebb [Hebb (1949)], a psychologist, created a hypothesis of learning based on the mechanism of neural plasticity, which is now known as *Hebbian Learning* (It is also called Hebb's rule, Hebb's postulate, or cell assembly theory). Hebbian learning is considered to be a 'typical' unsupervised learning rule and its later variants were the early models for long-term potentiation. These ideas started being applied to computational models in 1948 with Turing's B-type machines [Webster (2012)].

In 1954, Farley and Wesley [Farley and Clark (1954)] first proposed NN computational machines, called calculators, to simulate a Hebbian network at MIT. Other NN computational machines were created by Rochester et al. [Rochester, et al. (1956)] in 1956. In 1958, Rosenblatt [Rosenblatt (1958)] proposed the perceptron that an algorithm for pattern recognition is based on a two-layer computer learning network using simple addition and subtraction. With mathematical notations, Rosenblatt described circuitry not in the basic perceptron, such as the exclusive or circuit (XOR gate), but a circuit whose mathematical computation could not be processed until after the back-propagation algorithm was created by Werbos [Werbos (1974)] in 1974. This effectively solved the XOR problem.

In 1969, Minsky and Papert [Minsky and Papert (1969)] discovered two key issues with the computational machines that processed neural networks: (1) single-layer ANNs were incapable of processing the XOR problem, and (2) computers did not have sufficient processing power to effectively handle the long run times required by large ANNs.

Since the 1980s, research on ANNs has seen remarkable developments and have come to be widely applied. In 1986, parallel distributed processing became popular under the word 'connectionism'. The book by Rumelhart and McClelland [Rumelhart and McClelland (1986)] provided a full exposition on the use of connectionism in computers to simulate neural processes. In 1989, a second generation of neurons integrated with a nonlinear activation function facilitated increased interest in ANNs [Hornik, et al. (1989)ANN], therein allowing nonlinear problems to be solved.

Support vector machines and other methods, such as linear classifiers, gradually overtook NNs in popularity for machine learning. However, the advent of deep learning in the late 2000s sparked renewed interest in ANNs. Table 1.2 summarizes the work on ANNs.

Table 1.2 Summary of Research on Artificial Neural Networks [Ding, et al. (2013), Maarouf, et al.]

Year	Work	Reference
1943	The first paper on ANNs	McCulloch and Pitts [Mccllochw and Li (1943)]
1948	Alan Turing's unorganized machines and ANNs	Webster [Webster (2012)]
1949	Hebbian learning	Donald Hebb [Hebb (1949)]
1954	Computational machines for Hebbian networks	Farley and Wesley [Farley and Clark (1954)]
1954	Neural-analog reinforcement systems	Minsky [MinskyReinforcement (1954)]
1956	Cell assembly theory	Rochester et al. [Rochester, et al. (1956)]
1958, 1962	Perceptrons	Rosenblatt [Rosenblatt (1958), Rosenblatt (1962)]
1969	Computational machines to process neural networks	Minsky and Papert [Minsky and Papert (1969)]
1974	Back-propagation algorithms	Werbos [Werbos (1974)]
1982	Hopfield network	Hopfield [Hopfield (1982)]
1985	Multilayer feed-forward networks	LeCun [LeCun (1985)], Parker [Parker (1985)]
1986	Connectionism	Rumelhart and McClelland [Rumelhart and McClelland (1986)]
1989	The second generation of neurons	Hornik et al. [Hornik, et al. (1989)ANN]
2000s	Neural networks in machine learning	Russell [Russell (2012)]

• **Evolutionary Computation**

In the 1950s, researchers started to work on evolutionary systems with the idea that evolution could be utilized as a tool for engineering problem optimizations, where the overall idea was to evolve a population of candidate solutions to a given problem using evolutionary operators inspired by natural genetic variation and natural selection. As also mentioned in the Fogel's work [Fogel (2002)], a comprehensive series of simulations of evolutionary processes, such as simulation of genetic systems, varying effects of linkage, epistasis, rates of reproduction and additional factors on the rates of advance under selection, the genetic variability of a population and other statistics, were carried out by Fraser [Fraser (1957),

Fraser (1957), Fraser (1958), Fraser (1960a), Fraser (1960b), Fraser (1960c), Fraser (1962), Fraser (1968), Fraser (1970)] and collaborated with others in this area [Fraser and Burnell (1967), Fraser and Burnell (1967), Fraser, et al. (1966), Fraser and Hansche (1965), Barker (1958a), Barker (1958b), Gill (1963), Gill (1965), Gill (1965)Selection, Martin and Cockerham (1960), Crosby (1960), Crosby (1963), Crosby (1973), Justice and Gervinski (1968), Fogel and Fraser (2000), Wolpert and Macready (1997)], such as computer models in genetics. In 1958 and 1959, Friedberg and his colleagues [Friedberg (1958), Friedberg, et al. (1959)] evolved a learning machine with language code. In the mid-1990s, Fogel's work [Fogel (1998)] provided a historical review of the efforts in evolutionary algorithms from the early 1950s to the early 1990s.

Since then, evolutionary computation (EC), or evolutionary algorithms (EAs), have become popular for optimization, machine learning and solving design problems. EC utilizes simulated evolution to search for optimal solutions to determine optimal parameters of complex problems. In this book, we apply the term EC in all the chapters. There are many different types of evolutionary algorithms; and this book introduce 3 types: GAs, evolution strategies and evolutionary programming [De Jong, et al. (1997), Whitley (2001), Melanie (1998), Sumathi and Paneerselvam (2010)].

1. **Genetic Algorithms**

Genetic algorithms (GAs) are a family of computational methods inspired by Darwin's evolutionary theory and initialized by Bremermann [Bremermann (1958)] in 1958. One of the most popular works of GA was Holland's book *Adaptation in Natural and Artificial Systems* in the early 1970s [Holland (1975)]. His work originated with studies of cellular automata, which he conducted along with his colleagues and his students at the University of Michigan [De Jong (1975), Holland (1968), Goldberg (1989)]. Holland introduced a formalized framework for predicting the quality of the next generation of evolution, known as Holland's Schema Theorem [Holland (1975), Holland (1968)]. Research in GAs remained largely theoretical until the middle of the 1980s, when the first international conference on GAs was held in Pittsburgh, Pennsylvania, USA. Since then, more studies have focused on GAs [Deb and Kumar (1995), Nix and Vose (1992), Vose and Liepins (1991), Rogers and Pr*ddot*ugel-Bennett (1999)] and a few further developed GA algorithms have been proposed [Kotani, et al. (2001)], such as micro-GAs [Goldberg (1989), Krishnakumar (1989), Chen and Song (2012)] and Mendel-GAs [Chen and Zhang (2013)]. Table 1.3 provides a brief literature review on the research of Evolutionary Computation and GAs.

Table 1.3 Summary of Research on Evolutionary Computation and GAs

Year	Work	Reference
1957	Introduction to simulation of genetic systems and effects of linkage	Fraser [Fraser (1957), Fraser (1957)]
1958–1959	Learning machine	Friedberg et al. [Friedberg (1958), Friedberg, et al. (1959)]
1958	Monte Carlo analyses of genetic models	Fraser [Fraser (1958)]
1958	Initialization of GA	Bremermann [Bremermann (1958)]
1958	Selection between alleles at an autosomal locus	Barker [Barker (1958a)]
1958	Selection between alleles at a sex-linked locus	Barker [Barker (1958b)]
1960	5-linkage, dominance and epistasis	Fraser [Fraser (1960a)]
1960	Effects of reproduction rate and intensity of selection	Fraser [Fraser (1960c)]
1960–1973	Computer simulation in genetics	Crosby [Crosby (1960), Crosby (1963), Crosby (1973)]
1960	High speed selection studies	Martin and Cockerham [Fraser (1960c)]
1963–1965	Simulation of genetic systems	Gill [Gill (1963), Gill (1965), Gill (1965)Selection]
1965	Major and minor loci	Fraser and Hansche [Fraser and Hansche (1965)]
1967	Inversion polymorphism	Fraser and Burnell [Fraser and Burnell (1967)]
1967	Models of inversion polymorphism	Fraser and Burnell [Fraser and Burnell (1967)]
1968	Simulation of the dynamics of evolving biological systems	Justice and Gervinski [Justice and Gervinski (1968)]
1968	The evolution of purposive behavior	Fraser [Fraser (1968)]
1968	Schema theory	Holland [Holland (1968), Holland (1975)], Fogel and Ghozeil [Fogel and Ghozeil (1997)] Grefenstette [Grefenstette (1993)], Poli [Poli (2000)] Radcliffe [Radcliffe (1997)]
1970	Computer models in genetics	Fraser [Fraser (1970)]
1975	The first book on GAs	Holland [Holland (1975)]
1975	Further study of GAs, block hypothesis	DeJong [De Jong (1975)], Goldberg [Goldberg (1989)]
1989	Micro-GAs	Goldberg [Goldberg (1989)], Krishnakumar [Krishnakumar (1989)], Chen [Chen and Song (2012)]
1992	Modeling GAs with Markov chains	Nix and Vose [Nix and Vose (1992)]
1994	Co-operative co-evolutionary GAs	Potter and De Jong [Potter and De Jong (1994)]
1995	Simulated binary crossover	Deb [Deb and Kumar (1995)]
1997	No free lunch theorems for optimization	Wolpert and Macready [Wolpert and Macready (1997)]
1998	Evolutionary computation: the fossil record	Fogel [Fogel (1998)]
1999	Steady-state GAs	Rogers [Rogers and Prddotugel-Bennett (1999)]
2000	Running races with Fraser's recombination	Fogel and Fraser [Fogel and Fraser (2000)]
2001	Variable-length-chromosome GAs	Kotani et al. [Kotani, et al. (2001)]
2013	Mendel-GAs	Chen [Chen and Zhang (2013)]

Particularly, in 1992, Koza [Koza (1992), Koza (1994)], utilized GAs to evolve programs to perform certain tasks in a process called genetic programming (GP). GP is a technique of enabling a GA to search a potentially infinite space of computer programs rather than a space of fixed-length solutions to a combinatorial optimization problems. These programs often take the form of Lisp symbolic expressions, called *S-expressions*. The idea of applying GAs to S-expressions, rather than combinatorial structures, is due originally to the work of Fujiki and Dickinson [Fujiko and Dickinson (1987)] and was brought to prominence through the work of Koza [Koza (1992)]. Table 1.4 lists a few works on GP in recent years.

2. Evolution Strategy

In the 1960s, Rechenberg [Rechenberg (1965), Rechenberg (1973)] proposed the idea of evolution strategies (*ES, Evolutionsstrategie* in German), which were employed to optimize real-valued parameters for devices such as airfoils. This method was then further developed by Schwefel [Schwefel (1975), Schwefel (1977), Schwefel (1981), Schwefel, et al. (1991), Schwefel (1995)]. Since then, the field of ES has remained an active area of research and has been mostly developed independently from the other fields of EC; however, it has been interacting with the other fields of EC [Knowles and Corne (1999), Beyer and Schwefel (2002)]. Table 1.5 summarizes the works on ES.

3. Evolutionary Programming

In 1965, Fogel et al. [Fogel, et al. (1966)] proposed the technique of evolutionary programming (EP) in which candidate solutions were represented as finite-state machines and then evolved randomly by mutating state-transition diagrams and selecting the fittest solution. Since the middle of the 1980s, EP has been developed to solve more general tasks, including prediction problems, numerical and combinatorial optimizations and machine learning [Fogel and Ghozeil (1997), Schwefel (1981), Fogel (1991), Fogel and Atmar (1991), Fogel (1995)Machine, Fogel (1999)]. In 1992, the first annual conference on EP was held at La Jolla, CA. Since then, further conferences have been held annually. Table 1.6 summarizes the works on EP.

• Swarm Intelligence

In recent years, swarm intelligence has been attracting substantial attention of researchers and has been applied successfully in a variety of applications. In general, swarm intelligence addresses the modeling of the collective behaviors of simple agents interacting locally among themselves and their environment, leading to the emergence of a coherent functional global pattern [Kennedy and Eberhart (2001)]. A swarm can be defined as a group of agents cooperating with certain behavioral patterns to achieve certain goals. From the computational point of view, swarm intelligence models are computing

Table **1.4** Summary of Research on GP [GPbib (2017), GPorg (2007), Langdon and Gustafson (2010)]

Year	Work	Reference
1992	GP, the first paper on GP	Koza [Koza (1992)]
1994	GP II	Koza [Koza (1994)]
1998	Code growth	Soule [Soule (1998)], Langdon and Poli [Langdon and Poli (1998)], Langdon et al. [Langdon, et al. (1999)]
1999	GP III	Koza et al. [Koza, et al. (1999)]
2001	Genetic-based machine learning (GBML) and GP	Sette and Boullart [Sette and Boullart (2001)]
2003	GP IV	Koza et al. [Koza, et al. (2003)]
2008	A field guide to GP	Poli et al. [Poli, et al. (2008)]

Table **1.5** Summary of Research on ES [Melanie (1998), Schwefel, et al. (1991), Beyer and Schwefel (2002)]

Year	Work	Reference
1965	First proposed ES, (1+1)-ES, (μ+1)-ES	Rechenberg [Rechenberg (1965), Rechenberg (1973)]
1975	Further development, ($\mu+\lambda$)-ES	Schwefel [Schwefel (1975), Schwefel (1977), Schwefel (1981), Schwefel, et al. (1991), Schwefel (1995)]
1996	Derandomized mutation step	Hansen et al. [Hansen and Ostermeier (1996)]
1996	Mutation of ES using Markov chains	Rudolph [Rudolph (1996)]
1996	Hybrid of GA and ES	Smith et al. [Smith and Fogarty (1996)]
1999	ES for multi-objective optimization	Knowles [Knowles and Corne (1999)]
2001	$\frac{1}{5}$ Success rule	Rechenberg [Rechenberg (1965), Rechenberg (1973)], Rudolph [Rudolph (2001)]

Table 1.6 Summary of Research on EP

Year	Work	Reference
1966	The first book on EP	Fogel et al. [Fogel, et al. (1966)]
1980s	Further studies on EP	Schwefel [Schwefel (1981)], Fogel et al. [Fogel (1991), Fogel (1995), Fogel and Ghozeil (1997), Fogel (1999)]
1991	Self-adaptation	Schwefel [Schwefel (1981)], Fogel et al. [Fogel and Atmar (1991)], Liang et al. [Liang, et al. (2001)]
2008	Unbiased evolutionary programming (UEP)	MacNish and Yao [MacNish and Yao (2008)]

algorithms that are useful for addressing distributed optimization problems. As introduced in Table 1.7, a number of researchers have been investigating and proposing various models of swarm intelligence, among which the most widely known models include ant colony optimization (ACO) [Colorni, et al.(1992), Colorni, et al.(1992), Dorigo, et al. (1996)], particle swarm optimization (PSO) [Kennedy and Eberhart (1995), Del Valle, et al. (2008)] and artificial bee colony (ABC) [Seeley (1996), Teodorović, et al (2006)].

Table 1.7 Summary of Research on Swarm Intelligence [Lim and Jain(2009), Banks, et al. (2007)]

Year	Work	Reference
1992	Ant colony optimization	Colorni, Dorigo and Maniezzo [Colorni, et al.(1992), Dorigo, et al. (1996)]
1995	Particle swarm optimization	Kennedy and Eberhart [Kennedy and Eberhart (1995), Del Valle, et al. (2008)]
1996	Artificial bee colony	Seeley [Seeley (1996), Teodorović, et al (2006), Teodorović, et al (2006)]
1998	Modified particle swarm optimizer	Shi and Eberhart [Shi and Eberhart (1998)]
2000	Bacterial foraging	Passino [Passino (2000), Passino (2002)]
2000	Social insect behavior	Bonabeau [Bonabeau, et al. (2000)]
2002	Canonical particle swarm	Clerc et al. [Clerc and Kennedy (2002)]
2003	Artificial fish swarm algorithm	Li [Li (2003)]
2004	Fully informed particle swarm	Mendes [Mendes, et al. (2004)]
2005	Fuzzy bee system	Teodorovic [Teodorovi$acutec$ and Dell'orco]
2005	Virtual bee algorithms	Yang [Yang (2005)]
2008	Firefly algorithm	Yang [Yang (2008)]
2010	Bat algorithm	Yang [Yang (2010)]
2010	Firework algorithm	Tan and Zhu [Tan and Zhu (2010)]
2015	Artificial dolphin swarm algorithm	Chapter 7.7

- **Artificial Immune Systems**

  Immune systems are highly distributed, adaptive and self-organizing in nature; they also maintain a memory of past encounters and have the ability to continually learn about new encounters. From the computational point of view, the immune system has much to offer by means of inspiring researchers. Inspired by biological immune systems, artificial immune systems (AIS) have emerged during the last decade and are used by researchers to design and build immune-based models for a variety of applications. An AIS can be defined as a computational paradigm that is inspired by theoretical immunology, observed immune functions, principles and mechanisms [deCastro and Timmis (2003)].

  The AIS emerged in the middle of the 1980s through articles authored by Farmer et al. [Farmer and Packard (1986)] and Bersini and Varela [Bernstein and Vazirani (1991)] on immune networks. However, it was only in the middle of the 1990s that AIS became an independent field [deCastro and Timmis (2003), de Castro and Von Zuben (1999)]. Forrest et al. and Kephart et al. [Kephart (1994)] published their first papers on AIS in 1994 and Dasgupta conducted extensive studies on negative selection algorithms. Hunt and Cooke [Hunt and Cooke (1996)] began working on immune network models in 1995. The first book on AIS was edited by Dasgupta in 1999 [Dasgupta (1999)]. In 2000, Timmis et al. [Timmis, et al. (2000)] developed an immune network theory-inspired AIS based on work undertaken by Hunt and Cooke [Hunt and Cooke (1996)], in which the proposed AIS consisted of a set of B-cells, links between those B-cells and cloning and mutation operations performed on the B-cell objects. This AIS was tested on the Fisher Iris dataset

[Fisher (1936)], with some encouraging results. Timmis and Neal continued this work and made some improvements [Timmis and Neal (2001)]. In 2002, De Castro and Timmis proposed a framework for AIS [deCastro and Timmis (2002)] that provides a representation to create abstract models of immune organs, cells and molecules; a set of affinity functions to quantify the interactions of these 'artificial elements'; and a set of general-purpose algorithms to govern the dynamics of the AIS. In 2008, Dasgupta and Nino [Dasgupta and Nino (2008)] published a textbook on immunological computation that presents a compendium of up-to-date work related to immunity-based techniques and describes a wide variety of applications.

Currently, the new ideas along AIS lines, such as danger theory and algorithms inspired by the innate immune system, are also being explored. However, some believe that these new ideas do not yet offer any truly 'new' abstract over and above the existing AIS algorithms. This, however, is hotly debated and the debate provides one of the main driving forces for AIS development at the moment. Other recent developments involve the exploration of degeneracy in AIS models [Andrews and Timmis (2006), Mendao, et al. (2007)], motivated by its hypothesized role in open-ended learning and evolution [Edelman and Gally (2001), Whitacre (2010)]. Table 1.8 lists a few works on AIS.

Table 1.8 Summary of Research on AIS [deCastro and Timmis (2002), Timmis, et al. (2008), Timmis, et al. (2004), Al-Enezi, et al. (2010)]

Year	Work	Reference
1994	The first paper on AIS	[Kephart (1994)]
1994	Negative selection	[Forrest, et al. (1994)]
1996	Artificial immune network models	[Hunt and Cooke (1996)]
1999	First book on AIS	[Dasgupta (1999)]
1999	Clonal selection-based algorithms	[de Castro and Von Zuben (1999)]
2000	AIS consisting of a set of B-cells	[Timmis, et al. (2000)]
2000	Supervised learning system (Immunos-81)	[Carter (2000)]
2001	Sparse distributed memory (SDM) using immune system metaphor	[HartRoss and Ross (2001)]
2002	A framework for AIS	[deCastro and Timmis (2003), deCastro and Timmis (2002)]
2002	Multi-layered immune-inspired learning	[Knight and Timmis (2002)]
2003	Boolean competitive neural network using immunology	[de Castro and Von Zuben (2003)]

- **Quantum Computing**

  Quantum computing (QC) can be defined as a paradigm that exploits a computational model relying on the principles of quantum mechanics [Furia (2006)]. The introduction of quantum mechanical computational models was based on the computational difficulty of simulating quantum systems with a classical computer. The first works were performed by Manin [Manin (1980), Manin (1999)] and Feynman [Feynman (1982)] in the early 1980s. Table 1.9 gives a brief review of the research on QC.

The QC concerns theoretical computation systems that directly utilize quantum mechanical phenomena, such as superposition and entanglement, to perform operations on data. It seems obvious that QC will largely contribute to the engineering goals of CI by applying it to various CI systems to speed up the computational process. However, it is indeed very difficult to design quantum algorithms for solving certain CI problems that are more efficient than existing classical algorithms for the same purpose.

Table 1.9 Summary of Research on QC [Furia (2006), Nielsen and Chuang (2010), Ying (2010)]

Year	Work	Reference
early 1980s	The first work on QC	Manin [Manin (1980), Manin (1999)], Feynman [Feynman (1982)]
early 1980s	Quantum mechanical computational model	Feynman [Feynman (1982)] and Benioff [Benioff (1980), Benioff (1982), Benioffdissipate (1982)]
mid 1980s	Quantum mechanical computational devices	Feynman [Feynmancomputers (1985), Feynmancomputers (1986)], Deutsch [Deutsch (1985)]
mid 1980s	Applying quantum models to cryptography	Bennett and Brassard [Bennett and Brassard (1984)], Wiesner [Wiesner (1983)]
early 1990s	Deutsch-Jozsa algorithm	Deutsch and Jozsa [Deutsch and Jozsa (1992)]
1993	Quantum complexity theory	Bernstein and Vazirani [Bernstein and Vazirani (1993), Bernstein and Vazirani (1997)]
1994	Shor-type quantum algorithms	Shor [Shor (1994), Shor (1997)], Lenstra and Lenstra [Lenstra and Lenstra (1994)]
1990s	Fault-tolerant techniques	Shor et al. [Shorreducing (1995), ShorFault (1996), RobertShor and Shor (1996)], Steane [Steane (1996)], Gottesman [Gottesman (1996)], Preskill [Preskill (2001)]
1996	Grover-type quantum algorithms	Grover [Grover (1996), Grover (1997)]
1998	Implementations of quantum computing device	Chuang et al. [Chuang, et al. (1998)], Aaronson [Aaronson (2004)]

- **DNA Computing**

DNA (deoxyribonucleic acid) can be found in every cellular organism as the storage medium for genetic information. It is composed of units called nucleotides, which are distinguished by the chemical group, or base, attached to them. The four bases are *adenine, guanine, cytosine* and *thymine*, abbreviated as **A, G, C,** and **T**. The single nucleotides are linked together end-to-end to form DNA strands.

DNA computing (DNAC) is an area of natural computing based on the idea that molecular biology processes can be used to perform arithmetic and logic operations on information encoded as DNA strands [Kari, et al. (2010)]. As a new computing paradigm, DNAC has advantages for addressing complex problems: (1) DNA codes and DNA molecular operations are suitable for representing complex information, (2) DNAC can reduce computation time due to its parallel nature, and (3) DNA molecules have a huge storage capacity that provides a benefit for solving large-scale, multi-variable problems. However, in comparison with conventional approaches, DNAC has certain disadvantages, such as being inconvenient, un-scalable and expensive. Therefore, it is difficult to solve practical engineering problems using DNA molecules. Hence, with the help of digital computers, Garzon et al. simulated a virtual test tube and reproduced Adleman's experiment using electronic DNA [Garzon, et al. (1999)].

On the other hand, theoretical research on DNAC includes designing potential experiments for solving various problems through DNA manipulation. Descriptions of such experiments include the satisfiability problem [Lipton (1995)], breaking the data encryption standard [Boneh, et al. (1996)], expansions of symbolic determinants [Leete, et al. (1999)], matrix multiplication [Oliver (1996)] and the bounded post correspondence problem [Kari, et al. (2000)]. Table 1.10 summarizes the work on DNAC.

Table 1.10 Summary of Research on DNAC [Pisanti (1997), Xu and Tan (2007), Bakar and Watada (2008)]

Year	Work	Reference
1994	The first DNA computing experiment	Adleman [Adleman (1994), Adleman (1998)]
1995	DNA solution of hard computation problems	Lipton [Lipton (1995)]
1996	Solving problems by DNAC	Boneh et al. [Boneh, et al. (1996)], Leete et al. [Leete, et al. (1999)], Oliver [Oliver (1996)], Kari et al. [Kari, et al. (2000)]
1999	Virtual test tubes	Garzon et al. [Garzon, et al. (1999), Garzon, et al. (2004)]
1997	DNAC model	Kari [Kari (1997)], Bonizzoni et al. [Bonizzoni, et al. (2001)], Zimmermann et al. [Zimmermann (2002)], Daley et al. [Daley and Kari (2002)]
2004	Genetic algorithms in DNA computing	Li et al. [Li, et al. (2004)]

1.3 On the Way to Industry 4.0

The industrial revolution began in Great Britain and most of the important technological innovations were British. The earliest use of 'industrial revolution' was in a letter written on 6 July 1799 by French envoy Louis-Guillaume Otto, announcing that France had entered the race to industrialise [Francois (1996)]. Figure 1.2 shows the brief history of the industrial revolution.

- The first industrial revolution marked the transition to new manufacturing processes in the period from about 1760 to sometime between 1820 and 1840. The commencement of the first industrial revolution is closely linked to a small number of innovations, such as textiles, steam power, iron making and the invention of machine tools [Bond, et al. (2011)], beginning in the second half of the 18th century [Wikipedia (2018)].

- The first industrial revolution evolved into the second industrial revolution in the transition years between 1840 and 1870, when technological and economic progress continued with the increasing adoption of steam transport (telephone, light bulb, phonograph, the internal combustion engine, railways, boats and ships, etc.), the large-scale manufacture of machine tools and the increasing use of machinery in steam-powered factories [Wikipedia (2018)].
- The third industrial revolution is the digitisation of design and manufacturing in a sustainable era of distributed capitalism, ushering the technologies of the internet, green electricity and 3-D printing, etc. [Rifkin (2015)].
- The fourth industrial revolution, also known as Industry 4.0 (i4), is the current trend of automation and data exchange in manufacturing technologies, which include cyber-physical systems, the Internet of things, cloud computing, robotics, artificial intelligence, nanotechnology, biotechnology, etc. The i4 has begun and offers attractive opportunities to industrial companies [Wikipedia (2018)].

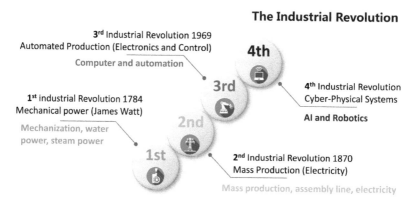

Fig. 1.2 The industrial revolution

Today, i4 refers to the industrial value chain and technological evolution towards smart manufacturing, with associated concepts of networked embedded systems, cyber-physical systems (CPS) [Sha and Gopalakrishnan (2009), Kim and Kumar (2012)], smart factory, the Internet of Things (IoT), the Internet of Services (IoS), 'Internet+', and '5G' telecommunications, to name but a few [Flores, et al. (2015)], as depicted in Figure 1.3. With i4, design and manufacture are currently shifting to such a new paradigm, targeting innovation, lower costs, better responses to customer needs, optimal solutions, intelligent systems and alternatives towards on-demand production. All these trends have in common the integration of several features at the same place as a response to challenges of computerised decision making and big data, which are proliferated by Internet and cloud com-

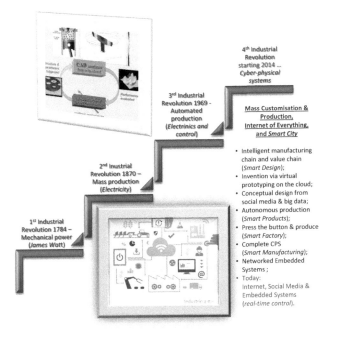

Fig. 1.3 On the way to industry 4.0

puting [Lee (2013)]. The state of the art of i4 research and development is mostly represented by manufacturers on smart manufacturing to elevate digital manufacturing, i.e., on the industrial Internet [Annunziata and Evans (2012)] of smart and connected production to enable the factory floor to become an automated innovation center [Siemens (2015)]. However, what are lacking at present are 'smart design' tools commensurate with i4 further up the value chain and 'market informatics' at the end of the value chain, both of which the factory-floor innovation needs to take into account.

1.4 Need for Computational Intelligence in Design and Engineering

Research shows that manufacturers using digital prototyping build half the number of physical prototypes as the average manufacturer, get to market 58 days faster than average and experience 48 per cent lower prototyping costs [Aberdeen (2006)]. This is achieved even without using the techniques of CI-assisted design and engineering (CIAD). This book introduces the conceptual CIAD frameworks and their multidisciplinary applications, with their objectives of shorter times and reduced costs of potentially 10–50 per cent. This book contributes to a driving force in the research portfolio in design and manufacture, as CIAD will be able to help industry and society maintain their competitive advantage as well as deliver improvements and impacts in multi-disciplinary fields such as engineering, science, health care, education and social services.

In the era of industry 4.0, reflecting one of the most innovative activities in businesses, design and creation are essential to the global economy and manufacturing industries. A rising demand of a smart design system is seen in increased requirements of customization and flexibility, design efficiency and responsiveness, and cost-effectiveness of products and their manufacturing. These requirements become urgent in the international race to the next 'industrial revolution', as highlighted by the German 'Industrie 4.0' initiative that aims at upgrading the entire manufacturing value chain by turning the factory floor into an innovation centre capable of mass customization.

"The term 'Industrie 4.0' was revived in 2011 at the Hannover Fair. In October 2012, the Working Group on Industry 4.0 presented a set of Industry 4.0 implementation recommendations to the German federal government. The Industry 4.0 workgroup members are recognized as the founding fathers and driving force behind Industry 4.0." [Wikipedia (2018)]

As this book spans both theoretical and applicable research, we foresee not only a clear potential impact on the ever-expanding needs of the intelligent design, engineering and manufacturing fields but also a methodological impact on a wide range of multi-disciplinary fields studied by universities and industrial partners such as science, technology, engineering and mathematics (STEM); life science; and social science. Beneficiaries in academia include researchers and students working on robotics, engineering design and manufacture, dynamics and control, big data systems and related professional circles in a variety of disciplines.

This book provides new insights into economic growth through the innovation of design and manufacture automation and into possible ways of increasing economic activities. It will contribute to developments in knowledge integration with industrial partners and end users, thus helping to maximize the impact of the research for knowledge sharing and transfer for potential long-term collaborations.

1.5 Terms and Definitions

- **Agents**:
  In CI, an agent is an autonomous entity that acts in a given environment; the agent observes through sensors, acts upon its environment using actuators, and directs its activity towards achieving goals through learning and pre-requisite knowledge [Poole and Mackworth (2010)].

- **Chromosome**:
  All living organisms consist of cells and each cell contains a set of chromosomes (strings of DNA) that serve as a 'blueprint' for an organism. In CI, the term

chromosome typically refers to a candidate solution to a problem, often encoded as a bit string.

- **Alleles**:
The different possible settings for a trait (e.g., blue, brown and hazel) are called alleles. In CI, an allele in a chromosome is either 0 or 1.

- **Gene**:
A chromosome can be conceptually divided into genes, each of which encodes a particular protein. Basically, we can consider a gene as the encoding of a trait, such as hair color. Each gene is located at a particular locus (position) on the chromosome. In CI, the 'genes' are either single bits or short blocks of adjacent bits that encode a particular element of the candidate solution. Each gene controls a particular characteristic of the individual; similarly, each bit in the string represents a characteristic of the solution.

- **Fuzzy Logic**:
A type of logic using graded or qualified statements rather than statements that are strictly true or false. The results of fuzzy reasoning are not as definite as those derived by strict logic, but they cover a larger field of discourse [Zadeh (1984a)].

- **Fuzzy Modifiers**:
Fuzzy modifiers are operations that change the membership function of a fuzzy set by spreading out the transition between full membership and non-membership, by sharpening that transition, or by moving the position of the transition region [Zadeh (1984a)].

- **Fuzzy Sets**:
Fuzzy sets are sets that do not have a crisply defined membership but rather allow objects to have grades of membership from **0** to **1** [Zadeh (1965a)].

- **Linguistic Variables**:
Linguistic variables are ordinary-language terms that are used to represent a particular fuzzy set in a given problem such as 'large', 'small', 'medium' or 'OK' [Zadeh (1984a)].

- **Ultra-fuzzy Sets**:
Ultra-fuzzy sets are those whose membership function is itself fuzzy. Here, an object in the set, rather than being given a membership grade between **0** and **1**, is assigned a range of possible membership grades, for example, **0.4** to **0.8**, instead of **0.55** [Zadeh (1984)].

- **Genotype**:
A genotype is the specific genetic make-up of an individual in the form of DNA. Together with the environmental variations that influence the individual, it codes

for the phenotype of that individual. In CI, a genotype is a basic data structure or type [Fogel (1995)].

- **Phenotype**:
  A phenotype is the total physical appearance and constitution or a specific manifestation of a trait, such as size, eye color, or behavior that varies between individuals. In CI, a phenotype represents a solution to a problem.

- **Genotypic Algorithms**:
  Genotypic algorithms operate on strings representing the system.

- **Phenotypic Algorithms**:
  Phenotypic algorithms operate directly on the parameters of the system itself. Both EP and Evolutionary Strategies are known as Phenotypic Algorithms, whereas the GA is a Genotypic Algorithm.

- **Industry 4.0**:
  "INDUSTRIE 4.0 is identified by the German government as one of ten 'Future Projects' as part of its High-Tech Strategy 2020 Action Plan. The INDUSTRIE 4.0 project (aka the fourth industrial revolution) began as a marketing opportunity for Germany to establish itself as an industry lead market and technology provider. It has now subsumed into the business lexicon as a catchall covering the automation of manufacturing, machine-to-machine and machine-to-product communication, the industrial internet and technology needed for mass customisation of production." [BDO Industry 4.0 Report - IMechE (2016)]

1.6 Specialized and Application Areas

Ever since the inception of CI research, CIAD have branched into many specialized and application areas such as those listed below (but not limited to):

- Energy [Chen, et al. (2013)]
- Drug Development [Xu, et al. (2012), Liu et al. (2012)]
- Economics and Finance [Chen and Zhang (2013)]
- Sustainable Development [Chen, et al. (2013), Chen, et al. (2012)]
- Engineering [Chen and Cartmell (2007), Chen, et al. (2012), Chen, et al. (2011), Chen, et al.(2011), Chen, et al.(2011), Chen, et al. (2012)]
- Societal Application [Chen and Song (2012)]

1.7 Information Sources

- **Journals**

  1. *Advanced Engineering Informatics*
  2. *Applied Mathematics and Optimization*
  3. *Applied Soft Computing*
  4. *Artificial Intelligence*
  5. *Artificial Intelligence in Medicine*
  6. *Annals of Mathematics and Artificial Intelligence*
  7. *Computer Sciences in Engineering*
  8. *Computers & Industrial Engineering*
  9. *Computers and Mathematics with Applications*
  10. *Computational Statistics and Data Analysis*
  11. *Engineering Applications of Artificial Intelligence*
  12. *Evolutionary Computation*
  13. *Evolutionary Intelligence*
  14. *Expert Systems*
  15. *Expert Systems with Applications*
  16. *Fuzzy Sets and Systems*
  17. *IERI Procedia*
  18. *International Journal of Approximate Reasoning*
  19. *International Journal of Intelligent Systems*
  20. *Information Sciences*
  21. *Information Processing and Management*
  22. *Information and Computation*
  23. *Journal of Automated Reasoning*
  24. *Journal of Information Science*
  25. *Journal of Machine Learning Research*
  26. *Journal of Statistical Planning and Inference*
  27. *Journal of Supercomputing*
  28. *Journal of Optimization Theory and Applications*
  29. *Knowledge-Based Systems*
  30. *Mathematics and Computers in Simulation*
  31. *Mathematical and Computer Modelling*
  32. *Machine Learning*
  33. *IEEE Transactions on Knowledge and Data Engineering*
  34. *IEEE Transactions on Neural Networks and Learning Systems*
  35. *IEEE Computational Intelligence Magazine*
  36. *IEEE Transactions on Evolutionary Computation*
  37. *IEEE Transactions on Fuzzy Systems*
  38. *IEEE Transactions on Autonomous Mental Development*
  39. *IEEE/ACM Transactions on Computational Biology and Bioinformatics*
  40. *IEEE Transactions on Computational Intelligence and AI in Games*
  41. *IEEE Transactions on Nano Bioscience*

42. *IEEE Transactions on Information Forensics and Security*
43. *IEEE Transactions on Affective Computing*
44. *IEEE Transactions on Smart Grid*
45. *IEEE Transactions on Pattern Analysis and Machine Intelligence*
46. *Natural Computing*
47. *Neural Computation*
48. *Neural Networks*
49. *Neurocomputing*
50. *Pattern Recognition*
51. *Parallel Computing*
52. *Theoretical Computer Science*

- **Conferences and Workshop**

  1. UK Workshop on Computational Intelligence
  2. Annual IEEE Congress on Evolutionary Computation
  3. IEEE International Conference on Fuzzy Systems (FUZZ-IEEE)
  4. IEEE Congress on Evolutionary Computation (CEC)
  5. International Joint Conference on Neural Networks (IJCNN)
  6. International Workshop on DNA-based Computers
  7. Genetic and Evolutionary Computation (GECCO) Conference
  8. International Conference on Artificial Immune Systems
  9. Annual ACM Symposium on the Theory of Computation

- **Books**

  1. Genetic Algorithm + Data Structures = Evolution Programs (3rd ed.) [Michalewicz (1996)]
  2. Evolutionary Multiobjective Optimization—Theoretical Advances and Applications Abraham and Goldberg (2005)
  3. Multi-objective Optimization using Evolutionary Algorithms [Deb (2001)]

- **Organizations**

  1. The International Fuzzy Systems Association (IFSA)
  2. The IEEE Computational Intelligence Society
  3. The Online Home of Artificial Immune Systems [The Online Home of Artificial Immune Systems (2015)]
  4. Quantum Artificial Intelligence Laboratory (QuAIL), NASA
  5. Ant Colony Optimization [Antorg (2015)]

1.8 How to Use This Book

This textbook can be used in many different areas, including computer science, control systems and various other STEM areas. The following list gives suggested contents for different courses at the undergraduate and graduate levels based on the syllabi of many universities around the world, such as the University of Glasgow.

- Teaching Suggestions
  The material in this book can be adapted for a one-quarter or one-semester course. The organization is flexible, therein allowing the instructor to select the material that best suits the requirements and time constraints of the class.

- Before You Use This Book
  A number of chapters in this book assume at least a basic knowledge of numerical simulation, programming, dynamics, control engineering, applied mathematics, engineering design and computational intelligence. If you are unfamiliar with MATLAB®, C/C++, pseudo-code and program flowcharts, we strongly recommend that the reader attempts to access the above-mentioned pre-requisite knowledge. Our book attempts to allow researchers who want more information on more specific topics and on the numerous new methods that are constantly being developed to take advantage of emerging research opportunities. The goal of this volume is to present the varied methods in short, practically-oriented chapters that will allow readers to integrate the methods into their own research.

- Overview of Book Structure
  This book is organized into four parts:
  Part I introduces three core technologies of computational intelligence: *fuzzy logic theory*, *ANNs* and *GAs*.
  Part II introduces some advanced computational intelligence algorithms, such as swarm algorithms as optimization tools and a few metric indices that can be employed to assess their performance.
  Part III introduces the conceptual framework of CIAD and its applications in science and technology, such as control systems and battery capacity prediction.
  Part IV introduces some CIAD applications in social science, for example, exchange rate investigation, electricity consumption and spatial analysis for urban studies.

- Problems, Tutorials, Laboratory and Coursework
  This book provides *Problems* in Chapter 1, *Tutorials* in Chapters in Part I, and *Laboratory and Coursework* in the Appendices.

- Web Companion
  There is a companion website to this book. Students and practitioners are encouraged to visit `http://www.i4ai.org/EA-demo/`. Additional resources

will be posted on this website as they become available, including more examples of CIAD definitions, case studies, related CIAD material, illustrations and useful links.

Problems

1. Write an essay on the history of CI.
2. Discuss the need for CI in smart design, smart manufacture and industry 4.0.
3. Discuss at least three well-publicized systems employing CI approaches and the reasons for their use.
4. What are the CI-related tasks performed during the smart design process?
5. Define the following terms: CI, soft computing and nature-inspired computing.

Part I
Hands-on Learning of Computational Intelligence

Part I introduces three core technologies of computational intelligence: *fuzzy logic theory*, *artificial neural networks* and *genetic algorithms*.

This section is about learning computational intelligence and its applicability to enabling design automation within a modern engineering context. The major aims are:

- Master the powerful technology of evolutionary computing that borrows the principles of natural evolution to help with global optimization through virtual generations of digital prototyping;
- Master neural computing technology that mimics how human learning and training can achieve optimal results;
- Master fuzzy systems techniques that human beings utilize in gathering information and making decisions;
- Understand the industrial relevance of the technology in solving many traditionally practically unsolvable engineering problems, in 'intelligently' refining the solutions, in optimizing engineering designs and in design automation.

Part I Learning Outcomes:

1. Knowledge: Methods of neural network and evolutionary computing, fuzzy logic and their applications in solving smart design and engineering optimization problems.
2. Understanding: The process of natural evolution of the human species, the process of human learning in the eye-brain and central nervous system and information gathering and inference in a fuzzy manner. Why evolutionary methods are usually more thorough but usually a slower adaptation and optimization tool than the neural learning methods. How evolutionary algorithms can be used to find solutions to practical engineering problems with a global optimality. Why neural networks can be used to store and retrieve information and mimic the behavior of many engineering systems. Training strategies. How neural networks can be used both to represent and control systems; how fuzzy systems to help with learning and decision making and neural and evolutionary computing help with devising the fuzzy systems. Use and abuse of computational intelligence techniques.
3. Skills: Programming evolutionary algorithms. Designing simple fuzzy logic systems; selecting neural network training strategies; genetic evolution of neural networks and optimizing engineering solutions.

2

Global Optimization and Evolutionary Search

2.1 Mimicking Natural Evolution

To meet the ever-growing demand in quality and competitiveness, engineers face a challenge in delivering within a limited period of time good design solutions that are globally optimal and that can meet mixed objectives. At present, a design is often tuned in a manual trial-and-error process by interacting with a computer-aided design (CAD) simulator repeatedly. In recent years, however, dramatic increase in processor power and affordability of computational redundancy has seen rapid development in, and applications of, soft-computing paradigms. These techniques aim at solving a practical problem with the flexibility of a population (or a network) of potential solutions (or elements) as opposed to using a rigid, single point of computing. Use of multiple points and redundancy, flexible learning and computational intelligence can be achieved.

Evolutionary computation, artificial neural network and fuzzy-logic-based soft-computing paradigms mimic human evolution, learning and decision-making and offer tractable solutions to real-world problems. For example, in evolutionary computation, the power of multiple trial-and-error based a posteriori 'intelligence' is utilized to test and discover possible solutions to many conventionally unsolvable problems. When interfaced with an existing CAD package, an evolutionary algorithm (EA) enables computer-automated design (CAutoD) by extending passive CAD simulation into active design search.

An EA enabled design automation process can start from a human designer's existing database or from randomly generated potential solutions. To evolve towards the next 'generation' of generally better solutions, the EA selects higher performing candidates from the current generation using 'survival-of-the-fittest' learning. It then exchanges some parameter values or search co-ordinates between two candidates through an operation called 'crossover' and introduces new values or co-ordinates through an operation called 'mutation'. This way, the evolutionary process makes use of the past trial information for global optimality in a similarly 'intelligent' manner to a human designer. Then the new generation is evaluated and better candidates are reproduced again with a higher probability of survival. This process repeats itself automatically and improves solutions generation by generation, leading to a number of finally evolved top-performing designs.

In summary, when coupled with an evaluator such as an existing CAD simulator, evolutionary computation can offer:

- A fully computerised design procedure
- A reduced design cycle
- Designs that take into account practical constraints (e.g., current, voltage or power saturation)
- Improved or novel designs with a globally optimal parameter set
- Novel configurations that could extend the present performance bounds
- Ability to react quickly to changing market conditions

Soft-computing techniques have been successfully applied to control engineering. In this course, basics of evolutionary and related techniques and some state-of-the-art developments in this area are to be presented with a focus on CACSD automation. Applications to controller design, LTI design unification, nonlinear and sliding-mode control, fuzzy and neural control, modelling and identification will be demonstrated.

2.1.1 Breading Engineering Solutions

1. Think about the process of designing a circuit, a filter, a control system or a radio.
2. The geometric shape of a car or Boeing 777, aerodynamics, ergonomics, etc.
3. Computers, assisted by initial human analysis and calculations based on physical principles and measurements, can nowadays simulate, analyze and evaluate almost any engineering systems. But,

   - Design is a revered process of simulation
   - Associated with the design are problems of multi-dimensionality, multi-modality and multi-optimality
   - Engineering constraints, geometric constraints, environmental constraints, etc., need to be considered:
     a. Most products are so far designed through repetitive trial-and-errors
     b. Today's designs are actually 'evolved' over many 'generations'
     c. In nature, why sexual reproduction?
     d. Involve many 'neural learning'
     e. Human interpretation of performance, beauty, etc., using fuzzy logic and reasoning

2.1.2 Conventional Computers

Conventional computers have to be explicitly programmed, i.e., given step-by-step instructions to follow in order to solve a problem; cannot automatically solve problems. Need soft computing and computational intelligence.

2.1.3 Genetic Evolution—A Way to Solve Complex Optimisation Problems

- The complexity of a problem lies in the complexity of the solution space that may be searched through. This complexity arises due to:

  1. size of the problem domain
  2. non-linear interactions between various elements (epistatis)
  3. domain constraints
  4. performance measure with dynamics and many independent and co-dependent elements
  5. incomplete, uncertain and/or imprecise information

- Systems of nature routinely encounter and solve such problems. Good examples include genetic evolution of species, which become better and better suited to the environment generation by generation
- A Darwinian machine or an evolutionary program emulates this process and does not require an explicit description of how the problem is to be solved. It tends to evolve optimal solutions automatically
- Evolutionary computing technology:

  1. Maps an engineering system directly onto a 'genetically' encoded (numerical or character) string of system parameters and structures
  2. Can convert an automatic design problem to a simulation problem to solve
  3. Solves a difficult design problem by intelligently evaluating performance and evolving optimal candidates based upon the evaluations
  4. A universal tool for design automation
  5. Also applicable to the design of ANNs, called Darwinian selective learning
  6. Such a nearly untapped powerful technology that can make a revolutionary impact on engineering design in the near future.

2.2 Nondeterministic Methods for Optimization and Machine Learning

With the dramatic increase in computer power in the past few decades, machine learning and optimization can afford increasingly bold and broad explorations.

Non-deterministic algorithms ease the restrictions on direction and guidance. These belong to a category of modern optimization and machine-learning algorithms.

Using such an algorithm, the next step or search is generated at random and requires no 'teacher' or 'supervisor'. As may be inferred, these methods must be a-posteriori because the correctness of the search is not reasoned a-priori and may only be assessed after the move, i.e., this is in effect a trial-and-error process. Note again that an a-posteriori comparison or selection does not require the existence of the gradient of the objective function.

In a nondeterministic algorithm, the search steps are not pre-determined. This means that if the search objective or cost function is the same and we repeat the optimization process again from the same starting point, the search pattern in every run is likely to be different. Therefore, to assess the performance of such an algorithm, many repeated runs may be necessary and the performance may only be judged by statistical means.

2.2.1 Nondeterministic Hill-Climbing

Let's start a usual product design or refinement process. When a human cannot guide his design by a-priori directions, he conducts trial-and-error a-posteriori. A first step towards machine intelligence would be to use a computer to automate this trial-and-error-based learning process when exhaustive search is impractical. Such a process may be tedious to human beings, but suits a computer just fine.

Nondeterministic a-posteriori Hill-Climbing (HC) is a local optimization and search technique that emulates the human trial-and-error learning process. If the performance index is 'well-behaved', this algorithm performs well.

(i) Heuristic Algorithm—Its strategy is simple. Start at a random point or a point known to be good. Then try several points around it and see if the trials lead to better solutions or not. Pseudo-C/Java code for an iterated steepest ascent hill-climber is shown below.

This hill-climber is similar to 'localised' random search and is thus a little more orderly than pure global random search. Its search effort is focused around the best found so far and this makes a little use of the past search information, while pure random search disregards this information (although it keeps the best one found so far).

Note:

a. N small short-sighted (too local); difficult to improve.
b. N large random 'exhaustive' search; time too long.
c. Fancier/modified hill-climber may improve performance, but cannot solve the problem at the root.

d. Another method of making use of search history is the so-called 'taboo search', which tends to prevent new search from repeating in or going back to known poor areas.

```
public hillclimber()
{
 declaration and initialisation;
  for (i=0; i<Number_of_repetition_required; i++)
  {
   local = FALSE;

   use a known or existing Pc as current
   (or generate it at random);

   evaluate Pc by f(Pc);
    do {
        select N new designs in
        the neighborhood of Pc;

        evaluate all N new sets;

        select the design Pi that
        has the largest f;

        if ( f(Pc) <  f(Pi) )
        // This means improvements possible
          Pc = Pi;
        else
        // This means no improvements made
          local = TRUE;
        }
        while ( !local )
     // Continue if local best not found
        output Pc;
     }
  // Finished the prescribed number of trials
  return;
}
```

(ii) Monte Carlo Perturbation as Mutation—If the parameters to be optimized are in binary codes, the method used to select neighboring points in the above hill-climber can be just to flip '0' to '1' or from '1' to '0'. Such a diversifying method is also known as 'mutation' in evolutionary computing.

If the parameters are more 'continuous' (e.g., being floating-point numbers as in general engineering design cases), the selection can be achieved by using a small amount of random perturbations as found in Monte Carlo simulation. We call such perturbations 'Monte Carlo Mutation' and one approach to this is to add to the current parameter value a 'small' random number with, for example, a Gaussian distribution or a hyperbolic tangent distribution:

$$\varepsilon = 0.1T \tanh\left(random[-\pi, \pi]\right) \tag{2.1}$$

where, ε is perturbation, T can be a constant or a function of number of iterations, which can be given by:

$$\tanh = \frac{e^x - e^{-x}}{e^x + e^{-x}} = \frac{1 - e^{-2x}}{1 + e^{-2x}} = \frac{2}{1 + e^{-2x}} - 1 \tag{2.2}$$

(iii) *Multi-modal   and   A-priori   Problems   Unsolved.*   Example   6.1 [Michalewicz (1996)].

Suppose the search space is a set of binary strings P_i of length 30. Three such strings are given by:

$$P_1 = 1101 \quad 1010 \quad 1110 \quad 1011 \quad 1111 \quad 1011 \quad 0110 \quad 11 \tag{2.3}$$

$$P_2 = 1110 \quad 0010 \quad 0100 \quad 1101 \quad 1100 \quad 1010 \quad 1000 \quad 11 \tag{2.4}$$

$$P_3 = 0000 \quad 1000 \quad 0011 \quad 0010 \quad 0000 \quad 0010 \quad 0010 \quad 00 \tag{2.5}$$

Here $n = 30$ and $1 \le i \le 2^{30} = 1,073,741,824$. The objective function to be maximized is given as:

$$f(P_i) = |11 \cdot one(P_i) - 150| \tag{2.6}$$

where, the function $one(P_i)$ returns the number of 1's in the string P_i. For example, the above three strings would evaluate to:

$$f(P_1) = |11 \cdot 22 - 150| = 92 \tag{2.7}$$

$$f(P2) = |11 \cdot 15 - 150| = 15 \qquad (2.8)$$

$$f(P3) = |11 \cdot 6 - 150| = 84 \qquad (2.9)$$

Can we calculate $\nabla f(P_i)$ easily? No, even if we square f. Note that this is a high-dimensionality problem and any exhaustive search will break down, as 1 billion mini-seconds ≈ 12.4 days. However, the objective is well-behaved and we can compare, for example, $f(P_i)$ with $f(P_i + 1)$. Thus HC may be applied easily to solve this problem.

Using HC, we shall still encounter multi-modal and a-priori problems. There is one local maximum being:

$$f(P_{local}) = f(000000000000000000000000000000) = |11 \cdot 0 - 150| = 150 \qquad (2.10)$$

and one global maximum being:

$$f(P_{local}) = f(111111111111111111111111111111) = |11 \cdot 30 - 150| = 180 \qquad (2.11)$$

Note that there are also many minima being

$$f(P_{min}) = |11 \cdot 14 - 150| = 4 \qquad (2.12)$$

If search point is at P_c, where $one(Pc) = 6$ (i.e., $f(P_c) = 84$), then any attempt of a neighboring P_i towards the global maximum, i.e., increase in the number of 1's, will result in a smaller fitness. For example, increasing the number of 1's by 2 to 8, the fitness will be decreased to 62 (< 84). The hill-climber will have to reverse the search direction by decreasing the number of 1's and will thus end up with the local maximum (without knowing there exists a global optimum). The chance of finding the global maximum is thus slim and is dependent upon the initial guess (i.e., the starting point base on a-priori information must be 'good').

(iv) Summary—A local hill-climber performs well for a single modal problem but encounters the following problems:

- Difficulties with multi-modality
- Performance is dependent on the starting point (a-priori problem)
- Zero-gradient problem (i.e., where to move next if $f(P_c) = f(P_i)$?)

2.2.2 *Simulated Annealing*

Globalising HC cannot necessarily lead to the global optimum when a new trial position is on the global hill but the performance at that particular position is poorer than the current one (which is, however, on a smaller hill). A possible solution towards the global optimum is thus not to discard the inferior search point (i.e., the design parameter set) altogether. This strategy is adopted in Simulated Annealing (SA), which allows some inferior neighboring positions to replace the current one for possible correct directions leading to global optimum. This means that the selection is now also nondeterministic. The probability of replacement is determined by Boltzmann's energy distribution and this selection mechanism is termed Boltzmann selection.

The algorithm is based on the annealing/cooling process in metallurgy. In this process, a solid piece of metal is heated to a temperature at which it melts and all particles of the metal re-arrange their positions randomly. This is then followed by slowly cooling through the temperature to reach a thermal equilibrium of solid metal. In such a process, all particles arrange themselves in the low energy ground state of a corresponding lattice. If cooling is too fast, however, defects can be 'frozen' into the solid and metastable amorphous structure may be reached, instead of the low energy crystalline lattice structure.

In SA, as 'temperature' decreases, the Boltzmann distribution concentrates on the states of lowest energy and, when the temperature approaches 'zero', only the minimum energy states have a non-zero probability of occurrence. Pseudo-code for simulated annealing is shown below.

The thermal-equilibrium loop is usually unnecessary. However, including this allows examining, at every temperature level, an additional number of points (usually 10~20). This is analogous to letting the metal settle at each temperature level in order to prevent flaws occurring due to rushing the cooling process. Such additional sub-process increases the computational time but does yield finer results than the conventional SA.

In the above code, N = 1 usually. Note that neighboring points are often generated by Monte Carlo mutation with a given probability distribution, such as that of (6.1) with T being the annealing temperature. This way, the size of perturbation is statistically proportional to the value of T and thus reduces with the decreasing temperature.

```
public double simulated_annealing()
{
declaration and initialisation;
// e.g., T = 20
initialise temperature T;
select a current set Pc at random;
evaluate Pc by calculating  f(Pc);
do
 do
    select N new sets in the neighborhood of Pc;
    evaluate all N new sets;
    select the string Pi with the largest f;
    if ( random(0, 1) < exp{[ f(Pi) -  f(Pc)]/T} )
    Pc = Pi;
    while {thermal-equilibrium not reached}
    // The above means to continue if there
    is improvement
    // \alpha < 1, to tighten up control
    T = \alphaT ;
    while {stop-criterion not met}
    return Pc;
 }
```

Example 6.2 [Michalewicz (1996)]

Consider again s similar problem with another binary string:

$$P_4 = \{1110 \quad 0000 \quad 0100 \quad 1101 \quad 1100 \quad 1010 \quad 1000 \quad 00\} \qquad (2.13)$$

This string has twelve 1's and evaluates to

$$f(P_4) = |11 \cdot 12 - 150| = 18 < 84 \qquad (2.14)$$

In HC, any 'short-sighted' attempt to improve it will result in decreased 1's, leading to the local maximum of 150.

However, SA would accept an 'inferior' string. For example, moving against the 'conventional wisdom' to thirteen 1's yields poorer performance as indexed by:

$$f = |11 \cdot 13 - 150| = 7 < 18 \qquad (2.15)$$

but such an 'inferior' replacement may be accepted with a probability of

$$p = \exp\left(\frac{f(P_i) - f(P_c)}{T}\right) = \exp\left(\frac{7 - 18}{20}\right) = 0.577 > 50\% \qquad (2.16)$$

if $T = 20$. This means that this inferior (but potentially better directional) string is more likely to be accepted for the next move than to be rejected.

2.2.3 Parallelism C—An Essential Step Towards Global Search

(i) Looking Beyond Search Algorithms

- Hill-climbing and simulated annealing are nondeterministic search algorithms
- Parameter encoding is unnecessary (implications of pros and cons)
- Existence and practical problems are overcome, if CAD evaluation exists
- Simulated annealing is one step closer to global optimisation
- Are they adequate in modern engineering environment?

(ii) Do We Need High-Performance Optimization in Real-world Applications?
'Several industrial oil applications have each seen 3 million of extra profit annually from process optimization of just one unit.'

Evolutionary computing and neural networks are both adaptive learning algorithms. What is learning? It is a process of self-adjustment and optimisation to adapt to the environment and to find better solutions. Thus the central role of evolutionary computing and neural networks is adaptive optimization.

(iii) Multiple-point Search
Clearly, search may be enhanced by a 'team', instead of by an individual. This increases the chance of finding the global optimum in a multi-modal space, i.e., increases the likelihood of finding the best available solution to the problem at hand. An example is Parallel (or multi-point) Hill-Climbing (PHC) or Parallel Simulated Annealing (PSA). In the latter case, it is also sometimes termed 'Segmented Simulated Annealing (SSA)'.
Parallelism would also help with better usage of a prior experience and allow some individuals to start the search from some known points. Such search paradigms may be implemented on multiple or parallel processors directly.

2.3 The Simple Genetic Algorithm

Theoretically and empirically, GAs have been proven to perform efficient and effective search in complex spaces. Compared with traditional gradient-based search methods, GAs are more powerful as in that they do not need the problem to be reformulated to search a non-linear and non-differentiable space. Another advantage of GAs is the flexibility of the fitness function formulation, which can be expressed as a proper function of the system's output, e.g., a polynomial function. Using a GA, optimal engineering criteria can be represented by fitness functions.

In this section, we shall study the most representative evolutionary computing paradigm, the Genetic Algorithm (GA). We shall also see that GA-based evolutionary computing is capable of meeting the challenges faced by a modern CAD environment.

2.3.1 Mutation and Crossover

Based on the survival-of-the-fittest Darwinian principle in natural selection, biological reproduction and mutation, the genetic algorithm has been in development for three decades [Goldberg (1989), Holland (1992), Srinivas and Patnaik (1994)]. A GA utilizes both 'crossover' and 'mutation' operations to diversify the search. This approach has proved to be particularly effective in searching through poorly understood and irregular spaces.

Example 6.3 [Michalewicz (1996)]

Two relatively poor strings:

$$P_5 = 1111 \quad 1000 \quad | \quad 0000 \quad 1101 \quad 1100 \quad 1110 \quad 1000 \quad 00 \qquad (2.17)$$

$$P_6 = 0000 \quad 0000 \quad | \quad 0001 \quad 1011 \quad 1001 \quad 0101 \quad 1111 \quad 11 \qquad (2.18)$$

both evaluate to 16. Combination of these two by crossover produces a much better offspring string. For example, a simple crossover at the 8th position yields:

$$P_7 = 1111 \quad 1000 \quad | \quad 0001 \quad 1011 \quad 1001 \quad 0101 \quad 1111 \quad 11 \qquad (2.19)$$

with a fitness value of

$$f(P_7) = |11 \cdot 19 - 150| = 59 \qquad (2.20)$$

2.3.2 Coding, Genes and Chromosomes

Importantly, a GA conducts search and evolution not by the parameters-to-be-optimized themselves, but by some code of the parameters, known as 'genetic code' in the form of 'genes'. This way, the search information is easily manipulated and passed onto the next generation of candidate solutions. This is analogous to nature: we only inherit genetic material from our parents, not all the cells. A complete set of genes form a 'chromosome' which represents an entire parameter set under study. A conventional GA generally uses coded strings (chromosomes) of binary numbers (genes) in the search, as seen in Examples 1.1 to 1.3. See the following optional section for more details about the human DNA coding sequence.

DNA (Deoxyribo Nucleic Acid) Double-helix and Information Coding (part of this section is courtesy of Bioelectronics IV Course Notes, Dr. G.R. Moores, Institute for Biomedical and Life Sciences, University of Glasgow).

The human genome is a biological map laying out the exact sequence of the estimated 3.5 billion pairs of chemicals that make up the DNA in each human cell. Those chemicals are arranged in specific ways to create the estimated 80,000 to 100,000 human genes, which in turn carry the instructions that determine individual characteristics of each person; the colour of eyes or skin, for example. Genes, arranged in tightly coiled threads of DNA organized into pairs of chromosomes in most cells of the body, also determine whether cells will function normally or not.

Four nitrogen containing bases:

A - Adenine
C - Cytosine
G - Guanine
T - Thymine

Maximum hydrogen bonding (basis only appears in complementary base-pairing):

A-T and T-A
C-G and G-C

There are four pairs as shown above and they are represented by 4 letters in one amino acid sequence. There are 20 different amino acids.

- DNA Replication (with high fidelity information coding): Requires 4 letters to represent 20 different amino acids. Thus 1 letter cannot be used to represent only 1 acid, because this way 4 letters can only represent 4 acids. Permutation of multiple letters needed. 2 out of 4 gives $4 \times 4 = 16$ representations, but 3 out of 4 gives $4^3 = 64 > 20$. So 3-letter (3-base) words (also called codons) are used in a DNA to specify the amino acids

- Genetic code: A segment of DNA containing coded information of polypeptide chains, with 3-base coding for every amino acid
- Chromosome: A DNA molecule, contains 80,000~100,000 genes, $\sim 3.5 \times 10^9$ base-pairs long
- Cell: Contains 46 (23 pairs of) chromosomes

The GA coding mechanism is based on an analogy to the genetic code in our own DNA structure, where the genetic information is coded by using 4 bases. Compare this with binary coding used in conventional GAs and computers, which have only 2 bases. A coded parameter is similar to a 3-base word having $4^3 = 64$ different values (In a human DNA, this is actually over-specified, as only 20 out of the 64 values are used to represent amino acids). Thus a coded chromosome is composed of many such words. This analogy inspired the development and use of Base-7 coding for fuzzy logic [NgPhd (1995)], decimal coding (Base-10) and other integer coding mechanisms [Li and Ng (1996a)] for general engineering systems.

For further terminology used in GAs and evolutionary computing, refer to the following table:

Table 2.1 Genetics Terminology and Evolutionary Computing

Genetics Terminology	GA Equivalent or Related
Fitness	Fitness, performance index, objective function or (inverse) cost function to be optimised
Genotype	Coded string structure (with coding)
Phenotype	Set of parameter values (without coding)
Letter	Binary bit
Word (3 letters)	Word (2 bytes or 16 bits)
Gene	Bit or digit of a parameter, parameter, feature or character
Chromosome, individual or string	A coded or floating-point parametric set (a set of genes)
Population	A collection of search points or chromosomes
Generation	Next iteration in the artificial evolution process
Schemata or similarity templates	Building blocks
Niche	Sub-domains of functions
Allele	Feature value
Locus	String position

2.3.3 The Working Mechanism

As shown in Figure 2.1, a schematic diagram of the genetic evolution for a coded search problem and its interface with a conventional CAD package are shown below. Usually, in the process of evolution, a fixed-size population of chromosomes are updated according to the relative individual fitness that reflects the evaluated performance index.

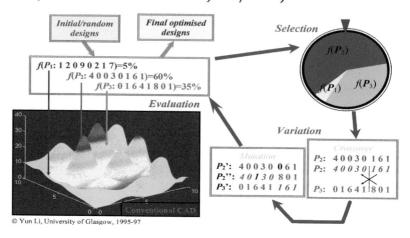

Fig. 2.1 Computer-automated design by artificial evolution

A GA-based learning cycle mainly involves three steps, namely, evaluation, selection and variation. The GA itself is run by three operators: reproduction (together with selection), crossover and mutation (both to realise variation). There exists another operator, inversion, which can be derived from crossover and mutation, and is thus not commonly used in a GA. However, this operator can sometimes be very useful in applications like travelling salesman's problem, where direct crossover and mutation are 'illegal'.

For adaptation to the evolving environment, the crossover operation exchanges information between the parental pair. Mutation changes the value of a gene, bringing new material and diversity to the population (see [Michalewicz (1996)] or [Goldberg (1989)] for detail). All these operators are applied in a probabilistic (i.e., non-deterministic) manner. More details on using GAs will be given in the following sections.

2.3.4 GAs Transform Exponential Problems to NP-Complete Problems

Solving a practical design problem in practical time would be the unique strength of CAD, but is such a problem solvable by numerical means first? Let us formalize the answer:

What is Polynomial Time?

Suppose x stands for the number of parameters that need to be determined (i.e., the size of the problem). Then polynomial time of the search (or optimization) process is described by:

$$t = O(x^M) \leq a_0 x^0 + a_1 x^1 + \cdots + a_M x^M \tag{2.21}$$

where $M < \inf$, and $a_i \in R \forall i$. If $M = \inf$, then the complexity of the problem is not polynomial, i.e., it requires exponential time given by:

$$t = O(e^x) \geq a e^x \tag{2.22}$$

Problem Classification

Before studying whether a problem is solvable, it is desirable to review problem-classification [Sedgewick and Wayne (2011)] used in computer science and algorithm engineering. This is depicted in Figure 2.2. The background area represents the set of *unsolvable problems* and the shaded areas represent *solvable problems*.

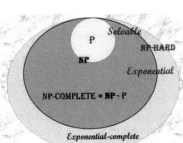

Fig. 2.2 Complexity of classification of numerical problems

The solvable problems are further classified into the following categories:

- **P** = {Problems that can be solved by a *deterministic algorithm* in *Polynomial time*}

- **NP** = {Problems that can be solved by a *Nondeterministic algorithm* in **P**olynomial time}
  = {Problems that can be solved in polynomial time}
  = **P** $\cup$ **NP-COMPLETE** $\supseteq$ **P**;

- $\overline{NP}$ = {Solvable problems that can **NOT** be solved by any means in polynomial time}
  = {Problems that can be solved in *exponential time*}

- **NP-COMPLETE** = {Problems that can be solved by a nondeterministic algorithm in polynomial time but cannot be solved by any deterministic algorithms in polynomial time}

- **NP-HARD** = {Solvable problems that cannot be solved by any deterministic algorithms in polynomial time}
  = {Solvable problems that are at least as **hard** as an **NP-complete** problem}
  = **NP-COMPLETE** $\bigcup \overline{NP}$

It is reported that the majority of science and engineering problems belong to the category of *NP*-hard problems [Sedgewick and Wayne (2011)]. A simulation (analysis and evaluation) problem is usually solvable by numerical means on a digital computer. Existing CAD packages are just developed to carry out this task.

Numerical methods used in CAD packages are conventionally based on calculus. They can perform *CAutoD*, if they are incorporated with a numerical optimization tool. Such a tool is usually based on conventional *gradient-guidance* or *a-priori techniques*. Some design problems could be solved in polynomial time by these techniques. Such a scenario does not, however, exist in practical systems.

Therefore, the advantage of CAutoD is that it will turn a CAD package from a passive simulation tool to a direct or automated design facility.

Solving a design problem by trial-and-error CAD, in exponential time

Using a CAD package based on conventional numerical techniques, a design engineer can solve a design problem by searching through or optimizing simulations. The engineer has first to input certain *a priori* system parameters, such as those obtained from some preliminary analysis and should then undertake simulations and evaluations using the package. If the simulated performance of the 'designed' system does not meet the specification, the designer would modify the values of the parameters randomly or by the engineer's real-time gained experience. The engineer would then run the simulations repeatedly until a 'satisfactory' design emerges.

Clearly, such a design technique suffers from:

- That the design process is not automated and can be tediously long
- That the design cannot be carried out easily, since mutual interactions among multiple parameters are hard to predict (i.e., *multi-dimensional problem*)
- That the resulting 'satisfactory' design does not necessarily offer the best or near-best performance (*multi-modal problem*) and room is left for further improvement
- That the manually designed system needs to be assigned an *a-priori* structure which may not best suit the application at hand (*structural problem*)

Since the analysis problem is solvable and existing CAD packages have been developed to guarantee so, one approach to achieving a solvable and possibly automated design could be to *exhaustively* evaluate in **S** all the possible design choices $\mathbf{P}_i \forall i$.

To illustrate this method, let $n = 8$ and suppose in **S** each parameter has 10 possible values. Every candidate design could then be encoded by a string of 8 decimal digits. Then there are a total of:

$$Max(i) = 10^n \qquad (2.23)$$

permutations of design choices. By *enumerating* all the digits of string one by one could span the entire quantized design space S. Now suppose that each evaluation by numerical simulation takes 0.1 second on an extremely fast computer, then the entire design process would take 0.1 second x 10^8 = 3.8 months non-stop to complete. **This is unacceptable in practice**.

In summary, such an *exhaustive/enumerative search* scheme does transform an unsolvable problem to a solvable one. However, the number of search points and thus the search time are $\overline{\text{NP}}$, i.e., increase exponentially with the number of parameters that need to be optimized (*domain size problem*). Even the highly regarded exhaustive scheme *dynamic programming* breaks down on problems of 'moderate' dimensionality and complexity [Goldberg (1989)].

Some specialized, or problem-dependent, numerical schemes work much more efficiently than the exhaustive search, but they are confined to a very narrow problem domain. One such example which is conventionally widely used is as follows:

Solving a design problem by CIAD, in polynomial time

During the past four decades, many heuristic and other Derivative-Free Optimizer (DFO) [Conn, et al. (2009)] have been developed. They have been successfully applied to a wide range of real-world problems. Coupled with a CAD package, a DFO can turn CAD into CAutoD or CIAD, thereby solving a design problem in polynomial time.

At present, there exist the following DFO algorithms:

- *Heuristics (a posteriori)*:

  - Random search
  - Bayesian optimization
  - Data-based Online Nonlinear Extremumseeker
  - Simulated annealing, a special case being heuristic hill-climbing
  - Swarm intelligence
  - Cuckoo search
  - Genetic algorithm (GA) and evolutionary computation

- *Direct Search (a priori)*:

  - Simplex search
  - Pattern search
  - Powell's conjugate search
  - Coordinate descent and adaptive coordinate descent
  - Multilevel coordinate search

Example of using a GA to train a neural network for XOR classification
The four supervising patterns for the exclusive-OR (XOR) problem are given in Table 2.2.

Table 2.2 The Four Supervising Patterns

Pattern no.	1	2	3	4
input: x1	0	1	1	0
input: x2	0	0	1	1
output: y	0	1	0	1

The fitness (performance index) function used is the inverse of sum of the four absolute errors, as given in Equation (2.24):

$$f = \left(0.001 + \sum_{k=1}^{4} |y_k - \hat{y}_k| \right) \tag{2.24}$$

Two-digit decimal encoding of the weights and thresholds, as given in Table 2.3:

Table 2.3 Two-digit Decimal Encoding of the Weights and Thresholds

Weights (phenotype)	-5.0	-4.9	-4.8	⋯	0.0	⋯	+4.7	+4.8	+4.9
Coding (genotype)	00	01	02	⋯	50	⋯	97	98	99

GA control parameters:

- population size: $s_p = 50$
- crossover rate: $p_c = 60$
- mutation rate: $p_m = 5$

The training process by a GA is characterized by decreasing learning errors through generations in evolution as shown in Figures 2.3 and 2.4.

The trained neural network is given in Figure 2.5 and the trained input/output behavior of the NN is shown in Figure 2.6.

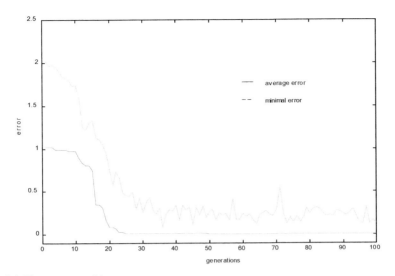

Fig. 2.3 The average and lowest errors

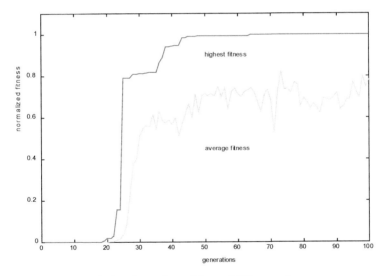

Fig. 2.4 The average and highest fitness for a trial run of 100 generations

The Schema Theory [Goldberg (1989), Holland (1992), Srinivas and Patnaik (1994)] implies that a genetic algorithm requires an exponentially reduced search time, compared with the exhaustive search that requires a total evaluation time of $O(p^n)$, where n is the number of parameters to be optimized in the search and p the number of possible choices of each parameter.

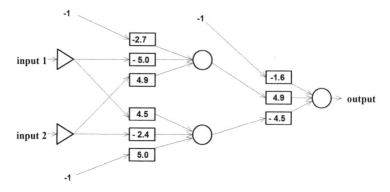

Fig. 2.5 The best learned NN with weights

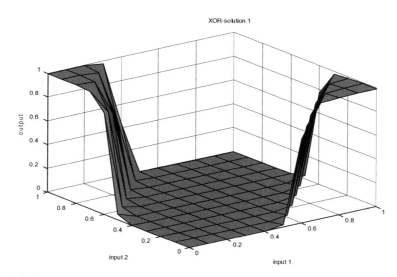

Fig. 2.6 Obtained input-output behavior

This means that an evolutionary search algorithm can be used to transform a non-polynomial (exponential) problem into an NP-complete problem. The experimental studies on an artificial problem [Smyth and Keane (1995)] have also shown that a GA, particularly when it is fine-tuned by Simplex, HC and SA [Li and Hauβler (1996), Li and Ng (1996b), Li, et al. (1996), Li, et al. (1997)], provides a much higher convergence rate and robustness than conventional optimization and search means.

Table 2.4 Typical Operators of a Genetic Algorithm

Operations	Chromosomes	Fitness
Initial population: Example of coded parameter sets forming an initial population with size 3. The performance of each parameter set is simulated and then assigned a fitness	P_1 : 12090217 P_2 : 40030161 P_3 : 01641801	$f(P_1) = 5$ $f(P_2) = 60$ $f(P_3) = 35$ (N.B. The above fitness values are examples)
Reproduction: A simple scheme is to allow the chromosomes to reproduce offspring according to their respective fitness. Thus P_1 has low probability of producing children, P_2 has a probability of producing two and P_3 one	P_2 : 40030 161 P_2 : 40030 161 P_3 : 01641 801	Evolution in progress (no need to re-calculate fitness here)
Crossover: Some portion of a pair of chromosomes is exchanged at the dotted position randomly specified	P_2' : 40030 061 P_2'' : 40130 801 P_3' : 01641 161	No fitness calculations needed here
Mutation: The decimal values of some genes of some chromosomes are varied. The value which has been changed as an example is highlighted by an underline	P_2' : 40030 **0**61 P_2'' : 40130 801 P_3' : 01641 161	A new generation is now formed and the fitness needs to be evaluated for the next cycle

2.4 Micro Genetic Algorithm

The term micro-GA (micro-GA or μGA) refers to a small-population GA with reinitialization, which can speed up the iteration procedure. This technique has been applied in numerous research fields [Coello and Pulido (2001), Lo and Khan(2004), Tam, et al. (2006), Davidyuk, et al. (2007), Andr, et al. (2009)].

The idea of the micro-GA was suggested through some theoretical results obtained by Goldberg [Goldberg (1989)], later being formally proposed by Krishnakumar [Krishnakumar (1989)]. He refers to a GA with a small internal population and reinitialization, according to which the population sizes of three were sufficient to achieve convergence, regardless of the chromosomic length. The population size used in a GA is usually in the order of tens to hundreds and sometimes thousands. With such a large number of individuals, the time needed to perform calculations of the fitness function can become formidable. It is, therefore, important to design a highly efficient GA for multi-objective optimization problems. A popular method in this direction is the micro-GA, which has a very small internal population (3 to 6 individuals).

Generally, as shown in Figure 2.7, there are two cycles in the micro-GA work flow, internal and external cycles, in which there are two groups of population mem-

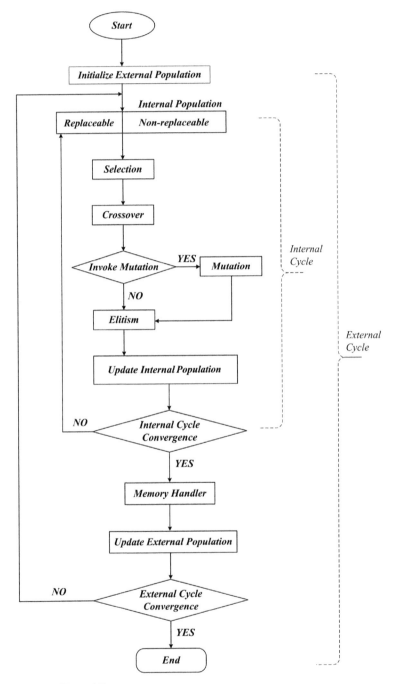

Fig. 2.7 Micro-GA workflow

ories: the internal population (P_I) and the external population (P_E). P_I is used as a source of diversity in the micro-GA internal cycle, and P_E is used to archive individuals of the Pareto optimal set. The 'internal cycle convergence' condition in this paper is the 'max internal generation' of the internal cycle.

As stated in Equation (2.25), the internal population P_I, which is used as the source of diversity in the internal loop, is composed of two sub-sections: the replaceable section P_{IR} and the non-replaceable section P_{IN}. Equation (2.26) defines the size of P_{IR} at generation t, which is calculated using the size of P_{IR} at generation $t-1$ plus ('+') the off-size of the replaceable section ΔP_{IR} at generation $t-1$, that is, the current generation P_{IR} decided by the last generation P_{IR} and ΔP_{IR}.

$$P_I(t) = P_{IR}(t) + P_{IN}(t) \tag{2.25}$$
$$P_{IR}(t) = P_{IR}(t-1) + \Delta P_{IR}(t-1) \tag{2.26}$$

The off-size replaceable population is proportional to the size of the original internal population P_I, as stated in Equation (2.27), where λ is a reproduction rate set by the user.

$$\Delta P_{IR} = \lambda P_I \tag{2.27}$$

In engineering applications, it is important to design a highly efficient GA for solving multi-objective optimization problems. A popular method in this direction is the micro-GA with a very small internal population ($3 \rightarrow 6$ individuals). Usually, P_I is divided into two parts, a replaceable part and a non-replaceable part, and the percentages of each part can be adjusted by the user. Because of the small internal population, mutation is an optional operator in micro-GAs, and Pareto-ranking methods will be taken into the elitism definition process inside the internal cycle, which may include Goldberg's method [Goldberg (1989)] or the Fonseca and Fleming's method [Fonseca and Fleming (1993)].

Basically, there are three parts to micro-GA optimization: fitness function definition, encoding and decoding definition and genetic operator definition, which includes selection, crossover and mutation (optional for μGA). Once the three parts have been well defined, the micro-GA can then create a population of solutions and apply genetic operators, such as mutation and crossover, to find acceptable results.

In this proposed method, there is also an 'adaptive block', which contains three sub-blocks: 'update population statistics', 'range adaptation' and 'elitist-random reinitialization'. In the evolutionary workflow, the VPμGA creates a population of solutions and applies genetic operators, such as selection and crossover, to find a better result that satisfies objective error limits.

2.5 Genetic Algorithm Using Mendel's Principles

Mendel's principles (some biologists refer to Mendel's 'principles' as 'laws') are always applied to genetically reproducing creatures in a natural environment.

Mendel's conclusions from his pea-plant experiments can be summarized through two principles [Mendel (1865), O'Neil (2009), Marks (2008)]:

- The principle of segregation: For any particular trait, the pair of alleles of each parent separate and only one allele passes from each parent to an offspring
- The principle of independent assortment: Different pairs of alleles are independently passed to their offspring

Since ancient time, farmers and herders have been selectively breeding their plants and animals to produce more useful hybrids. Knowledge of the genetic mechanisms finally came about as a result of careful laboratory breeding experiments conducted and published by Mendel in the 1860s [Mendel (1865), O'Neil (2009)]. Through selective cross-breeding of common pea plants over many generations, Mendel discovered that certain traits show up in offspring without any blending of parental characteristics. For example, pea flowers are either purple or white; intermediate colors do not appear in the offspring of cross-pollinated pea plants. Mendel's research only refers to plants; however, the underlying principles of heredity also apply to humans and animals because the mechanisms of heredity are essentially the same for all complex life forms on this planet. For example, Figure 2.8 shows daffodils with different hybrid characteristics of color (white or whitish, green, yellow, pink, orange, etc.), shape (flat, slim, wrinkled, etc.) and size.

According to Mendel's experiments on pea plant hybridization, the same hybrid forms (e.g., color, shape and size) always reappear whenever fertilization occurs among the same species. For example, in pea-plant experiments, the pea plants with dominant hybrid characteristics (e.g., green and smooth) present a better ability to increase their population compared to pea plants with recessive characteristics (e.g., yellow or wrinkled). By assigning a dominance factor to each gene of a specific chromosome and conducting genetic processes, similar to the pea-plant experiment, the chromosomes are guided by Mendel's principles implicitly in the GA process.

Fig. 2.8 Daffodils with different hybrid characteristics

In recent years, numerous studies have been conducted in the area of GAs using Mendel's principles [Mendel (1865), O'Neil (2009), Song, et al. (1999), Park, et al. (2001), Kadrovach, et al. (2001), Kadrovach, et al. (2002)]. In this section, a new GA method using Mendel's principles (Mendel-GA) is proposed, which includes the following differences to the standard GA method and to previous research:

(1) In this chapter, the Mendel operator is inserted after the selection operator, thus allowing the Mendel operator's local search ability to be utilized, as shown in Figure 2.9. In previous studies, the Mendel operator was inserted into the GA process after the mutation operator. Based on the mutation probability P_m, mutation may generate an unstable population, which will lead to polluted outputs for the whole evolutionary processes, biologically and mathematically [Ishibuchi, et al. (2002), Furió, et al. (2005)]. Typically, the mutation operator is a randomly introduced changing of a binary bit from '0' to '1' and vice versa. The basic method of mutation is able to generate new recombinations of improved solutions at a given rate; however, the possibility of damage to the dominant population, a loss of good solutions and convergence trends also exist [Zitzler (1999), the National Academy of Sciences (2004), Haupt and Haupt (2004)]. The Mendel operator will amplify such an unstable population with its local search ability from a micro evolutionary point of view.

(2) Mendel's principles are represented by the Mendel operator, which is easily synchronized with multi-objective GA processes, such as the multiple-objective genetic algorithm (MOGA) [Fonseca and Fleming (1993)], the niched Pareto genetic algorithm (NPGA) [NPGA, et al. (1994)], the non-dominated sorting genetic algorithm (NSGA) [Srinivas and Deb (1994)] and the non-dominated sorting genetic algorithm II (NSGAII) [Deb, et al. (2002)].

(3) The standard GA is based on Darwin's theory, which is represented by differential survival and reproductive success; in the Mendel-GA, Mendel's law is represented by the equal gametes, which unite at random to form equal zygotes and reproduce equal plants throughout all stages of the life cycle.

Figure 2.9 shows Mendel's principles embedded in the GA's overall evolutionary framework, which can accurately describe the heredity meiosis process from a micro evolution point of view in a GA process. A GA using Mendel's principles can be defined as a Mendel genetic operator introduced into the basic GA.

In this chapter, the binary encoding method is utilized for the Mendel-GA's encoding/decoding operation. Typically, all bits (genes) of a chromosome can be encoded using the two binary alphabets '1' and '0', in which '1' represents that a gene is active and '0' represents that it is inactive.

As shown in Figure 2.10, '0' and '1' are the two bit values of the binary encoded chromosome and the chromosome length is called Mendel's percentage (*MP*), which is a proportionality factor included to balance the bit length of the parental chromosome. Clearly, $MP = 0$ means that no bit is involved in the current generation of evolution and $MP = 1$ means that all bits of the parental chromosome are involved in the evolution, $MP \in [0,1]$ or $[0,100\%]$.

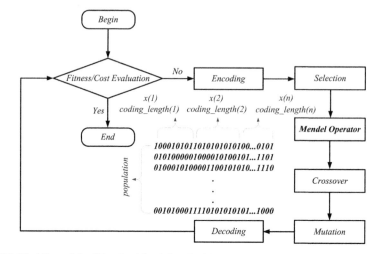

Fig. 2.9 Workflow of the GA using Mendel's principles

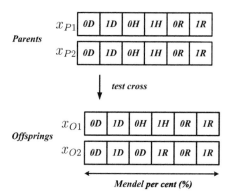

Fig. 2.10 Mendel operator: parents to offspring

The Mendel operator generates the offspring's chromosomes (x_{O_i}) from the parent's chromosomes (x_{P_i}) guided by every bit's attribute, which is described by the Punnet square. Two parental chromosomes x_{P_1} and x_{P_2} can generate one child chromosome x_{O_1}, as shown in Figure 2.10. To obtain the second child chromosome x_{O_2}, the productive process needs to be repeated.

As stated in Table 2.5, in the Punnet square for the Mendel operator, each chromosome bit is assigned an attribute, which indicates the gene's corresponding character. There are three types of attributes: D (Dominant, the pure and dominant gene), R (Recessive, the pure and recessive gene), and H (Hybrid, the hybrid gene). $attrP_1$ and $attrP_2$ are the attributes of a bit of the parental chromosomes x_{P_1} and x_{P_2}, and $attrO$ is the attribute of a bit of a child's chromosome x_{O_1} or x_{O_2}.

Table 2.5 Mendel Punnet Square for 1-bit Chromosome

		$attrP_1$		
$attrO$		D	H	R
$attrP_2$	D	100%D	50%D 50%H	100%H
	H	50%D 50%H	25%D 50%H 25%R	50%R 50%H
	R	100%H	50%R 50%H	100%R

From Table 2.5, it is clear that

$\langle 1 \rangle$ IF $attrP_1 = D$ AND
IF $attrP_2 = D$, THEN $attrO = 100\%D$;
IF $attrP_2 = H$, THEN $attrO = 50\%D$ or $50\%H$;
IF $attrP_2 = R$, THEN $attrO = 100\%H$;

$\langle 2 \rangle$ IF $attrP_1 = H$ AND
IF $attrP_2 = D$, THEN $attrO = 50\%D$ or $50\%H$;
IF $attrP_2 = H$, THEN $attrO = 25\%D$ or $50\%H$ or $25\%R$;
IF $attrP_2 = R$, THEN $attrO = 50\%R$ or $50\%H$;

$\langle 3 \rangle$ IF $attrP_1 = R$ AND
IF $attrP_2 = D$, THEN $attrO = 100\%H$;
IF $attrP_2 = H$, THEN $attrO = 50\%R$ or $50\%H$;
IF $attrP_2 = R$, THEN $attrO = 100\%R$;

According to Mendel's principles, hybridization is the process of combining different attributes of all bits of the chromosomes to create the offspring bits with Punnet-square-driven attributes. Compared with a crossover operator or a mutation operator, the Mendel operator is performed with the Punnet-square attributes.

Problems

1. Workflow of Micro-GA with variable population (Figure 2.11)
2. Unconstrained optimization problem

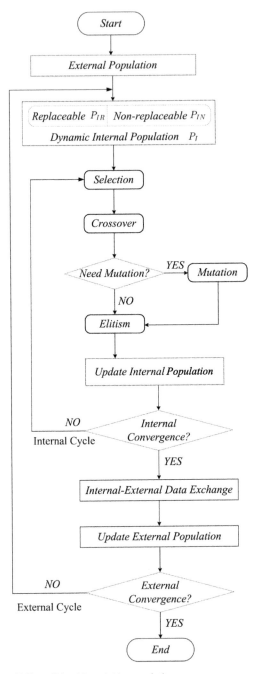

Fig. 2.11 Workflow of Micro-GA with variable population

$$\begin{cases} Max: f(x_1, x_2) = 21.5 + x_1 \sin(4\pi x_1) + x_2 \sin(20\pi x_2) \\ s.t. \quad -3.0 \le x_1 \le 12.1 \\ \qquad 4.1 \le x_2 \le 5.8 \end{cases} \tag{2.28}$$

2.6 Characteristics of Evolutionary Design Tools

2.6.1 Advantages of Evolutionary Techniques

Compared with natural evolution, the emulated process is more efficient, controllable and yet more flexible for artificial optimization. All these methods are probabilistic in nature and exhibit global search capabilities, thus making them attractive for almost all areas of human activity. Evolutionary algorithms accommodate all the facets of soft computing and other attractive features, namely,

- Overcoming all drawbacks of conventional optimisation techniques
- Domain-independent operation
- Robustness
- Nonlinearity dealt with
- Imprecision and uncertainty dealt with
- Adaptive capabilities
- Providing multiple optimal solutions
- Inherent parallelism

2.6.2 Preparation and Conditions of Use

- The system must be analyzable or may be simulated, i.e., the performance of candidate designs can be evaluated
- Selection of an index that can gauge the performance of candidate solutions against specification in an application. The values of the index should provide more information than a simple True-or-False answer, as otherwise reproduction will be completely polarized and the EA will become a purely random search
- Selection of the structure of candidate solutions for HC, SA, EP, ES and GA (encoding needed) or anticipated structures for mGA encoding or building blocks and terminals for GP encoding

2.7 Tutorials and Coursework

1. A 'performance index' in an application of the integer variable x is given by:

$$f(x) = |4 \cdot one(x) - dec(x)| \qquad (2.29)$$

where, the function $one(x)$ returns the number of 1s in the unsigned binary code of x and $dec(x)$ the decimal value of x. Find the integer or integers x_0 such that

$$f(x_0) = \sup_{\forall x \in [0,15]} f(x) \qquad (2.30)$$

and determine f(x0). How many global minima are there? Reason for the method that you use. Are there analytical or quicker numerical solutions?

2. Briefly describe the problem of design automation of engineering systems, assuming that a given engineering system can be analyzed by numerical simulations. Is this practical design problem solvable by deterministic numerical means under such an assumption and, if so, in what order of time would it be solved? What are NP-hard computing problems? In your opinion, why are the majority of engineering design problems widely recognized as NP-hard problems?

3. What difficulties encountered in the calculus-based optimization techniques are overcome by hillclimbing and simulated annealing techniques? Are they global search methods? How can global search be better achieved? By what means can the tractability of practical design problems be largely improved by slightly trading off accuracy?

4. Suppose the problem of Q4 is to be solved by a genetic algorithm with a search resolution of $\Delta x = 0.001$. What is the minimum number of bits that are needed in binary genotype coding? Obtain the codes for $x = 0.875$ and $x = -0.88375$. Without detailed calculations, give the corresponding decimal phenotype value of the code {00 1000 0000 0000}.

5. Estimate by graphical means the supremum of the function,

$$f(x) = 10(x-1)^4 \sin(10\pi x) \qquad (2.31)$$

for the domain interval [-1, 2] and also estimate the value of $x \in [-1, 2]$ at which the global peak occurs. Are there any end analytical solutions? Are there numerical solutions? Draw this 'amplitude-modulated' waveform and discuss whether the value of your estimated x should be increased or decreased in order to further approach the exact global peak?

2.8 Summary

In this chapter, we discussed three examples of simple evolutionary algorithms. There exist many such algorithms and handbooks for them. In order to get a bird's eye view, we outline a number of them here and use the summary as an example to guide the user as when to use which type of EC algorithms and as shown in Figure 2.12.

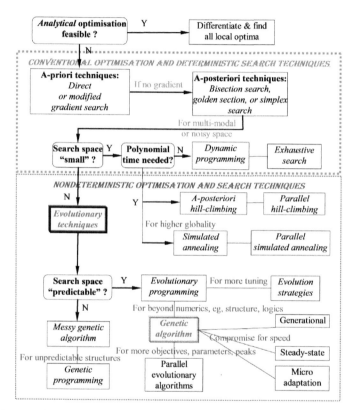

Fig. 2.12 Use and abuse of DFO and search techniques

In a nutshell:

- **Hill Climbing (HC)**
  = Trial-and-error based a posteriori selection and 'guidance'

- **Simulated Annealing (SA)**
  = HC + Boltzmann selection

- **Evolutionary Programming (EP)**
  = Multi-point search (phenotype) with Monte Carlo mutation + stochastic

tournament selection
$\approx$ SSA (PSA)
$\approx$ MHC (PHC)
$\approx$ Phenotype GA without crossover

- **Evolution Strategy (ES)**
  = Multi-point search (phenotype) with Monte Carlo mutation + recombination (in modern ES) + deterministic selection of best points + adaptive 'chromosome' variances and covariances (as strategic parameters of internal model controlling mutation and evolution process)
  $\approx$ MHC + recombination (in modern ES)
  $\approx$ SSA + recombination (in modern ES)
  $\approx$ Phenotype GA (with uniform crossover)

- **Genetic Algorithm (GA)**
  = Multi-point search with coded genotype + stochastic selection + mutation + crossover
  $\approx$ EP + coding
  $\approx$ ES - internal model

- **Classifier System (CS)**
  = GA for machine learning
  $\approx$ GA + NN (+ FS)

In general, preparations are necessary before using the chosen EC technique and conditions for use that should be observed, include:

- If the problem can be simulated including via a CAD package, then the performance of candidate designs can be evaluated and fed back to form the next potential better generation of designs, as shown in the CIAD framework
- Selection of an index that can gauge the performance of candidate solutions against specification in an application. The values of the index should provide more information than a simple True-or-False answer, as otherwise the reproduction will be completely polarized and the EA will become a purely random search
- Selection of the structure of candidate solutions for HC, SA, EP, ES and GA (encoding needed) or anticipated structures for mGA encoding or building blocks and terminals for GP encoding
- A termination condition (e.g., no further meaningful improvement or certain time limit is reached).

3

Artificial Neural Networks and Learning Systems

3.1 Human Brain and Artificial Neural Networks

3.1.1 Central Nervous System and Conventional Computer

Figures 3.1 and 3.2, give the comparison in features of the human brain and the conventional computer

- Electrical signals are transmitted and received by the neural cell body
- Signal is transmitted through the axon which acts as a distribution transmission line
- Relayed by synapse, the terminal bouton generates a chemical that affects the relay
- The chemical controls action and the amount of reception at dendrites of the receiving neuron
- The receiving neuron can respond to the total inputs aggregated within a short time interval (period of latent summation)
- If the total potential of its membrane reaches a certain threshold level ($\sim$40 mV) during the period of latent summation, an electrical response is generated by the neural body
- The generated response signal is of limited amplitude but usually saturated (i.e., very binary)
- The signal is then asynchronously transmitted to the next neurons through its axon fibers with directions
- After carrying a pulse, an axon fiber becomes unexcitable for an asynchronous refractory period ($\sim$ms)

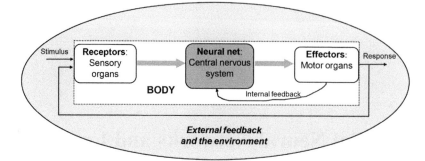

Fig. 3.1 Information flow in the human brain nervous system and picture of a biological neural cell

Fig. 3.2 Nervous impulse train (similar to that seen in ElectroEncephaloGrams or ECGs)

Table 3.1 Comparison of Features Between Human Brain and Conventional Computer

Features	Human Brain	Conventional Computer
Computing elements	10^{10} - 10^{11} neurons (PEs)	1 CPU
Computing means	distributed with central pattern	1 CPU with separated memory
Communication links	10^4 synapses per neuron 10^{10}	1 memo access per bus $\times$ 1 CPU
Communication means	asynchronous data-flow/wavefront	Synchronous clock
Pulse/clock/switch rate	10^3 /s	10^9 /s
Signal strength	$\sim$mV	$\sim$volts
Efficiency (J/s)	10^{-16} per operation	10^{-6} per operation
Sizes	$\sim$ 1 m circuits, $\sim$ 100 m neuron	$\sim$ 0.1 ?m tech, $\sim$ 1 cm CPU chip
Operations	Complex, nonlinear, parallel	ALU, linear, serial
Perceptual recognition	$100 \sim 200$ ms	hours $\sim$ days
How?	Learning by experience	By instruction series/programs

3.1.2 'Reproduce' the Human Brain by Artificial Neural Networks

A model of the neuron by a perceptron and artificial neural network (Figures 3.3 and 3.4) reflects these observations. Briefly, the major functions, are namely:

- weighting (i.e., memory) and weighted inputs
- summation (i.e., information gathering)
- activation (usually nonlinear with a threshold)

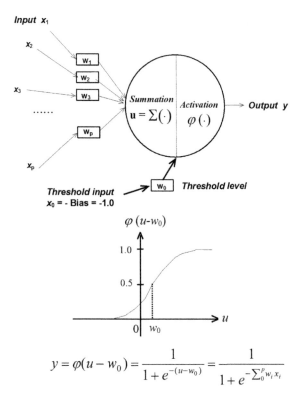

$$y = \varphi(u - w_0) = \frac{1}{1 + e^{-(u-w_0)}} = \frac{1}{1 + e^{-\sum_0^p w_i x_i}}$$

Fig. 3.3 A perceptron (an artificial neuron with summated inputs going through a, usually, nonlinear transfer)

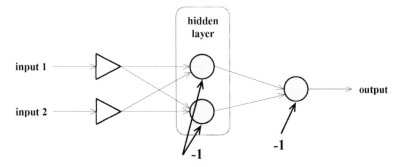

Fig. 3.4 An ANN, or simply, a neural network (NN)

Note that, however, the latent summation period is NOT included in the perceptron model, but may be included in the network (as a temporal network; details later).

Implemented by using either:

- electronic components (emulating Figure 3.3, Kirchhoff's circuit laws (KCL) + transistor)
- simulated in software on one digital computer
- parallel processing elements

In summary, an ANN is a machine that is designed to artificially emulate the way in which the human brain performs a particular task or function of interest, as given in Figure 3.4.

Definition of an ANN: A neural network is a massively parallel distributed processor that has a natural propensity for storing experimental knowledge and making it available for use. It resembles the human brain in three aspects:

- Knowledge is acquired through a learning process
- Inter-neuron connection strengths (known as synaptic weights) are used to store knowledge
- Information and knowledge are processed in a parallel manner

3.1.3 Mathematical Models and Types of ANNs

1. The Perceptron
   A mathematical model represents the descriptions of the three functions: weighting, summation and activation. Refer to Figure 3.3 for the functions and symbols.

   (a) Weighted Summation: Let subscript k represent the k-th neuron in the network. Suppose there are p inputs associated with this perceptron, then the weighted sum is:

$$u_k = \sum_{j=1}^{p} w_{kj} x_j \tag{3.1}$$

   Note that the threshold level of this neuron is represented by $\theta = w_{k0}$ and the threshold input by $x_0 = -1$.

   (b) Sigmoid Activation: The total sum going through the activation function is:

$$v_k = u_k - \theta_k = \sum_{j=1}^{p} w_{kj} x_j \tag{3.2}$$

   The most commonly used activation function is the sigmoid function. The unipolar sigmoid function is termed a logistic function, as given by:

$$\varphi(v) = \frac{1}{1 + e^{-av}}, \in (0, 1) \tag{3.3}$$

   where a is called the slope parameter.

> Q: What is the actual slope of this function
> at the origin?
> A: a/4.

Usually, the slope parameters of all neurons in a network are uniformly set as a = 1.

> Q: Is the scaling necessary if the weights
> can be trained?
> A: No.

The output of the neuron is thus given by:

$$y_k = \varphi(u_k - \theta_k) = \frac{1}{1 + \exp\left(-\sum_0^p w_{kj}x_j\right)} \tag{3.4}$$

Note that this sigmoid function yields only unipolar (positive) output signals. If the weights have only positive values, the inputs to the next neurons can only be positive. In many cases, this may be undesired and thus bipolar weights are used.

If only positive weights are allowed in an implementation, a bipolar sigmoid function must be used to generate bipolar input signals to the next neurons. This sigmoid is realized by the hyperbolic tangent function as given by:

$$\tanh(v) = \frac{\exp(av) - \exp(-av)}{\exp(av) + \exp(-av)} = \frac{1 - \exp(-2av)}{1 + \exp(-2av)} = \frac{2}{1 + \exp(-2av)} - 1, \in (-1, 1) \tag{3.5}$$

2. Other Types of Activation Functions

a. Threshold Function: For $a = \infty$, the logistic function becomes a threshold function:

$$\varphi(v) = \begin{cases} 1, v > 0 \\ 0, v < 0 \end{cases} \tag{3.6}$$

as shown in Figure 3.5.

b. Signum Function: For $a = \infty$, the hyperbolic tangent function becomes a signum function:

$$\varphi(v) = \begin{cases} 1, v > 0 \\ 0, v = 0 \\ -1, v < 0 \end{cases} \tag{3.7}$$

as shown in Figure 3.6.

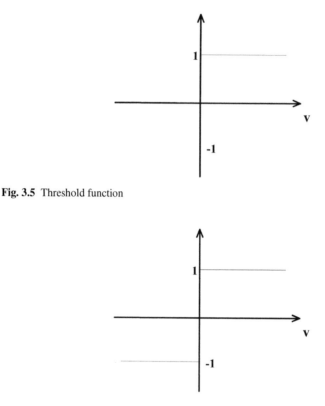

Fig. 3.5 Threshold function

Fig. 3.6 Signum function

c. Piecewise-Linear Function: A piecewise-linear function is a linear approximation of the nonlinear tanh function. This is shown in Figure 3.7 and is given
by:

$$\varphi(v) = \begin{cases} 1, v \geq 1 \\ v, -1 < v < 1 \\ -1, v \leq -1 \end{cases} \qquad (3.8)$$

Note that these types of hard activation functions have singular points in
their derivatives and thus may create difficulties in calculus or gradient-based
training.

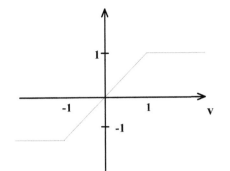

Fig. 3.7 Piecewise-linear function

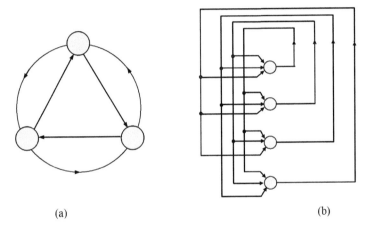

(a)   (b)

Fig. 3.8 Hopfield networks completely connected with both feedforward and feedback

3. Three Major Types of Architectures

(a) Hopfield Networks: A three-neuron architecture is shown in Figure 3.8(a) and a four-neuron network is rearranged in Figure 3.8(b).
  - Such an ANN is viewed as a complete digraph, i.e., every neuron may be connected to every other neuron in both directions
  - This means both feedforward and feedback information paths may exist in the network
  - These networks are sometime also called 'recurrent networks' because of feedback
  - Difficult to train and at least, more time is needed

(b) Layered Feedforward Networks: A two-layered architecture of such a network
    is shown in Figure 3.4.
    - Partitioned complete digraphs. Neurons are partitioned into a number of
      subgroups and every member of one subgroup may be completely con-
      nected to all the neurons in the next subgroup in one direction (i.e., no
      structural feedback)
    - Feedback may only be realized through training
    - The training (by, e.g., backpropagation) is relatively mathematically easy
    - This type of architectures are most widely used

(c) Layered Feedback Network
    - Circuit realization of a perceptron (nodal equation with KCL)
    - Realizing neurons in a digital computer requires time. Synchronisation is
      implemented by clocked sampling
    - This means the output of a neuron can only be observed when sampling
      period is behind the input, which is very harmful when feedback is involved
    - Thus a more accurate model that reflects such dynamics needs to include
      this delay, shown in Figure 3.9. Here the unit-delay operator z-1 represents
      a latched storage element
    - (Temporal/real-time) recurrent (lattice) networks

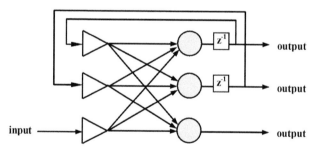

Fig. 3.9 A temporal feedback network

4. Important Features and Capabilities of ANNs

   ANNs are developed based on the hope that the flexibility and power of
   the brain can be reproduced by artificial means. The use of ANNs offers, or is
   expected to offer, the following capabilities:

   - Memory and knowledge representation through weights
   - Information gathering through summation and threshold
   - Nonlinearity in decision making through activation
   - Learning capability through training, with positive examples and/or negative
     examples

- Universal model by direct I/O mapping though the 'black-box'. No physical insight or other types of mathematical equations of the system being modelled are needed, for instance, pattern recognition; dynamic behavior of a car
- Integrated control where both model and controller are learnt by the same ANN. The design of a neural decision maker is based directly on real data and no prior model building is necessary, for instance, driving a car (fuzzy logic)
- Adaptivity in on-line learning. Integrated adaptive (model-reference and self-tuning) control, for instance, driving another car
- Evidential response means not only make response to an input but also knows how confident the decision is made. Can incorporate fuzzy logic easily
- Contextual information in which a neuron interacts with other neurons, i.e., its knowledge is embedded within the context of the network. The structure represents the global knowledge
- Parallelism with integrated design and implementation
- Uniformity and modularity for analysis and design
- VLSI implementability, because of modularity and locality
- Fault tolerance or 'graceful degradation' when fault
- Neurobiological analogy when engineers look to neurobiology for new ideas to solve real-world problems; Neurobiologists look to artificial neural networks for interpretation of neurobiological phenomena and a better understanding of their dynamics

5. Deep Learning

Recently, with proliferating availability of parallel and high-speed computing hardware, deep learning becomes feasible and has attracted increasing attention throughout the computational and artificial intelligence community. In a nutshell, deep learning involves an extended number of hidden layers of the Layered Feedforward Networks and offers a step upgrade of the important features and capabilities of ANNs listed above, except feature g-adaptivity, because of its demanding real-time speed.

With a deep cascade of multiple layers, a feedforward network structured for back-propagation provides hierarchical learning powerful for pattern analysis, feature extraction and transformation, where higher level features are derived from lower level features, corresponding to different levels of abstraction, for example. Learning can be driven by a large volume of data, which can be of high velocity and high variety, i.e., big data.

3.2 ANN Design and Learning

3.2.1 Three Steps in ANN Design

- Select an appropriate architecture or topology, with explicit input and output nodes
- Gain correct network functions through learning or training weight parameters
- Validate the trained network in a generalisation process by testing its recognition performance with data that have never been seen before

Note that all the three types of ANNs can be trained easily by an evolutionary algorithm, though training feedback networks may be difficult by conventional means.

Many engineering problems can be described by physical laws in a differential equation, such as

$$y = \frac{dx_1}{dt} + ax_2^2 \qquad (3.9)$$

Often an NN tends to approximate such a real-world problem universally. The vital weakness of an NN is thus its lack of direct mapping from the differential equation. This weakness poses a real design challenge to the development of a satisfactory solution by means of neural computing.

3.2.2 Knowledge Representation

Knowledge consists of two types of information:

- Prior information (a priori knowledge)
- Observations (measurements)

Noisy data, e.g., the optical character reader (OCR) problem, particularly for handwriting

General common sense rules for knowledge representation [Haykin (1994)]:

- Rule 1: Similar inputs from similar classes should usually produce similar representations inside the network and should therefore be classified as belonging to the same category
- Rule 2: Items to be categorized as separate classes should be given widely different representations in the network. (This rule is the exact opposite of Rule 1.)
- Rule 3: If a particular feature is important, then there should be a large number of neurons involved in the representation of that item in the network

- Rule 4: Prior information on the network and invariances should be built into the design of a neural network, thereby simplifying the network design by not having to learn them, for instance, a specialized/restricted structure, with a smaller number of independent (necessary connecting) parameters to train and thus a lower cost requires a small data set for training, learns faster and often generalizes better

3.2.3 Learning Process

Consider Perceptron k with p inputs and a threshold. Use the following symbols for variables and parameters at learning step n:

Threshold: $\theta_k(n)$
Inputs: $x_k(n) = [x_{k0}(n), x_{k1}(n), ..., x_{kp}(n)]^T = [-1, x_{k1}(n), ..., x_{kp}(n)]^T$
Weights: $w_k(n) = [w_{k0}(n), w_{k1}(n), ..., w_{kp}(n)]^T = [w_{k0}(n), w_{k1}(n), ..., w_{kp}(n)]^T$
Output: $y_k(n)$
Desired response: $d_k(n)$
Learning rate: $\eta(n) \in (0, 1)$

- Step 1: Initialization: Set $w_k(n)$ by Equation (3.10). Then perform the following computations for time $n = 1, 2, ...$

$$w_k(n) = [random(0, 1)] \tag{3.10}$$

- Step 2: Activation: At time n, activate the perceptron by applying continuous-valued input vector $x_k(n)$ and desired response $d_k(n)$
- Step 3: Computation of actual response: For the perceptron under training, compute:

$$y_k(n) = \varphi[w_k^T(n)x_k(n)] \tag{3.11}$$

- Step 4: Learning: Update the weight vector:

$$w_k(n+1) = w_k(n) + \Delta w_k(x_k, y_k, d_k, w_k, n) \tag{3.12}$$

Here the correction or learning is dependent upon the inputs, output, desired response, current weights and time. The amount of correction is also determined by a learning rate parameter that may be time-varying; for instance, moving towards the minimum error energy position. Thus the learning rate may have to decrease with time (i.e., coarse-learning at beginning and gradually change to fine-learning). This parameter is also used to ensure stability of the error adjustment process

- Step 5: More learning: To reach the steady-state of the learning curve, continue the learning process by incrementing time n by one unit and going back to Step 2

Clearly, the biggest question that we have not answered is, how do we decide the amount (as well as direction) of adjustment Δw_k?

3.2.4 Learning Methods

In a computer, these are realized by learning algorithms (small step-by-step rules) or paradigms (idealised conceptual models of software which dictate the approach taken by a programmer) (Recall the definition of programs). As can be inferred, there exist many learning algorithms that differ from one another in the way that decides the adjustment Δw_k. So they are not unique, each having its own advantages and disadvantages. The most representative algorithms are listed below:

- Hebbian learning
- Error-correction learning (most popularly realised in computers, such as back-propagation)
- Thorndike's law of effect
- Boltzmann learning
- Competitive learning
- Darwinian selective learning (such as that based on evolutionary algorithms)

3.2.5 Learning Paradigms

1. Supervised Learning: As shown in Figure 3.10, this is depicted in Figure 3.1. The central theme of this training paradigm is the external teacher, as in model reference control. Here the teacher has the knowledge of the environment that is represented by a set of input-output examples. The environment is, however, unknown to the neural network.

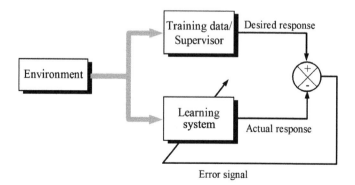

Fig. 3.10 Supervised learning (based on a reference model or training data set)

Supervised learning can be performed in an off-line or on-line manner. In the off-line case, a separate computational facility is used to design the supervised learning system. Once the desired performance is accomplished, the design is 'frozen', which means that the neural network operates in a static manner. On the other hand, in on-line learning the learning procedure is implemented solely within the system itself, not requiring a separate computational facility. In other words, learning is accomplished in real time, with the result that the neural network is dynamic.

The advantage of this learning paradigm is instructive feedback: when errors are found, they are to be corrected. Thus, one algorithm realizing this is the error-correction algorithm based on the gradient of the error-surface, such as the least mean squares based on backpropagation algorithm. A disadvantage of supervised learning, regardless of whether it is performed off-line or on-line, is the fact that without a teacher, a neural network cannot learn new strategies for particular situations as they are not covered by the set of examples used to train the network. This limitation may be overcome by the use of reinforcement learning.

2. Unsupervised Learning: In unsupervised learning (or self-organized learning), there is no external teacher. Thus, to guide the learning, self-assessment of performance is needed. The purpose of such learning is to discover significant patterns or features in the input data and to do the discovery without a teacher. The network usually starts from a weight set of locally optimal values and then self-organizes to form a globally optimal system. This also includes architectural learning, usually realized by means of growing the global network from the minimal network of only input and output neurons with local connections, hoping that the 'global order can arise from local interactions' [Turing (1952)].

The unsupervised learning combines features in supervised and reinforcement learning in a feedback environment. Note that in order to reinforce learning, the corresponding feedback must be positive. In order to correct an error, the corresponding feedback must be negative. Generally, there are three basic principles of self-organization [Haykin (1994)]:

- Modifications in synaptic weights tend to self-amplify
- Limitation of resources leads to competition among synapses and therefore, selection of the most vigorously growing synapses (i.e., the fittest) at the expense of others
- Modifications in synaptic weights tend to co-operate

There are mainly four types of such self-organizing neural networks (SONNs):

- Hebbian learning SONNs
- Competitive learning SONNs through feature mapping, where the locations of neurons correspond to intrinsic features of the input patterns
- Information theory-based SONN models, where the synaptic weights develop in such a way that the preserved amount of information is maximized. Here, according to Shannons information theory, the amount of information on an event is determined by the probability at which the event occurs, as defined by:

$$I(x_k) = \log \left[\frac{1}{P(x = x_k)} \right] = -\log [P(x = x_k)] \qquad (3.13)$$

whose units are nats, when the natural logarithm is used, and bits, when base 2 logarithm is used. The mathematical expectation, or the mean value of $I(x_k)$ is called entropy

- Darwin machine with cellular encoded GP or evolvable hardware

3. Reinforcement Learning: Reinforcement learning is on-line learning of an input-output mapping through a process of trial and error designed to maximize a scalar performance index called a reinforcement signal [Minsky (1961), Waltz and Fu (1965)]. However, the basic idea of 'reinforcement' has its origins in experimental studies of animal learning as stated in Thorndike's classical law of effect [Thorndike (1898), Thorndike (1911), Barto (1992), Sutton and Barto (1998)]: 'If an action taken by a learning system is followed by a satisfactory state of affairs, then the tendency of the system to produce that particular action is strengthened or reinforced. Otherwise, the tendency of the system to produce that action is weakened.'
Note that reinforcement learning can also be realized by 'survival-of-the-fittest' Darwinian selection.

Reinforcement learning addresses the problem of improving performance in a trial and error manner and thus learning on the basis of any measure whose values can be supplied to the system. Therefore, a reinforcement learning system may be viewed as an evaluative feedback system [Haykin (1994)].

In reinforcement learning, there is no teacher to supply gradient information during learning. The only piece of available information is represented by the reinforcement received from the environment; the learning system has to do things and see what happens to obtain gradient information. Although reinforcement is a scalar, whereas the gradient in supervised learning is a vector, the key point to note is that in reinforcement learning the information contained in reinforcement evaluates behavior but does not in itself indicate whether improvement is possible or how the system should change its behavior [Barto (1992)].

To gather information of a directional nature, a reinforcement learning system probes the environment through the combined use of trial and error and delayed reward; that is, the system engages in some form of exploration, searching for directional information on the basis of intrinsic properties of the environment, as seen in an evolutionary algorithm. In so doing, however, the reinforcement learning system is slowed down in its operation because a behavioral change made to obtain directional information is generally in conflict with the way in which the resulting directional information is exploited to change behavior for performance improvement, i.e., there is always a conflict between the following two factors [Barto (1992)]:

- The desire to use knowledge already available about the relative merits of actions taken by the system
- The desire to acquire more knowledge about the consequences of actions so as to make better selections in the future

This phenomenon is known as a conflict between:

- design and analysis
- exploitation and exploration [Thrun (1992), Holland (1975)]
- control and identification [Barto (1992)]

4. Self-Reproducing Networks: If the network is encoded by cellular encoding, the network representation may be executed as a 'program', such as by genetic programming. This way, the architecture may also be grown from initial input and output nodes to a full network with trained weights in an evolution process. When the network is realized in hardware, it is regarded as Darwin Machine and the hardware is also referred to as evolvable hardware by some researchers.

This means learning by evolution. A neural network is thus designed using evolutionary computing. If a genetic algorithm or genetic programming is used, its coding capability can allow the architecture of the network to be optimized together with weights. This means that this method does not impose pre-selection of network topology and offers a potential to arrive at an optimal one that may best suit the application on hand.

3.3 Learning Algorithms

3.3.1 Hebbian Learning

This was inspired by neuro-biological phenomenon. The basic rules are [Hebb (1949)]:

- If two neurons on both sides of a synapse are activated synchronously, then the strength of that synapse is selectively increased
- If two neurons on either side of a synapse are activated asynchronously, then that synapse is selectively weakened or eliminated

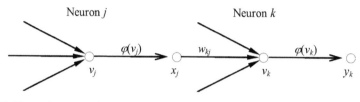

Fig. 3.11 Neuron j to neuron k

According to these rules, as shown in Figure 3.11, the weights are adjusted by:

$$w_{kj}(n) = F(y_k(n), x_j(n)) \qquad (3.14)$$

where,

- $x_j(n)$ = presynaptic activity (i.e., input)
- $y_k(n)$ = postsynaptic activity (i.e., output)
- F = function of both postsynaptic and presynaptic activities (i.e., both inputs and outputs)

A simple learning rule is the activity product rule as given by:

$$w_{kj}(n) = \eta y_k(n) x_j(n) \qquad (3.15)$$

where, η is the learning rate. With this learning rule, through repeated application of the input signal will lead to an exponential growth that finally drives the synapse's weight into saturation. To avoid this, a limit on the growth can be imposed by, for example, a nonlinear forgetting factor as follows:

$$\Delta w_{kj}(n) = \eta y_k(n) x_j(n) - \alpha y_k(n) w_{kj}(n) = \eta y_k(n)[x_j(n) - (\frac{\alpha}{\eta}) w_{kj}(n)] \qquad (3.16)$$

where α is a new positive constant, representing the forgetting rate. This learning rule is sometimes referred to as a generalized activity productivity rule. This implies that the weights at time $n+1$ will decrease if $\alpha w_{kj}(n)$ is greater than $\eta x_j(n)$.

In Hebbs postulate, it is also sensible to use statistical terms to determine the amount of adjustment. One way of doing this is the activity covariance rule, as given by:

$$
\begin{aligned}
\Delta w_{kj}(n) &= \eta cov[y_k(n), x_j(n)] \\
&= \eta E[y_k(n) - \bar{y}_k)(x_j(n) - \bar{x}_j)] \\
&= \eta \left(E[y_k(n)x_j(n)]\right) - \bar{y}_k\bar{x}_j)
\end{aligned}
\tag{3.17}
$$

where, E is the statistical expectation operator and 'cov' stands for the covariance function. Here $\bar{y}_k$ and $\bar{x}_j$ are the mean values of the presynaptic and postsynaptic activities, respectively.

3.3.2 Error-Correction Learning

Clearly, the actual response of neuron k differs from the desired response before and during training. The difference between the target response and the actual response is defined as an error signal:

$$
e_k(n) = d_k(n) - y_k(n)
\tag{3.18}
$$

This signal may be used to guide the learning process by minimizing a cost function based on the error.

A commonly used cost function is the means-square-error:

$$
J = E\left\{\frac{1}{2}\sum_k e_k^2(n)\right\}
\tag{3.19}
$$

where J stands for the (statistical average) effective error energy and the summation is over all the neurons in the output layer of the network. The factor $\frac{1}{2}$ is used to represent bipolar fluctuations (e.g., sine wave type of errors) and to simplify the derivative expressions. Note that minimization of the cost function J with respect to the network parameters leads to the so-called method of gradient (descent) guidance. A plot of J against weight in the multidimensional space is termed the error (performance) surface.

1. Least-Mean-Square Algorithm-Based Error-Correction: The theory of linear adaptive filters has been applied successfully in such diverse fields as communications, control, radar, sonar and biomedical engineering. It is also easily applied to a single linear neuron model [Haykin (1994)]. This concerns the linear optimum filtering problem. The solution to this is the highly popular least mean-square (LMS) algorithm. The design of the LMS algorithm is very simple, yet a detailed analysis of its convergence behavior is a challenging mathematical task.

   - Cost Function: Let $x = [x_1 x_2 \ldots x_p]^T$ be the individual signals provided by p input neurons with a corresponding set of weights $w = [w_1 w_2 \ldots w_p]^T$. These inputs are associated with an output neuron with its signal described by y. There is no threshold nor activation applied to the weighted summation of inputs. The requirement is to determine the optimum setting of the weights $w = [w_1 w_2 \ldots w_p]^T$ so as to minimize the difference (error) between the system output y and some desired response d in a mean-square sense. The solution to this fundamental problem lies in the Wiener-Hopf equations. The error signal is thus defined by:

$$e = d - y = d - \sum_{k=1}^{p} w_k x_k = d - x^T w \qquad (3.20)$$

A performance measure or cost function is the mean-squared error defined by:

$$J = E\left[e^2\right] = J(w) \qquad (3.21)$$

where E is the statistical expectation operator. The factor $\frac{1}{2}$ is included as a convenience for derivation. The solution of this filtering problem is referred to in the signal processing literature as the Wiener filter, in recognition of the pioneering work done by Wiener [Haykin (1994)].

Expanding the above terms, we get

$$\begin{aligned}
J &= \frac{1}{2}E\left[d^2\right] - E\left[\sum_{k=1}^{p} w_k x_k d\right] + \frac{1}{2}E\left[\sum_{j=1}^{p}\sum_{k=1}^{p} w_j w_k x_j x_k\right] \\
&= \frac{1}{2}E\left[d^2\right] - \sum_{k=1}^{p} w_k E\left[x_k d\right] + \frac{1}{2}\sum_{j=1}^{p}\sum_{k=1}^{p} w_j w_k E\left[x_j x_k\right]
\end{aligned} \qquad (3.22)$$

where double summation is used to represent the square of a summation and the weights are treated as constants and therefore taken outside the expectations.

• Wiener-Hopf Equations: If the cost function has well-defined or differentiable bottom or minimum points (Condition 1), the error surface will be either bowl-shaped (in which case a minimum point exists) or harness-shaped (where an inflection point, not extremum point, exists). In engineering applications, however, Condition 2 is usually not tested to prevent prolonging of the calculation time. The existence of minimum points is judged by engineering sense instead.

To determine the optimum condition, we differentiate the cost function J with respect to the weight w_k and then set the result equal to zero for all k. The derivative of J with respect to w_k is called the gradient of the error surface with respect to that particular weight. To find w_0, let

$$\frac{\partial J}{\partial w_k} = -E\left[x_k d\right] + \sum_{j=1}^{p} E\left[x_j x_k\right] w_j = 0 \qquad (3.23)$$

Then solving the simultaneous equations:

$$E\left[x_k d\right] = \sum_{j=1}^{p} E\left[x_j x_k\right] w_j \tag{3.24}$$

will lead to the possible optimal point w_0. This system of equations is known as the Wiener-Hopf equations and the filter whose weights satisfy the Wiener-Hopf equations is called a Wiener filter.

– Recursive Descent Method
   To solve the Wiener-Hopf equations to tap weights of the optimum filter, we basically need to compute the inverse of a p-by-p matrix. We may avoid the need for this matrix inversion by using the method of recursive descent. According to this method, the weights of the filter assume a time-varying form and their values are adjusted in an iterative fashion as given by:

$$w_k(n+1) = w_k(n) + \Delta w_k(n) \tag{3.25}$$

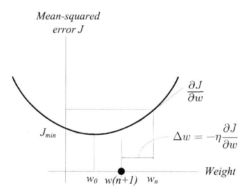

Fig. 3.12 w_k adjustment

This iteration is taken along the error surface with the aim of moving them progressively towards the optimum solution. The method of recursive descent has the task of continually seeking the bottom point of the error surface of the filter. Such an adjustment is illustrated in the case of a single weight.

It can be seen from Figure 3.12 that if the gradient $\frac{\partial J}{\partial w} > 0$, $\Delta w_k(n)$ is adjusted downwards, i.e., $w_k(n+1)$ should be decreased. It is wise to set the amount of decrease proportional to the gradient of the error surface. Thus the amount of correction in a time-varying form is given by:

$$\Delta w_k(n) = -\eta \frac{\partial J(n)}{\partial w_k}$$
$$= \eta \left\{ E[x_k d] - \sum_{j=1}^{p} E[x_j x_k] w_j \right\}, k = 1, 2, \cdots, p \tag{3.26}$$

This correction rule is called the delta rule or the Widrow-Hoff rule [Widrow and Hoff (1960)]. When the filter operates in an unknown environment, however, the correlation functions are not available, in which case we are forced to use estimates in their place.

– Least Mean-Square Algorithm
  The LMS algorithm is based on the use of instantaneous estimates of the autocorrelation function and the cross-correlation function. These estimates are made by:

$$E[x_k d] = x_k(n)d(n) \tag{3.27}$$

$$E[x_j x_k] = x_j(n)x_k(n) \tag{3.28}$$

which means

$$\Delta w_k(n) = \eta \left[x_k(n)d(n) - x_k(n) \sum_{j=1}^{p} w_j x_j(n) \right]$$
$$= \eta x_k(n)[d(n) - y(n)] \tag{3.29}$$
$$= \eta x_k(n)e(n)$$

This forms the LMS-based error-correction learning rule. Compare this with the activity product Hebbian learning rule.

Note that the LMS algorithm can operate in a stationary or nonstationary environment. In a nonstationay environment, the algorithm has the task of not only seeking the minimum point but also tracking it when it moves. Clearly, the smaller the learning rate η, the more accurate the tracking. However, small η will usually take a longer convergence time, but a large η may result in oscillations (unstable learning).

Note also that the cost function J formed by mathematical expectation may also be replaced by an instantaneous estimation as given by:

$$J = \frac{1}{2}e^2(n) \tag{3.30}$$

This will yield the same learning rule (see Tutorial 2).

2. Backpropagation for Multilayer Perceptrons: For a multilayered network, there are usually multiple output perceptrons at the output layer.

Thus use the cost function [Werbos (1974), LeCun (1985), Parker (1985), RumelhartHintonWilliams (1986), Rumelhart and McClelland (1986)].

$$J = \frac{1}{2} \sum_{j \in O} e_j^2(n) \qquad (3.31)$$

where, the set O includes all the neurons in the output layer.

$$e_j(n) = d_j(n) - y_j(n) \qquad (3.32)$$

Let $w_{ji}(n)$ represent the synaptic weight that connects neuron i to neuron j. Before activation, the weighted sum of inputs with a threshold to neuron j is given by:

$$v_j(n) = u_j(n) - \theta_j(n) = \sum_{i=0}^{p} w_{ji}(n) y_j(n) \qquad (3.33)$$

and the output of this neuron is given by:

$$y_j(n) = \varphi(v_j(n)) \qquad (3.34)$$

Similarly, the learning rule is describes by:

$$w_{ji}(n+1) = w_{ji}(n) + \Delta w_{ji}(n) \qquad (3.35)$$

where,

$$
\begin{aligned}
\Delta w_{ji}(n) &= -\eta \frac{\partial J(n)}{\partial w_{ji}} \\
&= -\eta \frac{\partial J(n)}{\partial y_j(n)} \frac{\partial y_j(n)}{\partial w_{ji}} \\
&= -\eta \frac{\partial J(n)}{\partial y_j(n)} \varphi'(v_j(n)) y_j(n)
\end{aligned}
\qquad (3.36)
$$

The last equation was obtained from Equations (3.19) and (3.20).

Case 1: If neuron j is in the output layer, then the error gradient over y_j is

$$\frac{\partial J(n)}{\partial y_j(n)} = -e_j(n) \qquad (3.37)$$

This means we can calculate the weight update easily using this and Equation (3.22).

Case 2: If the next layer to neuron j is the output layer k, then (see Tutorial 2 for proof) the 'local error gradient' is given by:

$$
\begin{aligned}
\frac{\partial J(n)}{\partial y_j} &= \Sigma_k \frac{\partial J(n)}{\partial y_k} \varphi_k' (v_k(n)) w_{kj}(n) \\
&= -\Sigma_k e_k(n) \varphi' (v_k(n)) w_{kj}(n)
\end{aligned}
\tag{3.38}
$$

Case 3: If neuron j is separated by layer k from the output layer, then iterate the above one more time and so forth.

Note that in BP based learning, it is required that local error gradients and $\varphi_k'(n)$ exist.

Summary of BP Algorithm:

- Initialization by setting: $w_{ji}(1) = random(0,1)$, for all i and j.
- Forward passing. For time $n = 1, 2, \cdots$, compute $y_j(n)$. Note that, for the input terminal layer, $y_j(n) = xj(n)$ no computation is needed.
- Backward passing. Then start at the output layer by passing the error signals and iterating the local error gradient layer by layer through the entire network to the input layer. This recursive process permits the synaptic weights of the network to undergo changes in accordance with the delta rule. Note that for the presentation of each training example, the input pattern of the input-output data set is fixed throughout the entire round-trip process.
- Training for a complete epoch. Repeat Steps 2 and 3 for all pieces of the input-output data. One complete cycle of training through the entire set of data is call an epoch.
- Termination. Repeat Steps 2, 3 and 4 for many epochs for convergence of the training process. It is to be hoped that the training process will converge and the training can be terminated if absolute values of all error gradients (usually measured by a norm of the gradient vector) become very small.

Thus the above mode of iteration is called the pattern mode (or incremental learning), where weight updating is performed after the presentation of each training example. If weight updating is performed after presentation of all the training examples that constitute an epoch instead, the learning mode is called the batch mode (or batch learning).

For real-time (or on-line) applications, the pattern mode may be preferred because it requires less local storage for each synaptic connection. Further, the pattern-by-pattern updating of weights makes the search stochastic in nature, which may make the algorithm less likely to be trapped in a local minimum. However, the use of batch mode provides a more accurate estimate of the gradient vector, which is particularly useful for final local tuning.

Improvement Using Learning Momentum
A weight update can be modified to include an amount of change made at the previous step multiplied by a momentum factor α to:

$$\Delta w_{ji}(n) = -\eta \frac{\partial J(n)}{\partial y_j} \varphi'_j(v_j(n)) y_i(n) + \alpha \Delta w_{ji}(n-1) \qquad (3.39)$$

This is intended to increase the speed of learning without increasing η (which may result in unstable learning). This method is called the generalized delta rule.

3.3.3 Competitive Learning

A limit is imposed on the 'strength' of each neuron, such as:

$$\sum_i w_{ji} = 1, \forall j \qquad (3.40)$$

Winner takes-all: A mechanism that permits the neurons to compete for the right to respond to a given subset of inputs, such that only one output neuron, or only one neuron per group, is active (i.e., 'on') at a time.

The standard competitive learning rule adjusts the synaptic weights by:

$$\Delta w_{ji} = \begin{cases} \eta(x_i - w_{ji}), & j+ \\ 0, & j- \end{cases} \qquad (3.41)$$

in which, $j+$, if neuron j wins the competition; $j-$, if neuron j loses the competition.

3.3.4 Darwinian Selective Learning and Darwin Machine

One way to realize competitive and reinforcement learning for error-correction is Darwinian selective learning. In tournament, proportionate and deterministic ranking selections, weight candidates are selected in a competitive way. In a tournament, roulette wheel, ranked roulette wheel, stochastic ranking and elitist strategy selections, fit candidates are reinforced. Note that, however, both learning mechanism are fundamentally related by the 'survival-of-the-fittest' Darwinian principle.

Through chromosome coding and genetic algorithms, the architecture of an ANN may be pruned (tailored to an appropriate sub-network) from a parent network, i.e., the architecture and weights can be trained simultaneously (for details, see [Li and Haußler (1996)]).

Clearly, other evolutionary algorithms can also be used in Darwinian selective training, which may not require coding of the weights but they may not be able to provide architectural training.

3.4 Tutorials and Coursework

1. Why can a neural network be used to classify many (usually unseen) patterns and to approximate the behavior of many 'systems' and decision makers? The ideas of neural networks were first developed and reported in the 1940's. In your opinion, briefly, what were the major reasons that only in the late 1980s neural computing began to be widely accepted and applied? You may compare this with similar situations involving parallel processing or evolutionary computing.

2. For a single linear perceptron j with N inputs, the cost functions used in error-correction training is a quadratic error. Show that the error-correction learning rule is also given by:

$$\Delta w_{kj}(n) = \eta e_j(n) x_j(n) \tag{3.42}$$

Calculate the update if the perceptron is subject to sigmoid activation; also calculate the update for an activation given by:

$$\varphi(v) = \frac{2}{\pi} tan^{-1}(v) \tag{3.43}$$

3. Use the backpropagation tool provided in the MATLAB® Neural Network Toolbox to train a two-layered feedforward network to solve the XOR problem in both pattern and batch modes. Change the learning and momentum rates and plot a family of convergence traces of the training processes.
   Then use the FlexTool (GA) [FlexToolGA] Toolbox to train the same network. Compare the trained weights and convergence with those obtained in the BP training process, as well as with those obtained by a genetic algorithm (Darwinian selective learning) as given in the example shown in the lecture notes.

4. In backpropagation-based error-correction training, the cost function is often the sum of quadratic errors. Let subscript k represent the output layer and j the layer immediately adjacent to the output layer. Show that the 'local error gradient' is given by Equation (3.38), where $\varphi(v)$ is the activation function.

Table 3.2 Data Table

Pattern No.	1	2	3	4
input: x1	0	1	1	0
input: x2	0	0	1	1
output: y	0	1	0	1

5. Consider the encoding problem in which a set of orthogonal input patterns, diag [1, 1, 1, 1, 1, 1], are mapped with the same set of orthogonal output patterns through a two-layered ANN. Essentially, the problem is to learn an encoding of a p-bit pattern into a $\log_2 p$-bit pattern, and then learn to decode this representation into the output pattern. Use a backpropagation algorithm in the MATLAB Neural Network Toolbox to train the network for identity mapping.

6. Use a simple evolutionary or any algorithm in the MATLAB Neural Network Toolbox to design and train an artificial neural network to mimic the XOR logic.

4

Fuzzy Logic and Fuzzy Systems

4.1 Human Inference and Fuzzy Logic

4.1.1 Human Inference and Fuzzy Systems

With the development of computational intelligence, human inference-oriented fuzzy systems (FS) and fuzzy logic control (FLC) have received increasing attention world-wide. For example, a fuzzy controller incorporates uncertainty and abstract nature inherent in human decision-making into intelligent control systems. It tends to capture the approximate and qualitative boundary conditions of system variables (as opposed to the probability theory that deals with random behavior) by fuzzy sets with a membership function. Such a system flexibly implements functions in near human terms, i.e., IF-THEN linguistic rules, with reasoning by fuzzy logic, which is a rigorous mathematical discipline. Hence, it is termed a type of expert systems that handles problems widespread with ambiguity. It is well known for its capability in dealing with non-linear systems that are complex, ill-defined or time-varying. In addition, fuzzy systems are reliable and robust and are straightforward to implement [Ng, et al. (1995)].

Successful applications, such as robot control, automotive system, aircraft, spacecraft and process control have shown that they offer potential advantages over conventional control schemes in:

- natural decision making and uses natural languages
- learning capability
- less dependency in quantitative models
- a greater degree of autonomy
- ease of implementation
- friendly user interface

4.1.2 Fuzzy Sets

When controlling a process, human operators usually encounter complex patterns of quantitative conditions, which are difficult to interpret accurately. The magnitude of the measurements is vaguely described as 'fast', 'slow', 'big', 'small', 'high', 'low', etc. To represent such inexact or qualitative information in a quantitative way, an approach called 'fuzzy set theory' was developed by Zadeh in 1965 [Zadeh (1965a)].

Fuzzy set theory, a generalization of classical set theory, reflects the observation that the more complex a system becomes, the less meaningful are the low-level details in describing the overall system operation. Therefore the acquisition for precision in a complex system becomes a difficult task and often unnecessary. However, one of Zadeh's basic assumptions is that a system's operational and control laws can be expressed in words. Hence, fuzzy logic, based on the fuzzy set theory incorporated into the framework of multi-valued logic, is developed. Although such fuzzy approach of expression in words seems to be inadequate, it can actually be superior to or easier than a more mathematical approach. The main advantage of fuzzy set theory is that it excels in dealing with imprecision.

In classical set theory, an element is classified either as a part of a set or not and there is no in-between relationship. In other words, partial membership is not allowed at all in classical set theory and the membership function can only restricted to either 1 or 0 which are classified as either 'true' or 'false' respectively. Fuzzy logic, on the contrary, allows partial set memberships and gradual transitions between being a full member of the set and full non-member of the set. Having a capability of flexible membership classification, fuzzy logic can be used to implement a crisp system. Most of the action occurs in transition—the partial membership regions of a set.

Fuzzy logic can be easily demonstrated in the following example. A fuzzy subset A with an element x has a membership function of $\mu_A(x)$, which is a degree of certainty or fulfilment expressed in the interval between 0.0 and 1.0. If $\mu_A(x)$ is 1.0 for example, the element is a full member of the set. As opposite, if $\mu_A(x)$ is 0.0, it is not a member at all. Considering now a fuzzy subset A with five elements, which has the membership function of 0.3, 0.7, 1.0 , 0.7 and 0.3. Then the element with a membership function of 1.0 is confidently a full member of A, whereas the others are only a partial member. The membership function, therefore, determines the degree to which the element belongs to the subset. If a fuzzy set A is defined as integers 'around 10' on the scale from 8 to 12, it might be described by the following,

$$A = \left\{ \frac{0.3}{8}, \frac{0.7}{9}, \frac{1.0}{10}, \frac{0.7}{11}, \frac{0.3}{12} \right\} \tag{4.1}$$

where 0.3, 0.7, 1.0, 0.7 and 0.3 are the degrees of memberships and 8, 9, 10, 11 and 12 are called the universe of discourse.

Fuzzy set theory involves several complicated theorems. The following three definitions would sufficiently form the basis for the decision table that is widely used in engineering where decision-making is necessary.

(1) Union of two sets: $A \cup B$, corresponds to the OR operation and is defined by

$$\mu_{A \cup B}(x) = max(\mu_A(x), \mu_B(x)) \quad (4.2)$$

(2) Intersection of two sets: $A \cap B$, corresponds to the AND operation and is defined by

$$\mu_{A \cap B}(x) = min(\mu_A(x), \mu_B(x)) \quad (4.3)$$

(3) Complement of a set: $\bar{A}$, corresponds to the NOT operation and is defined by

$$\mu_{\bar{A}}(x) = 1 - \mu_A \quad (4.4)$$

The fuzzy operations used in the rule inference in this thesis are adopted from the works of Mamdani [Mamdani (1974)] and therefore, the max-min operators are used throughout this research. To illustrate the application of these definitions, consider two qualitative statements, "Big" and "Small", with the following membership functions:

$$\mu_{Big} = \{0.0, 0.3, 0.7, 1.0\} \quad (4.5)$$

$$\mu_{Small} = \{0.2, 0.7, 1.0, 0.8\} \quad (4.6)$$

The three definitions can be applied directly to the two membership functions of the qualitative statements by the following:

$$\mu_{(Big \cup Small)}(x) = \{max(0.0, 0.2), max(0.3, 0.7), max(0.7, 1.0), max(1.0, 0.8)\}$$
$$= \{0.2, 0.7, 1.0, 1.0\} \quad (4.7)$$

$$\mu_{(Big \cap Small)}(x) = \{min(0.0, 0.2), min(0.3, 0.7), min(0.7, 1.0), min(1.0, 0.8)\}$$
$$= \{0.0, 0.3, 0.7, 0.8\} \quad (4.8)$$

$$\mu_{\bar{A}}(x) = 1 - \mu_A \quad (4.9)$$

$$\mu_{\overline{Big}}(x) = \{(1.0 - 0.0), (1.0 - 0.3), (1.0 - 0.7), (1.0 - 1.0)\}$$
$$= \{1.0, 0.7, 0.3, 0.0\} \quad (4.10)$$

4.1.3 Membership Functions

Having formulated the fuzzy control laws and identified the input and output decision variables, the next consideration is to define the membership functions of the linguistic sets. Terms like **Positive Big**, **Positive Medium**, **Positive Small**, **Zero**, **Negative Small**, **Negative Medium** and **Negative Big** are commonly used linguistic sets in any control environment and they are also termed as fuzzy sets. The shape of the fuzzy set is quite arbitrary and depends on the user's preference. Shapes that are commonly used are bell-shaped, quadratic, trapezoidal and triangular. Symmetrical type membership functions are very common and can be implemented easily. Non-symmetrical and piece-wise types can also be used.

After the selection of a preference shape, it is necessary to define the number of fuzzy sets along the universe of discourse. In illustration, a triangular type is used for simplicity as shown in the following Figure 4.1.

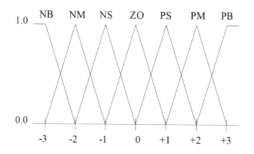

Fig. 4.1 Symmetrical membership functions of the seven elements

where -3 to +3 is the universe of discourse and seven fuzzy sets used are termed **NB**, **NM**, **NS**, **ZO**, **PS**, **PM**, **PB** corresponding to the above statements such as **Positive Big**, **Positive Medium**, etc. These fuzzy sets are later being used to describe the magnitude of measurements of the input and output of a control system which have to be quantized and mapped to the range of the universe of discourse.

4.2 Fuzzy Logic and Decision Making

The application of fuzzy logic techniques to control is tied together with the concept of linguistic control rules. A fuzzy controller consists of a set of control rules and each rule is a linguistic statement about the control action to be taken for a given process condition given by the following rule structure as given in Equation (4.11), in which, the <condition> is termed as the antecedent and the <control action> is the consequence.

$$IF < condition > THEN < controlaction > \qquad (4.11)$$

An example of the statements on <condition> could read like 'water level is high', 'opening of outflow is small' and etc. Likewise, statements on <control action> might read as 'increase water inflow' or 'increase the opening of outflow'.

The environment can be easily realized from the control of the water-level in a twin-tank system. The key items in the control rule are terms like 'high', 'small', 'increase'. In linguistic approximation by fuzzy logic, each of these terms are represented by a preference fuzzy membership function so as to establish a value in the interval [0, 1] for a given process condition. In fuzzy logic, the calculated value of a condition is taken to be any value of the interval [0, 1].

To establish a fuzzy controller, it is necessary to interpret rules that are based on experience or expert's knowledge so as to form a decision table that gives the input and output values of the controller corresponding to situation of interest. Beside the rules for the decision table, the membership functions are to be chosen by the fuzzy controller developer to represent the human's conception of the linguistic terms. However, in the construction of the rules, the choice for the input decision variables of the fuzzy controller is dependent on the developer's preference. One can use the controlled variables as the inputs to the fuzzy controller or use error related variables developed from the synthesis of a conventional controller, for example, PID type.

A schematic of a fuzzy control system developed from the conventional methodology is given in Figure 4.2:

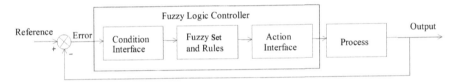

Fig. 4.2 A block diagram of fuzzy logic control systems

The fuzzy logic controller relates significant and observable variables to the control actions [Mamdani (1974)] and consists of a fuzzy relationship or algorithm. The time sequenced crisp variables are first converted to fuzzy variables and serve as conditions to the rule-base. Examples of crisp variables are $e(k)$ and $\dot{x}(k)$, where $e(k)$ is the error = reference - output, $\dot{x}(k)$ is the rate of change of error = $(e(k) - e(k-1))/\Delta t$ and Δt is the sampling period. In this case, the control action can be inferred from the interpretation of the amount of $e(k)$ and $\dot{x}(k)$. The process of this conversion is termed as fuzzification. These fuzzified variables ($X_e(k)$ and $X_{\dot{e}}(k)$) will then be used in the control rules evaluation using the compositional rule of inference, and the appropriate control action will be computed. The 'AND' logic is naturally most commonly used in forming the rule-base in the following form:

$$IF < condition, X_e(k) > AND < condition, X_{\dot{e}}(k) > THEN < controlaction >$$

$$(4.12)$$

A follow-up defuzzification technique will be employed to reconvert the control action to crisp value to regulate the process. So, the essential steps in designing fuzzy controllers include:

- defining input and output decision variables for fuzzy controllers
- specifying all the fuzzy sets and their membership functions defined for each input and output decision variables
- converting the input decision variables to fuzzy sets by a fuzzification technique
- compilation of an appropriate and sufficient set of control rules that operate on these fuzzy sets, i.e., formulating the fuzzy rule-base which are used as an inference engine
- devising a method that computes for a single resultant fuzzy control action
- devising a transformation method for converting fuzzy control action to crisp value

4.2.1 Formation of Fuzzy Decision Signal

The proportional plus integral plus derivative (PID) control algorithm is the most popular control scheme used in the industry. Today, some 95 per cent of industrial control systems are realized in various PID forms. As discussed earlier, the fuzzy logic controller is based upon $e(k)$ and $\dot{e}(k)$ inputs. This is similar to PD control and its control law is given by:

$$u_{PD}(k) = u_P(k) + u_D(k) = K_P \cdot e(k) + K_D \cdot \frac{\Delta e(k)}{\Delta t} \qquad (4.13)$$

The fuzzy PD controller is produced by using fuzzified error, $e(k)$ and the rate of change of error, $\frac{\Delta e(k)}{\Delta t}$ as input decision variables. In some cases, $e(k)$ and $\Delta e(k)$ variables are used. The use of $\Delta e(k)$ has its disadvantage such that when the sampling time, Δt is changed, the performance of the fuzzy controller may deteriorate. This is due to the inconsistent change of $\Delta e(k)$ for different Δt. On the contrary, the use of $\frac{\Delta e(k)}{\Delta t}$ would provide consistency for different Δt. However, both of them are the most commonly used fuzzy logic control schemes which reflect natural predictions. A two-dimensional rule-base is, however, required in this fuzzy controller. The control response produced by such fuzzy PD controller is similar to its conventional PD counterpart where both of them have suffered a similar problem of steady-state error. But due to the piece-wise nonlinearity of the rule-base, the fuzzy controller has excelled in performance in terms of robustness.

4.2.2 Fuzzy Rule Base

Based upon the fuzzy PD controller developed in Section 4.2.1, the two-dimensional rule-base of such fuzzy controller is designed with a phase plane in mind, in which the fuzzy controller acts as a switching operation. The tracking boundaries in the phase plane are related to the input variables. Usually, a phase plane diagram is partitioned according to the number of fuzzy sets used for the control actions. Graphically, the partitioning according to the number of fuzzy sets used for the control action is depicted in Figure 4.3. Note the arrangement of fuzzy sets for the control actions such that the trajectory of the tracking system is driven towards the 'ZO' region. Within this 'ZO' region, the tracking trajectory slides towards the origin which corresponds to the desired reference of the control system.

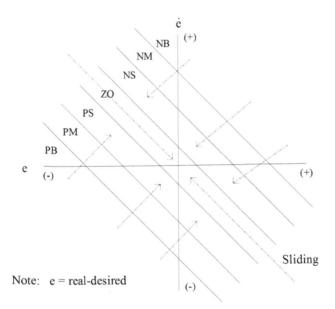

Fig. 4.3 Phase plane diagram

As shown in the above figure, the $(e, \dot{e})$ based inference is most commonly used in fuzzy decision making. This inference engine utilizes a rule-base formulated from a collection of rules in which the $\{X_e, X_{\dot{e}}\}$ pair produces a joint control action. The rules are written in the form of:

$$Rule_i : IF < X_e(k) = A_i > AND < X_{\dot{e}}(k) = B_i > THEN < X_O(k) = C_i > \quad (4.14)$$

where $i = 1, .., n$ and n is the number of rules. The rule-base can also be restructured in a 2-dimension lookup table, for example, a -3 to +3 universe of discourse for both input decision variables. The graphical representation showing the input and output decision variables defined in their respective universal discourse are also provided as shown in Figure 4.4. The phase plane partition of tracking boundaries can be used to assist the formation of the initial lookup table manually. Note in manual designs, the control actions are diagonally symmetrical through the centre of the table.

$X_{\dot{e}}$		-B / -3	-M / -2	-S / -1	ZO / 0	+S / +1	+M / +2	+B / +3
+3	+B	ZO	-S	-M	-B	-B	-B	-B
+2	+M	+S	ZO	-S	-M	-B	-B	-B
+1	+S	+M	+S	ZO	-S	-M	-B	-B
0	ZO	+B	+M	+S	ZO	-S	-M	-B
-1	-S	+B	+B	+M	+S	ZO	-S	-M
-2	-M	+B	+B	+B	+M	+S	ZO	-S
-3	-B	+B	+B	+B	+B	+M	+S	ZO

X_e

Fig. 4.4 A lookup table for an ideal FLC

In the lookup table, each control rule will prescribe a definite action to be taken for each particular condition existing in the physical environment. Therefore, the control actions are acquired through human knowledge or experience or by some form of rule acquisition techniques. The limitation of the ideal rule-base in Figure 4.4 is that the symmetrical properties of the lookup table may be insufficient for an asymmetrical system in the real world. An asymmetrical type will, therefore, be more appropriate in this case. The later section shows how this and other design tasks can be achieved systematically and automatically using genetic algorithm techniques.

In case of the lookup table, fuzzy control actions can be computed using the min-max operation as discussed in Section 4.1.2. For example, Figure 4.5 shows the graphical computation of min-max operations in a fuzzy control system. In this example, three activated rules are given and the min operation is performed on the antecedents of the rule where it computes for each consequence contributed by each corresponding rule. Max operation is later performed on these consequences to determine the final contribution of control actions. These final control actions are then used in the defuzzification process where a crisp executable value is computed. Such de-fuzzification will be discussed in the later section.

Min Operations :

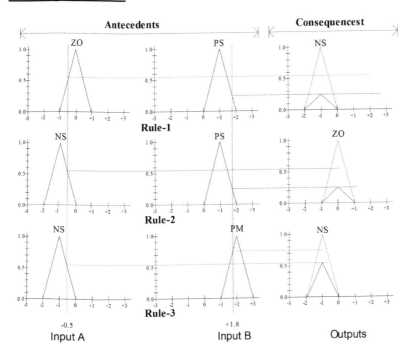

Max Operations :

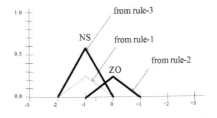

Note: The darken line is the final contribution of control actions.

Fig. 4.5 Example of min-max operations in a fuzzy control system

4.2.3 Measurements with Fuzzification and Scaling

From Figure 4.2, the condition interface includes:

(a) scaling factors used to map the range of the values of the controlled variables into a pre-defined universe of discourse;

(b) a quantization procedure to assist in the above mapping if the discrete membership functions are used and therefore, a discrete universe of discourse is used;

(c) an estimator is included to determine the rate of change of the controlled variables. This rate of change is normally treated by the controller as other input decision variables.

The mapping (fuzzification process) of a crisp value to a fuzzy terminology is used in the following equations for a universe of discourse of range -3 to +3 (representing -Big and +Big, respectively) in constructing a fuzzy PD controller given below. Two input scaling factors correspond to the error and the rate of change of error pair are to be defined. The scaling factor for the error, S_e, is obtained in

$$X_e(k) = \frac{3}{e_{max}} e(k) = S_e e(k) \tag{4.15}$$

where e_{max} is usually the magnitude of the step reference signal. Structurally, Equation (4.15) is also similar to the conventional control type:

$$u_P(k) = K_P e(k) \tag{4.16}$$

Similarly, in

$$X_{\dot{e}}(k) = \frac{3}{\dot{e}_{max}} \frac{\Delta e(k)}{\Delta t} = S_{\dot{e}} e(k) \tag{4.17}$$

the scaling factor for the rate of change of error, $S_{\dot{e}}$, is defined. The choice of $\dot{e}_{max}$ is determined empirically depending on the process dynamic. Here X_e and $X_{\dot{e}}$ are scaled and then used as fuzzified inputs to the fuzzy inference engine. Note, in fuzzy control implementations, the roles of the scaling factors are not identical to that of gains in conventional control as they perform a different task.

The choice of scaling factors has a significant effect on the controller performance in terms of the controller sensitivity, steady-state errors, stability, transients and number of rules being activated. In particular, the error scaling factor, S_e, affects the region of steady-state errors and the number of rules that are activated; and a large S_e will increase the sensitivity to the raised in the error signal. The rate of change of error scaling factor, $S_{\dot{e}}$, has an effect on the convergence of the system response, i.e., the transient response. Poor convergence will occur if too small a value is used.

4.2.4 Defuzzification and Output Signals

From Figure 4.2, the action interface has included the following:

(a) a procedure to convert the fuzzy information supplied by the controller into one unique control action called the combination and defuzzification process;
(b) scaling factors to map the action values into real process input values.

The combining of control actions is essential as more than one rule may be activated for any given set of inputs. In addition, the resulting single action (combined with the actions of activated rules) must be transformed from a fuzzy output value to a crisp and executable value. There are currently three popular defuzzification techniques used in the fuzzy control community, namely,

- the maximizer technique
- the centroid or the centre of gravity technique
- the singleton technique

The maximizer technique takes the maximum degree-of-membership value from the various activated rules to be its corresponding single control action. If, for example, given the consequences of three activated rules are as stated below:

$$
\begin{aligned}
\mu_{PS} &= 0.75 \\
\mu_{PM} &= 0.3 \\
\mu_{PB} &= 0.1
\end{aligned}
\tag{4.18}
$$

The corresponding control action with will $\mu_{PS} = 0.75$ be chosen, because the value of 0.75 is larger than the other two values of μ. However, there is some chance that two control actions with the same value of μ will occur, for which some form of conflict-resolving technique is necessary. One possible solution is to take the average of the corresponding control actions and another is to select the control action associated with a rule's location in the rule-base. The maximizer technique is the simplest approach of all but it suffers from the ignorance of important and potential control actions.

The 'centroid or the center of gravity' technique computes the resultant control action from the center of mass of the outputs of activated rules. The single executable output is a weighted average of the centroids of each degree-of-membership function. The approach is illustrated in Figure 4.6. In a general discrete case, the resultant control action can be written as follows [Lee (1990)]:

$$
X_O = \frac{\sum_{i=1}^{n} \mu_i U_i}{\sum_{i=1}^{n} \mu_i}
\tag{4.19}
$$

where μ is the degree-of-membership functions, U_i is the centroid of each membership function determined along the universe of discourse, n is the number of memberships used and X_O is the resultant fuzzy output action value. By avoiding the usage of the edges of the output membership functions, the resultant control action always corresponds to a single output value. Unfortunately, this approach is computationally intensive and suffers from an additional shortcoming. This shortcoming is the requirement of overlapping of input membership functions for a smooth operation over the entire output range such that several (at least two) rules must activate at each system iteration. Despite this shortcoming, the centroid method is currently the best technique for combination and defuzzification.

The singleton technique is a special case of centroid method. It represents each fuzzy output set as a single output value by using a weighted average to combine multiple actions [Brubaker (1992)]. This approach requires much less computational effort than the centroid method, but it still requires small overlapping for

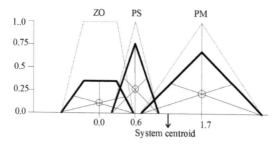

Fig. 4.6 The centroid method for deriving a single executable action

input membership functions to avoid discontinuities in the output. In the design of output membership functions, the designer only needs to determine the position of the membership function along the universe of discourse. The shape and base-length of output membership functions are made redundant. Due to its simplicity and less computational effort, it is used throughout in this thesis in the design of fuzzy controllers.

After combining and defuzzification process, the resultant fuzzy output action value is output to the plant through the output scaling process. The output scaling factor has an effect on the overall gain of the system (Linkens and Abbod 1992). A large value of the factor will cause the controller to operate in a bang-bang manner driving the system to saturation, whilst a small value will reduce the overall system gain giving a sluggish transient. The scaling equation is given by (Linkens and Abbod 1992):

$$x_O = \frac{x_{Omax}}{X_{Omax}} X_O = S_O X_O \tag{4.20}$$

where x_O the crisp output value to the plant, x_{Omax} the maximum plant input, X_O the resultant fuzzy output action value and X_{Omax} the maximum fuzzy output determined from defuzzifying a positive big fuzzy set with a given universe of discourse.

4.3 Tutorial and Coursework

1. A room temperature regulation system is to be built using fuzzy logic, where the decision on how to adjust the temperature is to be made using the observed temperature levels and not the internal thermodynamics of the room. For this, 20^oC is regarded as truly 'comfortable' and other temperature levels are classified as 'warm', 'hot', 'very hot', 'cool', 'cold' and 'very cold'. Let

$$A = \left\{ \frac{0.5}{15}, \frac{0.7}{17}, \frac{0.9}{19}, \frac{1.0}{20}, \frac{0.9}{21}, \frac{0.7}{23}, \frac{0.5}{25} \right\} \tag{4.21}$$

represent a fuzzy set defined around the 'comfortable' temperature level. Explain what the numbers 15, 17, 19, 20, 21, 23 and 25 mean and what the numbers 1.0, 0.9, 0.7 and 0.5 mean. Sketch two simple membership functions to demonstrate how fuzzy sets A and

$$B = \left\{ \frac{0.5}{10}, \frac{0.7}{12}, \frac{0.9}{14}, \frac{1.0}{15}, \frac{0.9}{16}, \frac{0.7}{18}, \frac{0.5}{205} \right\} \tag{4.22}$$

are fuzzified. What is the degree of membership of 17^oC belonging to both A and B and what is that of 18^oC belonging to either A or B?

Derive a decision table that may be used reasonably to adjust the temperature by both the observed temperature level and its trends. Explain whether conventional optimization techniques can be used to design such a fuzzy regulation system. Briefly explain how an artificial neural network or evolutionary algorithm may be used for the design and how a designer's existing knowledge may be directly incorporated to help the design process.

2. (a) A process temperature regulation system is to be built using fuzzy logic, where the decision on how to adjust the temperature is to be made, using the observed temperature levels and not the internal thermodynamics of the process. As given in Figure 4.7, for this case, 40^oC is regarded as the 'right' level and other temperature levels are classified as 'too warm', 'too hot', 'extremely hot', 'too cool', 'too cold' and 'extremely cold'. Let

$$A = \left\{ \frac{0.5}{12}, \frac{0.7}{17}, \frac{0.9}{30}, \frac{1.0}{40}, \frac{0.9}{45}, \frac{0.7}{50}, \frac{0.5}{60} \right\} \tag{4.23}$$

represent a fuzzy set defined around the 'right' temperature level.

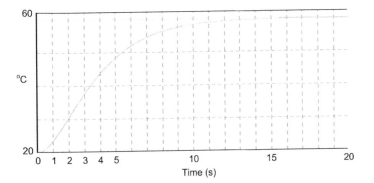

Fig. 4.7 Temperature regulation system

Explain what the numbers 12, 17, 30, 40, 45, 50 and 60 mean and what the numbers 0.5, 0.7, 0.9 and 1.0 mean.

(a) Within the 'universe of discourse' of the above system, it is known that the voltage applied to control the heater that regulates the temperature can be

varied from 0V to 200V. A typical step change of voltage from 50V to 150V
will result in a temperature change as observed in the following figure.
Sketch simple membership functions that may be used for 'fuzzification' and
'defuzzification' of the fuzzy decision-making system. Derive a decision table
that may be used reasonably to adjust the temperature to the 'right' level by
both the observed temperature level and its trends. What methods would you
suggest to be used to optimise or automate such decision-making?

3. The motion of the satellite is assumed to be confined to a plane passing through
the center of the earth. The relevant differential equations are:

$$\frac{dr}{dt} = f_r(r, v, \omega) = v$$

$$\frac{dv}{dt} = f_v(r, v, \omega) = \omega r^2 - \frac{MG}{r^2} \tag{4.24}$$

$$\frac{d\omega}{dt} = f_\omega(r, v, \omega) = \frac{1}{r}(u - 2\omega v)$$

where r is the radius of the orbit, v is the radial velocity component, ω is the angular velocity and u is the transverse force per unit mass exerted by the satellite's
thruster. M and G are constants where M is the mass of the earth 5.3×10^24 kg
and G is the gravitational constant of 6.7×10^{-11} Nm^2/kg^2.
The satellite dynamics sub-system should have one input, u and three outputs, r,
v and ω. Use MATLAB®/Simulink to design a fuzzy system to generate u so as
to maintain the orbit, as hinted below in Figure 4.8.

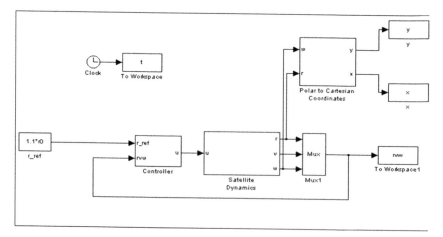

Fig. 4.8 The satellite dynamics sub-system by Simulink

Part II
CIAD and Advanced Computational Intelligence Tools

Part II introduces some advanced CI algorithms, such as swarm algorithms as optimization tools and a few metric indices that can be employed to assess their performance.

5

CIAD–Computational Intelligence Assisted Design

5.1 Introduction

Computational intelligence offers a wealth of nature-inspired approaches to complex problem solving, such as competitive and smart designs. Compared with traditional optimization methods, CI does not need to reformulate a problem to search a non-linear or non-differentiable space. Another advantage of CI is its flexibility in formulating the fitness function, with either single or multiple objectives, which can be assessed from the output of conventional CAD within the CIAD framework, as shown in Figure 5.1. This feature is particularly appealing if an explicit objective function is difficult to obtain.

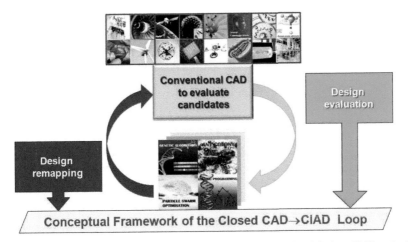

Fig. 5.1 Conceptual framework of the computational intelligence-assisted design, CIAD—closing the loop of CAD

A flowchart of computational intelligence-assisted design and engineering is given in Figure 5.2. It normally involves four phases: CI-assisted design (CIAD), CI-assisted engineering (CIAE), CI-resultant virtual prototype (CIVP) and CI-assisted manufacturing (CIAM). Because the CIAD framework provides an expanded capability to accommodate a variety of CI algorithms, it can accommodate a wide range of multidisciplinary applications, yielding three abilities: (1) mobilizing computational resources, (2) utilizing multiple CI algorithms, and (3) reducing computational costs.

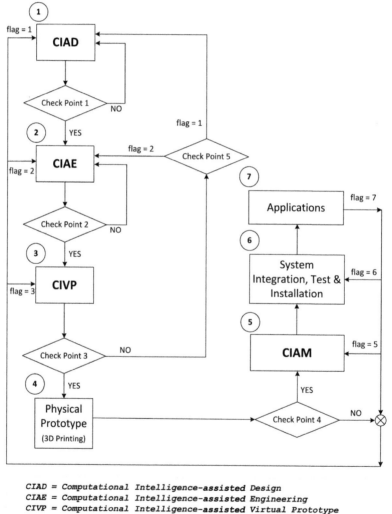

CIAD = Computational Intelligence-assisted Design
CIAE = Computational Intelligence-assisted Engineering
CIVP = Computational Intelligence-assisted Virtual Prototype
CIAM = Computational Intelligence-assisted Manufacturing

Fig. 5.2 Conceptual framework of computational intelligence-assisted design

5.2 Optimization, Design and Intelligent Design Automation

"Engineering design is the process of devising a system, component, or process to meet desired needs. It is a decision-making process (often iterative), in which the basic sciences, mathematics, and the engineering sciences are applied to convert resources optimally to meet these stated needs." [ABET (2017)]

5.2.1 Optimal Engineering Design

The task of designing an engineering system is mainly to optimise its 'parameters' so that the resultant system can best meet design specifications, or objectives, or can outperform existing systems. These parameters may also include structural representations and therefore, new configurations and new designs may be created or invented. Without loss of generality, a candidate design can be represented conveniently by a parameter set of the system, and the design tasks include:

- Better design of products for sustainability/circularity; new design approaches, i.e., design for re-use, repair, maintenance and remanufacture, e.g., products manufactured to allow easier separation of materials, reducing complexity
- Design of products to influence consumer behavior; for instance, opportunities with the retailer, household
- Getting consumers involved in design
- Design of supply chains
- Service design

DEFINITION 1
A candidate design of an engineering system can be represented by a uniform parametric vector given by:

$$\mathbf{P}_i = \{p_1, p_2, \cdots, p_n\} \in \mathbf{R}^n \tag{5.1}$$

where i stands for the ith possible design candidate, n the number of parameters required to determine the engineering system, $p_j \in \mathbf{R}$ the jth parameter of the ith design candidate with $j \in \{1,...,n\}$, and $\mathbf{R}^n$ the n-dimensional real Euclidean space.

NB. The term, 'parameter', may represent a 'design structures' or a 'design concept', and is not necessarily restricted to numerals. For the convenience of notation, it is assumed that the number of potential design candidates is 'countable', although there may exist potentially countless new and good designs.

DEFINITION 2

The *solution space* of the design problem is given by:

$$\mathbf{S} = \{P_i, \forall i | p_j \in \mathbf{R} \ and \ j \in \{1,...,n\}\} \subseteq \mathbf{R}^n \qquad (5.2)$$

NB. This space may not be confined within a known range. It includes all possible and potentially creative designs. Using a digital computer, the space is interpreted as a discrete domain (i.e., a sampled and quantized world, as opposed to a continuous and infinitely accurate world).

DEFINITION 3

The *objective* (*performance index*, or *fitness*) of a design is represented by a function, as given in Equation (5.3),

$$f(\mathbf{P}_i) : \mathbf{R}^n \longrightarrow \mathbf{R} \qquad (5.3)$$

with respect to design requirements or specifications, where R is the (*1*-dimensional) real space. If multiple objectives are required, a vector of performance indices, as given in Equation (5.4), should be used.

$$f(\mathbf{P}_i) : \mathbf{R}^n \longrightarrow \mathbf{R}^m \qquad (5.4)$$

NB. In many practical cases, maximizing the performance index is equivalent to minimizing a fitting, learning or search error. In this case a *cost function* that is to be minimized is used to replace the *fitness function* that is to be maximized.

For example, the performance index of a control system needs to reflect the following design criteria in the presence of practical system constraints; for more details, see [Feng and Li (1999)]:

- Good relative stability (e.g., good gain and phase margins)
- Excellent steady-state accuracy (e.g., minimal or no steady-state errors)
- Excellent transient response (e.g., minimal rise-time, settling-time, overshoots and undershoots)
- Robustness to the environment (e.g., maximal rejection of disturbances)
- Robustness to plant uncertainties (e.g., minimal sensitivities to parametric and structural variations)

DEFINITION 4

The problem of design is equivalent to the problem of finding a $\mathbf{P}_o$ such that:

$$\mathbf{P}_o : \{\mathbf{P}_o \in \mathbf{S} | f(\mathbf{P}_o) = \sup_{\forall \mathbf{P}_i \in \mathbf{S}} f(\mathbf{P}_i)\} \qquad (5.5)$$

Question

How do we define the problem of design if there are multiple objectives (which cannot be reconciled)? An example of addressing this problem is found in Economics, known as *Pareto optimality*.

Summary

The engineering design problem concerns first, finding the best design within a known range (i.e., learning) and second, a new and better design beyond the existing ones (creation). This is equivalent to an optimization (or learning or *search*) problem in an, almost certainly, multidimensional and multi-modal (multivariate) space with a combinational design objective or multiple objectives.

If the objective function (or, inversely, *cost function*) is differentiable under practical engineering constraints in the multidimensional space, the design problem may be solved easily by setting its *vector derivative* to zero. Finding the parameter sets that result in zero first-order derivatives which satisfy the second-order derivative conditions revealing all *local* optima. Then comparing the values of the performance index of all the local optima, together with those of all boundary parameter sets, lead to the *global* supremum, whose corresponding 'parameter' set will thus represent the best design.

For this, let us begin with conventional optimization and search methods. Then we shall move on to more modern methods and machine learning. We attempt to establish how a machine may be programmed to learn and, from this, may be programmed to create.

5.2.2 Difficulties with Conventional Optimization Methods

Design solutions may be found by computational techniques. A design problem may be solved by numerical optimization means. However, a-priori methods are dependent upon gradient information, which may be difficult to obtain.

The a-posteriori methods studied so far are deterministic though they do not require derivatives. Both they and a-priori methods are prone to noise and lack the ability to find the best solution globally if there exist multiple local optima. The drawbacks of conventional optimization techniques are summarized below.

- *Existence Problem*
  Gradient guidance can adjust $\mathbf{P}_i$ only when $\nabla f(\mathbf{P}_i)$ (and in some cases a monotonic second order derivative) exist or the objective functions have well-defined smooth slopes [Goldberg (1989)]

- *Practical Problem*
  Conventional techniques are almost impossible to work with due to constraints found in practical applications. These constraints include direct domain constraints (such as parameter range requirements and fixed relationships) and indirect inequalities (such as voltage or current limits and other hard nonlinearity). Further, in practical applications, performance information may include noise, or be discontinuous, incomplete, uncertain and/or imprecise

- *Multi-modal Problem*
  A-priori guidance usually leads to a local optimum, although parallelism may overcome this to a certain extent. Conventional parallelism, however, means no effective mechanism to exchange information among parallel search points

- *Multi-Objective Problem*
  Conventional optimization techniques can usually deal with one objective at a time. In engineering practice, there are usually multiple design objectives that may not best be weighted to form a single composite objective

- *A-priori Problem*
  It is difficult to incorporate knowledge and experience that a designer may have on the design

It can be concluded that, therefore, a practical system design problem can hardly be the solution through conventional numerical means.

5.2.3 *Converting a Design Problem into a Simulation Problem*

An engineering design problem is often unsolvable by analytical means because the derivative of the design objective can hardly be obtained. This is mainly due to the complexity and (explicit and implicit) constraints of the physical system, such as bandwidth (frequency-domain) limits and actuator saturation (time-domain) limits. Hence, analytical optimization is seldom adopted in practice.

A simulation problem is 'passive' and is thus much easier to solve than a design problem that is usually 'active' or creative. A simulation (or evaluation) problem is usually solvable by numerical means on a digital computer, despite constraints. Existing CAD packages, such as **SPICE**, have been developed to carry out this type of tasks. However, most existing CAD packages only provide a passive simulation tool for practical engineers and accommodate few direct, automated or creative design facilities.

5.2.4 Manual Design through Trial and Error

Using an existing CAD package, a design engineer solves a design problem by heuristic simulations. The engineer has first to supply by trial certain a-priori system parameters, such as those obtained through preliminary analysis. Then simulations and evaluations are undertaken by using the package. If the simulated performance of the 'designed' system does not meet the specification, the designer will modify the values of the parameters somewhat randomly or by his/her existing or real-time experience. Then the simulation and trial process is repeated until a 'satisfactory' design emerges. Clearly, such a design technique suffers from the following deficiencies:

- The design cannot be carried out easily since mutual interactions among multiple parameters are hard to predict (*multi-dimensionality problem*)
- The resulting 'satisfactory' design does not necessarily offer the best or near-best performance (*multi-modality problem*) and there may be room for further improvements
- The design process is not automated and can be tediously long.

5.2.5 Automate Design by Exhaustive Search?

Since existing CAD packages provide a means of analysis, simulation and evaluation, one approach to solving the design problem and achieving automated designs could be to *exhaustively* evaluate all the possible candidate designs.

To illustrate this method, let the number of parameters be $n = 8$. Suppose in the entire design space **S** each parameter has 10 candidate values, then there are a total of 10^n permutations of design choices. Now suppose that each design evaluation by CAD simulations takes 0.1 second, then evaluating all candidate designs will take 0.1 second $\times 10^8 = 3.8$ months to complete. This would be unacceptable in practice.

In summary, such an exhaustive or enumerative search scheme does transform the unsolvable design problem to the solvable simulation problem. However, the number of search points and search time increase exponentially with the number of parameters that need to be optimized. Even the highly regarded exhaustive scheme, called dynamic programming, breaks down on problems of 'moderate' dimensionality and complexity [Bellman (1957), Goldberg (1989)].

5.2.6 Further Requirements on a CAD Environment

In addition, a modern CAD paradigm should be an *open system* to meet the following design challenges:

- Deal with complexity of practical systems

- High quality and accuracy of design
- Speed of design with shortened time to market
- Robustness, reliability and safety arising from the design
- Competition with available design tools (e.g., ease of use)

It is found that many CAD systems do not yet meet these challenges easily mainly due to the limitations of conventional optimization techniques. We shall see how computational intelligence that simulates human intelligence could help achieve these. For problems of 'Algorithm and Problem Classification', please read Section 2.3.4.

5.3 Linking Intelligent Design with Manufacture

Further the design and manufacture chain from physical prototyping or 3D printing is virtual prototyping. Digital virtual prototyping 'gives conceptual design, engineering, manufacturing and sales and marketing departments the ability to virtually explore a complete product before its built' [Wikipedia (2015a) , Bullinger and Fischer (1998)]. Research shows that 'manufacturers that use digital prototyping build half the number of physical prototypes as the average manufacturer gets to market 58 days faster than average, and experience 48 per cent lower prototyping costs' [Lubell et al. (2012)] and this is achieved even without CAutoD.

However, digital prototyping can only unleash its power when it is optimized or fully extends the boundary of existing products or designs, the problem of which is unsolvable in practice. For example, consider a design process involving setting 10 parameters of a potential invention, where each parameter presents 8 possible choices.

For example, if every CAD run takes 1 second to complete, the total time required to complete the task will take 810 seconds = 12,428 days = 34 years. With CAutoD, however, such an exhaustive CAD process is transformed into an intelligent design automation process, which completes in polynomial (as opposed to exponential) time via a biologically-inspired environmentally-interactive learning technique [Wikipedia (2015b)]. Digital prototyping with CAutoD as an enabling technology will thus break down the conventional discipline barriers and change the traditional product development cycle from

$concept \longrightarrow build \longrightarrow test \longrightarrow fix$

to

$concept \longrightarrow optimiseorinvent \longrightarrow build.$

This way, not only can costly and prolonged physical prototyping be saved at the design stage, but innovative digital prototypes can also be created through CAutoD, meeting multiple design objectives and substantially reducing development gearing and time-to-market [Li, et al. (2004)].

Outcomes of a scientific breakthrough in this programme will, therefore, transform and bridge the gap between fundamental research at a low technology readiness level (TRL) (which underpins digital prototyping and futuristic exploration) and applied innovation at a high TRL with a shortened time-to-market. As depicted in Figure 5.3, this approach also aims to work seamlessly with different industrial sectors, whereby smart designs for a step change in creativity can be achieved for smart manufacturing and market-targeted manufacturing.

The state of the art of i4 research and development is mostly represented by manufacturers on smart manufacturing to elevate digital manufacturing, i.e., on the industrial Internet [Annunziata and Evans (2012)] of smart and connected production to enable the factory floor to become an automated innovation centre [Siemens (2015)]. However, what are lacking at present are smart design tools commensurate with i4 further up the value chain and market informatics at the end of the value chain, both of which the factory-floor innovation needs to take into account.

The technological heart of this work is based on the nature-inspired CAutoD technique, which will revolutionize the way that designs are created and machines are built. The use-inspired CAutoD approach will elevate the traditional CAD-based rapid prototyping and 3D printing for i4 [Bickel and Alexa (2013)], with knowledge-intensive design and integrated value chain connected with the market demand for a step improvement in industrial efficiency, performance and competitiveness [Ang, et al. (2005)].

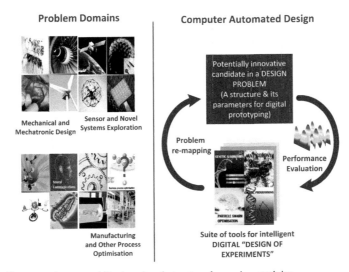

Fig. 5.3 Computer-Automated Design aimed at a step change in creativity

5.4 Computational Intelligence Integrated Solver

To utilize a diverse range of CI approaches, the CIS has been developed. As shown in Figure 5.4, the architecture of the CIS consists of three modules—the data input, the integrated solver and the result output, which can be described as follows:

- Module 1: Data Input (point A). This module prepares the data input for the integrated solver. It collects, filters, stores and pre-processes data originating from various sources such as statistical yearbooks, research analyses and government reports.
- Module 2: Integrated Solver. In this module, a set of nature-inspired computational approaches are integrated into one solver to optimize complex real-world problems, which primarily involve one or more of the following methods: GAs, simulated annealing (SA), artificial immune algorithms and quantum computation. In this chapter, the swarm dolphin algorithm (point C) is embedded in this solver and the details of this algorithm are introduced in Chapter 7.7.
- Module 3: Result Output (point B). This module reports the final results from module 2 and conducts all post-'Integrated Solver' tasks such as result visualization and data storage, and can be utilized for applications, such as artificial neural networks (ANNs) and fuzzy logic (FL).

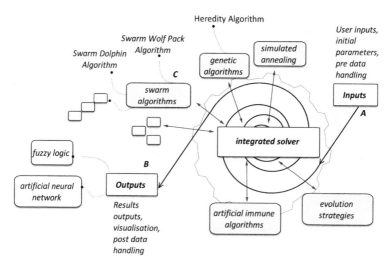

Fig. 5.4 The architecture of the CI integrated solver

5.5 CIAD

- *Phase 1*: CIAD, as also shown in Figure 5.5, can be deployed in 5 steps as follows:

  - *Step 1* is quantitative modeling under specific conditions for engineering ap-
    plications using CAD technologies.
  - *Step 2* is parameter optimization, which optimizes the CAD models via the
    pre-defined fitness functions according to the design objectives.
  - *Step 3* is a CIS that optimizes the parameters for the fitness function; the de-
    tails are given in Figure 5.4.
  - *Step 4* is the validation step, which is labeled as 'validation 1' for the CIAD
    phase. In particular, performance criteria are employed to assess the opti-
    mal results and then decide whether the optimization process should continue
    (*NO*) or be terminated before moving on to step 5 (*YES*).
  - *Step 5* produces the final results and completes the post-processing tasks.
    More specifically, this step reports the optimal solution, analyzes and visu-
    alizes the results, and presents the recommendations to the next phase, i.e.,
    CIAE.

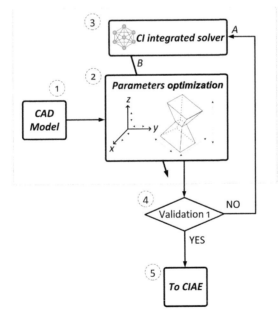

Fig. 5.5 CIAD phase

5.6 CIAE

- *Phase 2*: In CIAE, as shown in Figure 5.6, 5 steps are defined in this phase:

    - *Step 1* is to import data from CIAD and build models to satisfy the require-
      ments for engineering applications using computer-aided engineering (CAE)
      technologies, for example, the finite element method (FEM).
    - *Step 2* is parameter optimization for CAE models using the pre-defined fitness
      functions of engineering objectives.
    - *Step 3* is CIS solver work.
    - *Step 4* is the validation step, labeled as 'validation 2' in the CIAE phase. Sim-
      ilarly, based on the criteria, the decision of *NO* or *YES* should be made in this
      step.
    - *Step 5* passes the results to the next phase, i.e., CIVP.

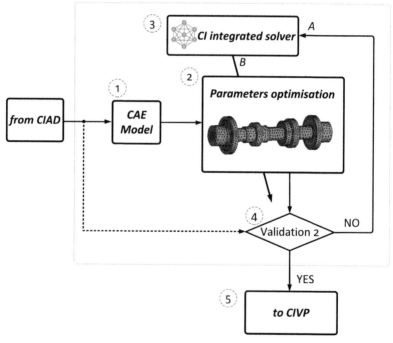

Fig. 5.6 CIAE phase

5.7 Intelligent Virtual Prototypes

- *Phase 3*: For CIVP, Figure 5.7 presents the five steps of this phase. The virtual prototype is built, for example, the dynamic model with an intelligent controller under the given virtual environments, in which co-simulations for the dynamic models and their controllers are performed with optimal parameters with the CIS solver. The data are imported from the CIAE phase when this phase starts.

  - *Step 1* is to design a control with optimal parameters determined by the CIS solver.
  - *Step 2* is parameter optimization for virtual prototype (VP) models of multi-body dynamics (MD) determined by the CIS solver, which could include rigid body systems, flexible body systems and multidisciplinary mechanical/mechatronic systems.
  - *Step 3* is the CIS solver that optimizes the parameters for the controllers and dynamic models.
  - *Step 4* is the validation step, similar to the validation steps of the two phases above and is labeled as 'validation 3' in the CIVP phase.
  - *Step 5* presents the results of the CIVP phase and conducts the analysis, visualization and data transfer to the next phase.

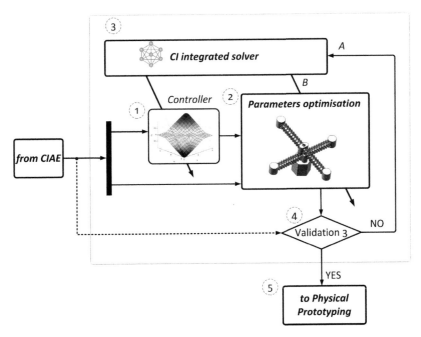

Fig. 5.7 CIVP phase

5.8 Physical Prototyping

- *Phase 4*: Based on the results and data imported from CIVP, this step performs the transformation of VPs into physical prototypes (PPs), where engineers can build a full-scale replica of an experimental system through rapid prototyping services. For example, to provide a rapid design, test and manufacture of mechanical, electro-mechanical and electronic products, 3D printing technology could be employed.

5.9 CIAM

- *Phase 5*: In CIAM, as also shown in Figure 5.8, validated by physical prototypes, digital models/data are imported in phases. The CIAM phase can be defined as consisting of five steps. CIAM is a phase subsequent to the CIAD, CIAE and CIVP phases. The models that were built, validated, optimized and refined in previous phases can be imported to the CIAM phase, which performs the computer-aided manufacturing (CAM) process. This process controls the machine tools and related machinery in manufacturing. In addition, it refers to the use of CIS systems for assisting all operations of a manufacturing plant, which may include planning, management, transportation, storage and other product lifecycle management (PLM) phases.

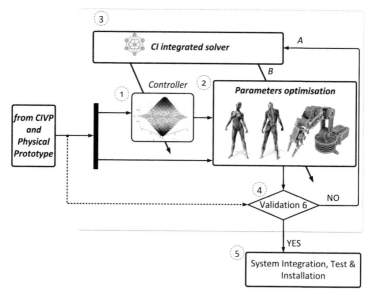

Fig. 5.8 CIAM phase

- *Step 1* is where a CIS-optimized controller is built and employed to drive computer numerically controlled (CNC) machines through a few numerical control (NC) programs using data imported from previous phases.
- *Step 2* is parameter optimization for manufacturing automation using the commands from the controller, which may include the use of computer applications to define a manufacturing plan for tool design, CAD model preparation, direct numerical control (DNC), NC programming, tool management, coordinate measuring machine (CMM) and execution, CNC machining, inspection programming, machine tool simulation, or post-processing.
- *Step 3* is the CIS that optimizes the CNC manufacturing process.
- *Step 4* is the validation step, which is labeled as 'validation 6' in the CIAM phase.
- *Step 5* summarizes the results of the CIAM phase and sends data to the next phase.

5.10 System Integration

- *Phase 6*: In system integration, testing and installation phase, all components are integrated, tested and installed to form a fully functional system or product.

5.11 Applications

- *Phase 7*: Multi-disciplinary applications supported by the previous phases, which have been demonstrated in some of our previous work in various diverse areas are: Terahertz spectroscopic analysis for public security [Chen, et al. (2011)], applied energy [Chen, et al. (2013)], sustainable development [Chen, et al. (2013), Chen, et al. (2012), Chen and Song (2012)], engineering modeling and design [Chen, et al. (2012)], aerospace engineering [Chen and Cartmell (2007)], automotive engineering [Chen, et al. (2012)], economics and finance [Chen and Zhang (2013)] and new drug development for public health care [Xu, et al. (2012), Liu et al. (2012)], etc.

5.12 Cyber-Physical Design Integration for Industry 4.0

Largely driven by technological innovation and intellectual ingenuity, smart design and manufacture play a key role in enhancing economic growth and competitiveness. A significant challenge to the current process is that it undergoes multiple phases, including: concept planning → prototyping → testing → correction → refinement → manufacture → end-user. Improving this process, digital technology promises Industry 4.0 improved design efficiency, short product's time-to-market, enhanced design and manufacturing flexibility and customization, reduced costs, and elevation of the manufacturing value chain, as illustrated in Figure 5.9.

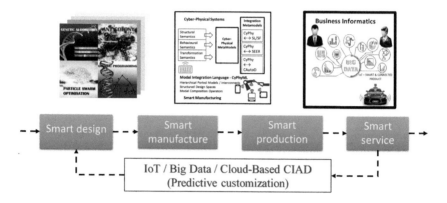

Fig. 5.9 Intelligent integration of Industry 4.0 value chain

Through the CIAD framework, a 'push button' solution can be achieved with cyber-physical integration, shortening the design and manufacture process to mining demand data → cyber-physical push button solutions → end-user. This helps increase the manufacturing competitiveness by an order of magnitude. It will also transform the business process, producing socio-economic benefits (e.g., machine/people safety and efficient industry value chains), mobilizing computational resources, utilizing multiple CI algorithms, reducing overall computational costs, facilitating knowledge integration between industrial partners and initiating international collaborations with a win-win outcome.

Looking forward, the CIAD-enabled integration of Industry 4.0 value chain will also facilitate the integration of internet of things and services, the cloud and big data technology and transform CI algorithms from the state of the art to state of practice. Overall, artificial intelligence improves production systems, nurtures the next generation of design and manufacture professionals and makes a step change in addressing societal challenges.

6

Extra-Numerical Multi-Objective Optimization

6.1 Introduction

Many real-world optimization problems involve extra-numerical and multiple objectives. CI approaches are able to encode the extra-numerical structure of the problem being tackled and the numerical parameters associated with the structure. Often the multi-objective nature presents objectives that cannot be reduced into one weighted composite objective, or multiple de-coupled independent objectives. Generally, a multi-objective (MO) optimization problem can be mathematically formulated as shown in Equation (6.1), subject to equality constraints $G_i(\mathbf{x})$ and inequality constraints $H_i(\mathbf{x})$, as given in Equations (6.2) and (6.3). Here, J is the number of objective functions given by $F_i: \Re^n \to \Re$; M is the number of equality constraints; I is the number of inequality constraints; $\mathbf{x}$ is the decision variable vector, as given in Equation (6.4); and K is the number of variables.

$$Minimize : F_i(\mathbf{x}) = [f_1(\mathbf{x}), f_2(\mathbf{x}), \cdots, f_J(\mathbf{x})], i = 1, 2, \cdots, J \qquad (6.1)$$

$$G_i(\mathbf{x}) = 0, i = 1, 2, \cdots, M \qquad (6.2)$$

$$H_i(\mathbf{x}) \leq 0, i = 1, 2, \cdots, I \qquad (6.3)$$

$$\mathbf{x} = [x_1, x_2, \cdots, x_K] \qquad (6.4)$$

Usually, the objectives in Equation (6.1) often conflict with each other, that is, improvement of one objective may lead to deterioration of another. An overall solution that can optimize all objectives simultaneously does not exist. Instead, the best trade-off solutions, called *Pareto* optimal solutions, are important to decision makers. A Pareto optimal set is a set of Pareto optimal solutions that are non-dominated with respect to each other. While moving from one Pareto solution to another, there is always a certain amount of sacrifice in one or more objectives to achieve a certain amount of gain in the other objective(s).

The term *multi-objective* refers to a search solution that provides the values of all the objective functions acceptable to decision makers [Coello Coello, et al. (2007)]. Thus, decision makers need to search for '*trade-offs*' rather than a single solution, which leads to a different solution in terms of '*optimality*' under multi-objective situations. The most widely used concept is *Pareto optimality*—an engineer can make a trade-off within this set under practical requirements by focusing on the set of *Pareto-front* choices. The *Pareto front* provides a visualized demonstration of the *Pareto-optimal* solutions but with an unclear indication of optimal diversities for decision making. A fast approach for Pareto-optimal solution recommendation has been proposed to address this issue, as introduced in Chapter 10, Section 10.6.

As shown in Figure 6.1, the best solution means a solution that is not worse in any of the objectives and at least better in one objective than the other solutions. An optimal solution is a solution that is not dominated by any other solution. Such an optimal solution is called a *Pareto-optimal* solution and the entire set of such optimal trade-off solutions is called a *Pareto-optimal set*. As evident, in a real-world situation, a decision-making (trade-off) process is required to obtain the optimal solution. Although there are several ways to approach an MO optimization problem, most efforts are concentrated on approximation of the Pareto set. Pareto optimal solution sets are usually preferred to single solutions because they can be practical when considering the real-life problems because the final solution for the decision maker is always a trade-off. Generally, an effective Pareto-based MO algorithm will converge on a solution set with the following properties: (1) the solutions are 'good' (i.e., close to the Pareto front); (2) the solutions are 'evenly spread' along the Pareto front; and (3) the solutions are 'widely spread' (i.e., they represent a good range). Chapter 10 will introduce a few metric indices used to address this topic.

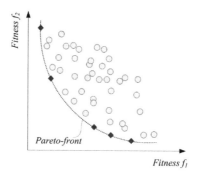

Fig. 6.1 Concept of Pareto-front

All vectors satisfying Equations (6.2) and (6.3) are called the solution set $\mathfrak{F}$, in which the decision variable set of $\mathbf{x}^* = [x_1^*, x_2^*, \cdots, x_K^*]$ yields the optimum values of all the objectives. The vector of decision variables $\mathbf{x}^* \in \mathfrak{F}$ is Pareto optimal if *there is no feasible vector of decision variables $\mathbf{x} \in \mathfrak{F}$ that increases some criterion without* *causing a simultaneous decrease in any other criterion.* The vectors $\mathbf{x}^*$ corresponding to the solutions included in the Pareto-optimal set are called *non-dominated vectors*. The curve of the Pareto-optimal set under the objective functions is called the *Pareto front* [Edgeworth (1881), Pareto (1896), Stadler (1988)], as shown in Figure 6.1.

6.2 History of Multi-Objective Optimization

The first MO evolution algorithm was introduced by Schaffer [Schaffer (1984)PhD, Schaffer (1985)] in the mid-1980s and is known as *the Vector Evaluated Genetic Algorithm (VEGA)*. In this method, the mating pool is divided into several parts, where each part is evaluated by a single objective fitness function. Although this algorithm has its limitations, it is typically used as a reference for benchmarking new algorithms. Since then, the growing importance of MO research has been reflected by a significant increase in the papers in international conferences and peer-reviewed journals, books, special sessions at international conferences and interest groups on the Internet [Coello Coello (2015)].

Typically, MO algorithms can be divided into three types [?]:

- Aggregating approaches (**AGG**)
- Population-based approaches (**POP**)
- Pareto-based approaches (**PAR**)

6.2.1 Aggregating Approaches

As one of the most straightforward approaches for addressing multiple objectives, 'aggregating function' techniques aggregate (or combine) all the objectives into a single objective using either an addition, multiplication or other combinations of arithmetical operations. An example of an AGG approach as a linear sum of weights of the form is given in Equation (6.5), where $\omega_i \geq 0$ are the weighting coefficients representing the relative importance of the J objective functions with the assumption given by Equation (6.6).

$$Minimize : F\left(\mathbf{x}\right) = \sum_{i}^{J} \omega_i f_i\left(\mathbf{x}\right) \qquad (6.5)$$

$$Minimize : \sum_{i}^{J} \omega_i = 1 \qquad (6.6)$$

Aggregating functions can be defined as linear or nonlinear and are employed in a wide range of evolutionary algorithms and related applications, for example, in game theory [Rao (1987)], goal programming [Deb (1999)] and the 'min-max' algorithm [Hajela and Lin(1992)].

In practice, the AGG optimization method would return a single solution rather than a set of solutions that can be examined for 'trade-off'. It can be very difficult to precisely and accurately select these weights, even for someone familiar with the specific applications. Aggregating functions have been largely studied by MO researchers because of the drawbacks of linear aggregating functions, namely, they cannot generate non-convex portions of the Pareto front regardless of the utilized weight combination [Das and Dennis (1997)]. The MO community tends to develop new algorithms based on non-AGG approaches. Therefore, decision makers often prefer a set of good solutions considering the multiple objectives.

6.2.2 Population-Based Approaches

In a POP approach, the population is employed to diversify the evolutionary search. As mentioned in Section 6.2, the first well-known POP approach was Schaffer's VEGA, proposed in 1984 [Schaffer (1984)PhD, Schaffer (1985)].

As shown in Figure 6.2, the VEGA consists of a simple GA with a modified selection mechanism. At each generation, some sub-populations are generated in turn by performing proportional selection according to each objective function.

Given the population as an $m \times n$ matrix, as shown in Figure 6.3, for a problem with J objectives, J sub-populations each of size m/J are generated. These sub-populations are then shuffled together to obtain a new population of size $m \times n$, to which the VEGA applies the crossover and mutation operators as a standard GA.

A few researchers have proposed variations of the VEGA and other similar population-based approaches in the last few decades [Venugopal and Narendran (1992),   Sridhar and Rajendran (1996), Norris and Crossley (1998), Rogers (2000), Coello Coello (2000)]. The main drawback of the VEGA is that Pareto dominance is not directly incorporated into the selection process, which is opposed to the concept of Pareto dominance. For example, if there is an individual that encodes a good compromise solution for all the objectives but it is not the best solution based on any of the objectives, it will be discarded.

6.2.3 Pareto-Based Approaches

Historically, Pareto-based approaches could be categorized into two generations [?], characterized as follows:

- The first-generation PAR using fitness sharing or niching techniques, combined with Pareto ranking [Goldberg (1989)]. Examples include the Non-dominated Sorting Genetic Algorithm (NSGA), Niched-Pareto Genetic Algorithm (NPGA) and Multi-objective Genetic Algorithm (MOGA).

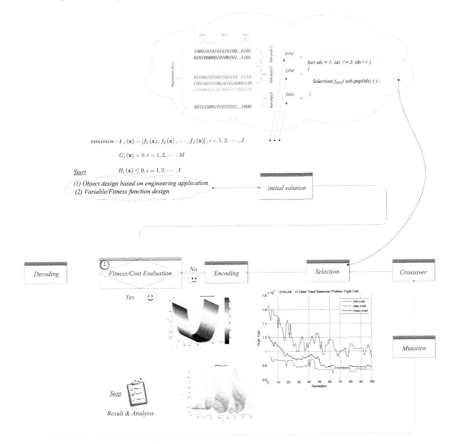

Fig. 6.2 Vector Evaluated Genetic Algorithm [Schaffer (1984)PhD, Schaffer (1985)]

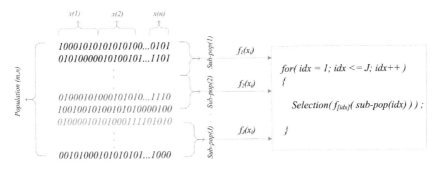

Fig. 6.3 A number of sub-populations for the VEGA

- The second-generation PAR using the mechanism of elitism. In this context of MO optimization, elitism usually refers to the use of an external population (also called the secondary population) to retain the non-dominated individuals. Examples include the Strength Pareto Evolutionary Algorithm (SPEA), Strength Pareto Evolutionary Algorithm 2 (SPEA 2), Pareto Archived Evolution Strategy (PAES), Non-dominated Sorting Genetic Algorithm II (NSGA-II), Niched Pareto Genetic Algorithm 2 (NPGA 2), Pareto Envelope-based Selection Algorithm (PESA) and Micro-Genetic Algorithm (μGA).

Table 6.1 Summary of Research on MO Optimization [?, Deb (2001), Coello Coello (2015), Tan, et al. (2002), Kim and de Weck (2005), Zhou (2011), Deb (2004)]

Year	Works	Type	Reference
1984	Vector Evaluated Genetic Algorithm (VEGA)	AGG	Schaffer [Schaffer (1984)PhD, Schaffer (1985)]
1989	Multi-objective Genetic Algorithm (MOGA)	PAR	Goldberg [Goldberg (1989)]
1992	Weight-based Genetic Algorithm (WBGA)	AGG	Hajela and Lin [Hajela and Lin(1992)]
1993	Niched Pareto Genetic Algorithm (NPGA)	PAR	Horn [Horn and Nafpliotis (1993)]
1994	Non-dominated Sorting Genetic Algorithm (NSGA)	PAR	Srinivas and Deb [Srinivas and Deb (1994)]
1995	Random Weighted Genetic Algorithm (RWGA)	AGG	Murata and Ishibuchi [Murata and Ishibuchi (1995)]
1999	Strength Pareto Evolutionary Algorithm (SPEA)	PAR	Zitzler and Thiele [Zitzler and Zitzler (1999)]
2000	Pareto-Archived Evolution Strategy (PAES)	PAR	Knowles and Corne [PAES and Corne (2000)]
2000	Pareto Envelope-based Selection Algorithm (PESA)	PAR	Corne et al. [PESA, et al. (2000)]
2001	Improved SPEA (SPEA2)	PAR	Zitzler et al. [Zitzler, et al. (2001)]
2001	Region-based Selection in Evolutionary multi-objective (PESA-II)	PAR	Corne et al. [PESAII, et al. (2001)]
2001	Micro-GA	PAR	Coello Coello and Toscano Pulido [Coello Coello and Toscano Pulido (2001)a, Coello Coello and Toscano Pulido (2001)b]
2002	Fast Non-dominated Sorting Genetic Algorithm (NSGA-II)	PAR	Deb et al. [Deb, et al. (2002)]
2002	Rank-Density Based Genetic Algorithm (RDGA)	PAR	Lu and Yen [Lu and Yen (2002)]
2003	Dynamic Multi-objective Evolutionary Algorithm (DMOEA)	PAR	Yen and Lu [Yen and Lu (2003)]
2005	Many-objective problems	PAR	Fleming et al. [Fleming, et al. (2005), Hughes (2005)]

Note that, although there are many variations of multi-objective GAs in literature, the above-cited GAs are well-known and credible algorithms that have been used in many applications, and their performances have been tested in several comparative studies [Konak, et al. (2006)]. Table 6.1 summarizes the work on MO while more details on the MOGA will be discussed in Section 6.4.

Most MO optimization problems found in literature have 2 or 3 objectives. In recent years, there has been a growing interest in the area of optimization wherein the problems might have more than 3 objectives, e.g., 4 to 20 objectives, which are referred as *many-objective* problems [Fleming, et al. (2005), Hughes (2005)].

6.3 Theory and Applications

Many real-world applications require the simultaneous optimization of several objectives, which are mostly competing against each other, such as a product's quality and its price.

An effective Pareto-based MO approach will converge on a solution set with the following properties:

- **Proximity**: Solutions are *good* - close to the Pareto front
- **Pertinency**: Solutions are *evenly spread* along the Pareto front
- **Diversity**: Solutions are *widely spread* in a good range along the Pareto front

To seek solution sets with the above properties, a few MO approaches have been proposed, some of which will be introduced in Section 6.4.

Current MO approaches have been widely utilized in scientific, engineering and industrial applications. Examples of MO applications are given, but not limited to, as follows:

1. **Engineering**
   - Electrical Engineering
   - Mechanical Engineering
   - Structural Engineering
   - Aeronautical Engineering
   - Robotics
   - Control Engineering
   - Telecommunications
   - Civil Engineering
   - Transport Engineering

2. **Industrial**
   - Design and Manufacture
   - Scheduling
   - Management

3. **Scientific**
   - Chemistry
   - Physics

- Medicine
- Computer Science
- Social Science

6.4 Multi-Objective Genetic Algorithm

Various widely used algorithms, such as the non-dominated sorting genetic algorithm II (NSGA-II) [Deb, et al. (2002)], have been developed to solve MO formulations.

GAs are global, parallel and stochastic search methods, founded on Darwinian evolutionary principles by Holland in 1975 [Holland (1975)]. GAs work with a population of potential solutions to a problem, and each individual within the population represents a specific solution to the problem and is expressed as some form of genetic code. Since then, GAs have frequently been applied as optimizers for different engineering applications. Practical problems are often characterized by several non-commensurable and often competing measures of performance, or objectives, with some restrictions imposed on the decision variable. Trade-offs exist between some objectives, where advancement in one objective will cause deterioration in another. It is very rare for problems to have a single solution. These problems usually have no unique or perfect solution; rather, they possess a set of non-dominated, alternative solutions, known as the Pareto-optimal set. The concept of Pareto optimality is only a first step towards solving an MO problem. The choice of a suitable compromise solution from non-inferior alternatives is not only problem dependent but also dependent on the subjective preferences of a decision agent, the decision maker. Thus, the final solution to the problem is the result of both an optimization process and a decision process.

The MOGA was proposed by Fonseca and Fleming in 1993 [Fonseca and Fleming (1993)]; it is an algorithm that applies Pareto ranking and sharing to the fitness objects' values. In this algorithm, an individual's rank corresponds to the number of individuals in the current population by which it is dominated. Non-dominated individuals are assigned the same rank, whereas dominated individuals are penalized according to the population density in the corresponding region of the trade-off surface. Fitness is assigned by interpolation, e.g., for a linear fitness function, from the best to worst individuals in the population, assigned values are averaged between individuals with the same rank.

The NPGA was proposed by Horn et al. [Horn and Goldberg (1994)]. This GA uses a tournament selection scheme based on Pareto dominance. Two randomly chosen individuals are compared against a subset of the entire population. When both competitors are either dominated or non-dominated, the result of the tournament is decided through fitness sharing in the objective domain (a technique called equivalent class sharing).

The NSGA was proposed by Srinivas and Deb [Srinivas and Deb (1994)]. It is based on several layers of classifications of individuals. Non-dominated individuals obtain a certain dummy fitness value and then are removed from the population. The process is repeated until the entire population has been classified. To maintain the diversity of the population, classified individuals are shared with their dummy fitness values.

The NSGA-II was proposed by Deb et al. [Deb, et al. (2002)]. It is a new version of NSGA that is more efficient (computationally speaking); it uses elitism and a crowded comparison operator that maintains diversity without specifying any additional parameters.

7
Swarm Intelligence

7.1 Introduction

The fundamental principle of swarm intelligence hinges on probabilistic-based search algorithms. All swarm intelligence models exhibit a number of general properties. Each entity of the swarm is composed of a simple agent. Communication among agents is indirect and short. Cooperation among agents is realized in a distributed manner without a centralized control mechanism. These properties make swarm intelligence models easy to be realized and extended such that a high degree of robustness can be achieved. In other words, the entire swarm intelligence model is simple in nature. However, the collective colony-level behavior of the swarm that emerges out of the interactions is useful in achieving complex goals [Bonabeau, et al. (2000), Poli, et al. (2007)].

7.2 Particle Swarm Optimization

Particle Swarm Optimization (PSO) is a population-based stochastic optimization technique developed by Eberhart and Kennedy [Kennedy and Eberhart (1995)] in 1995, inspired by the social behavior of bird flocking.

As shown in Figure 7.1, a swarm of 'particles' fly in the n-dimensional search space of the optimization problem looking for optimal or near-optimal regions. Given N particles, let us select any two particles, namely, particle $\bullet$ i and particle $\blacktriangle$ j. Each particle will be guided by two information sources: (1) the cognitive information based on the particle's own experience and (2) the social information based on observations of neighbors, that is, each particle communicates with its neighboring particles and is affected by the best point found by any member of its topological neighborhood.

We assume that the position vector is $X(t) = \{x_1(t), \ldots, x_i(t), \ldots, x_j(t), \ldots, x_N(t)\}$ and that the velocity vector is $V(t) = \{v_1(t), \ldots, v_i(t), \ldots, v_j(t), \ldots, v_N(t)\}$. The position $x_i(t)$ of particle i represents a solution candidate itself at time t, and the velocity $v_i(t)$ represents

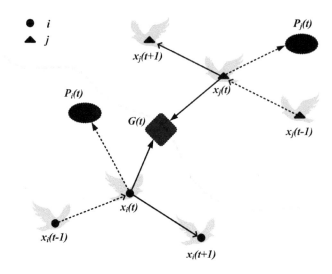

Fig. 7.1 The positions and velocities of particle i and particle j

information about the direction and changing rate of particle i at time t. For particle i, $x_i(t-1)$, $x_i(t)$ and $x_i(t+1)$ are the positions at time $t-1$, t and $t+1$, respectively. $P_i(t)$ is the best position found by particle i at time t, and $G(t)$ is the best position found by the best neighbor (particle j in this case) at time t.

Similarly, $x_j(t)$ and $v_j(t)$ represent the position and velocity of particle j; $x_j(t-1)$, $x_j(t)$ and $x_j(t+1)$ are the positions of particle j at time $t-1$, t and $t+1$, respectively. $P_j(t)$ is the best position found by particle j at time t.

The workflow of PSO is given in Figure 7.2 and can be summarized as follows:

- Initialize all the variables and parameters, e.g., the positions $X(t)$, the velocities $V(t)$ and the population
- Evaluate fitness functions and then update each of the positions x_i and velocities v_i from t to $t+1$ using the updating rules, as given by Equations (7.1) and (7.2), respectively. Here, w is the inertia weight, which is responsible for the scope of exploration of the search space. Larger values of w promote global exploration and exploitation, whereas smaller values of w lead to local search. This provides a common approach for balancing global and local search; r_1 and r_2 are random numbers between 0 and 1, and the coefficients c_1 and c_2 are given acceleration constants towards P and G, respectively
- Obtain the current global best positions $P(t)$ and $G(t)$ and keep their historical records. Compare the local and global solutions and update the solutions

$$x_i(t+1) = x_i(t) + v_i(t+1) \tag{7.1}$$

$$v_i(t+1) = wv_i(t) + c_1r_1(P_i(t) - x_i(t)) + c_2r_2(G_i(t) - x_i(t)) \tag{7.2}$$

- The iteration continues until terminal conditions are reached.

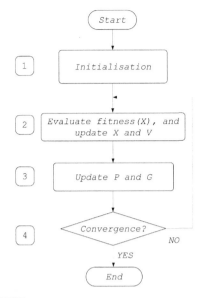

Fig. 7.2 The workflow of PSO

7.3 Ant Colony Optimization

Ant algorithms are iterative, probabilistic meta-heuristic for finding solutions to combinatorial optimization problems. They are based on the foraging mechanism employed by real ants when attempting to find a shortest path from their nest to a food source. When foraging, the ants communicate indirectly via pheromone their respective paths which they use to mark and which attracts other ants.

In the ant algorithm, artificial ants use virtual pheromone to update their path through the decision graph, i.e., the path that reflects which alternative an ant chose at certain points. The amount of pheromone an ant uses to update its path depends on how good the solution implied by the path is in comparison to those found by competing ants in the same iteration. Ants of later iterations use the pheromone markings of previous good ants as a means of orientation when constructing their solutions; this ultimately results in the ants focusing on promising parts of the search space. For additional applications of ant colony optimization (ACO), please refer to [Dorigo, et al. (1999)].

A swarm of such ants is capable of finding the shortest path connecting the nest to the foraging area containing the food, as the experiment shows. As shown in Figure 7.3, initially, all ants are located at the nest site. A number of ants start out from the nest in search of food, each laying pheromone on its path, and reach the first fork at point A. Since the ants have no information on which way to go, i.e., no ant has walked before and there is no pheromone of pre-requisite knowledge for their reference, the ants choose the left way (to point B) and right way (to point G) with equal probability.

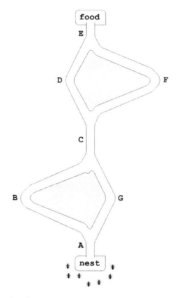

Fig. 7.3 Ants start to look for food

$$\tau_{i,j}(t+1) = \rho\,\tau_{i,j}(t) + \Delta\tau_{i,j}(t,t+1) \tag{7.3}$$

To describe the pheromone updates, let $\tau_{ij}(t)$ be the intensity of the trail on path or edge $L_{i,j}$ at time t, as given by Equation (7.3), where

- ρ is an evaporation coefficient. To avoid unlimited accumulation of trails, its value should be $\in [0,1]$. $(1-\rho)$ represents the evaporation of trails
- $\Delta\tau_{i,j}(t,t+1)$ is the individual update value for each ant, as given by Equation (7.4)

$$\Delta\tau_{i,j}(t,t+1) = \sum_{k=1}^{m} \tau_{i,j}^{k}(t,t+1) \tag{7.4}$$

- $\tau_{i,j}^{k}(t,t+1)$ is the quantity per unit of length of trail substance (pheromone) laid on $L_{i,j}$ by the k-th ant between times t and $t+1$, which can be defined by Equation (7.5), where Q is a constant and L^{k} is the tour length of the k-th ant

$$\tau_{i,j}^{k}(t,t+1) = \begin{cases} \dfrac{Q}{L^{k}} & \text{if } k\text{-th ant uses edge (i,j) in its tour} \\[2mm] 0 & \text{otherwise} \end{cases} \tag{7.5}$$

As shown in Figure 7.4, initially, one-half of the foraging ants are on the shorter route ($L_8 \rightarrow L_7$), with the remainder on the longer route ($L_1 \rightarrow L_2$) to intersection C. The ants that are on the shorter track will reach intersection C first and have to decide which way to turn. Again, there is no information for the ants to use as orientation

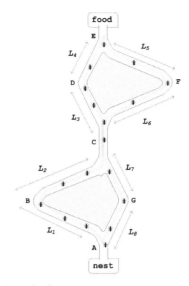

Fig. 7.4 Ants search for paths to food

information; thus, half of the ants reaching intersection B will turn back towards the nest, while the remainder will continue towards the foraging area containing the food.

As the search-for-food tour proceeds, the ants form a data structure, called a 'tabu list' (TBL), which is associated with each ant to avoid ants visiting a town more than once. The TBL memorizes the towns already visited up to time t, and ants are forbidden from visiting them again before they have completed a tour. When a tour is completed, the TBL is set to empty and every ant is again free to choose its way. We define S^k as a vector containing the k-th ant and $S^k(i)$ is the i-th element of S^k, which denotes the i-th point visited by the k-th ant in the current tour.

Equation (7.6) gives the transition probability $p_{i,j}^k(t)$ of the k-th ant on edge $L_{i,j}$ at time t by the 'random proportional transition rule' including heuristic information, which is a trade-off between visibility and trail intensity, where $\eta_{i,j}$ is visibility and α and β are parameters that allow a user to control the relative importance of trail $\tau_{i,j}$ versus visibility $\eta_{i,j}$. Equation (7.7) shows the j choosing approach with heuristic information perceived by the ant.

$$p_{i,j}^k(t) = \begin{cases} \dfrac{[\tau_{i,j}(t)]^\alpha [\eta_{i,j}]^\beta}{\sum_{h \in S^k}[\tau_{i,j}(t)]^\alpha [\eta_{i,j}]^\beta} & j \in S^k \\ 0 & \text{otherwise} \end{cases} \tag{7.6}$$

$$j = argmax[\tau_{i,j}(t)]^\alpha [\eta_{i,j}]^\beta \tag{7.7}$$

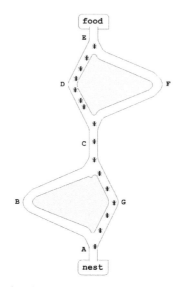

Fig. 7.5 Ants find the best path to food

Figure 7.5 shows the best path to food that the ants found. Figures 7.3 to 7.5 demonstrate the mechanism on how the ants search for food by marking paths with pheromones, which can be exploited to construct algorithms capable of solving highly complex combinatorial optimization problems.

7.4 Swarm Fish Algorithm

7.4.1 Swarm Fish Algorithm with Variable Population

In this section, a swarm fish algorithm with a variable population (AFVP) is introduced. Inspired by the swarm intelligence of fish schooling behaviors, the AFVP is an artificial intelligence algorithm that simulates the behavior of an individual artificial fish (AF) and then constructs an AF school.

The pseudocode of the AFVP is given in Figure 7.6 and the workflow of the AFVP is given in Figure 7.7. Similar to the basic fish algorithm [Li, et al. (2002), Shen, et al. (2011)], the AFVP includes 5-step operations: (1) behavior selection, (2) searching behavior, (3) swarming behavior, (4) following behavior, and (5) bulletin and update.

Behavior Selection: In the AFVP, the behavior selection step takes the searching behavior as the default or initial behavior for each AF. According to food density, the number of companions and visual conditions, the AF school selects their behaviors, including searching behavior, following behavior and swarming behavior.

```
Begin (1)

   t = 0 ;
   Initialise P(0);

      While ( Not termination-condition) do

        Begin (2)
        {
        t      = t + 1;
        flag = Evaluation P(t);

        switch( Behaviour Selection( flag ) )
        {
          state 1: swarming behaviour;
          state 2: following behaviour;
          default: searching behaviour;
        }

        bulletin P(t)= P_R + P_N;
          }

        End (2)

End (1)
```

Fig. 7.6 The pseudocode of the AFVP

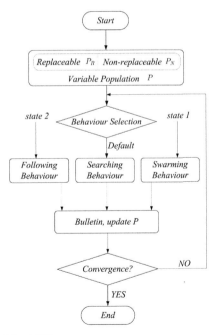

Fig. 7.7 The workflow of the AFVP

Searching Behavior: For a certain individual AF k, $S_k = \{s_1, \ldots, s_M\}$ is its finite state set, where there are M states that an AF can perform. Within the AF's visual field, suppose that the current state of this AF is S_i and that the next state is S_j. The AF moves from S_i to S_j randomly and checks the state updating conditions, as given by Equations (7.8) and (7.9). As demonstrated in Figure 7.8, $r_{ij} = \|S_j - S_i\|$ is the distance between the ith and j^{th} individual AF; $F = f(S)$ is the food density for this AF, where F is the fitness function; δ is the iteration step, and υ is the AF visual constant.

$$
S_{i+1} = \begin{cases} S_i + RAND \cdot \delta \cdot \dfrac{S_j - S_i}{\|S_j - S_i\|} & (\text{if } F_j > F_i) \\[4mm] S_i + RAND \cdot \delta & (\text{else}) \end{cases} \tag{7.8}
$$

$$
S_j = S_i + RAND \cdot \upsilon \tag{7.9}
$$

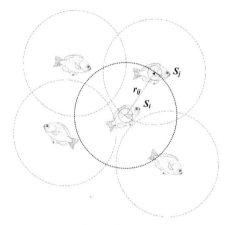

Fig. 7.8 The state distance between the ith and j^{th} individuals

Swarming Behavior: Suppose that the number of neighbors of this AF is α, the central state is S_c, the food density is $F_c = f(S_c)$, and η is the crowd factor. Within its visual field ($r_{ij} < \upsilon$), if $F_c/\alpha > \eta F_i$ and $\eta \geq 1$, the AF acts on the central state-driven step; otherwise, when $F_c/\alpha \leq \eta F_i$ or $\eta = 1$, the AF will continue the search behavior, as expressed by Equation (7.10).

$$
S_{i+1} = \begin{cases} S_i + RAND \cdot \delta \cdot \dfrac{S_c - S_i}{\|S_c - S_i\|} & \left(\text{if } \dfrac{F_c}{\alpha} > \eta F_i\right) \text{ and } (\eta \geq 1) \\[4mm] Equation (7.8) & \left(\dfrac{F_c}{\alpha} \leq \eta F_i\right) \quad or \quad (\eta = 0) \end{cases} \tag{7.10}
$$

Following Behavior: When the AF's companions reach the 'max' state S_{max} with the number α within the neighborhood, the food density reaches F_{max} in the mean time. As stated in Equation (7.11), under the same conditions as Equation (7.10), the AF updates its state in the highest food density area; otherwise, the AF will continue with the search behavior, as expressed by Equation (7.10).

$$S_{i+1} = \begin{cases} S_i + RAND \cdot \delta \cdot \dfrac{S_{max} - S_i}{\|S_{max} - S_i\|} & \left(\text{if } \dfrac{F_{max}}{\alpha} > \eta F_i\right) \text{ and } (\eta \geq 1) \\[3ex] Equation(7.8) & \left(\dfrac{F_{max}}{\alpha} \leq \eta F_i\right) \quad or \quad (\eta = 0) \end{cases} \tag{7.11}$$

Bulletin: The bulletin operation is a step used to compare each AF's current state S_i with historical state data. The bulletin data will be replaced and updated only when the current state is better than the last state, as described by Equation (7.12).

$$S_{j+1} = \begin{cases} S_j \; (\text{if } F_j > F_i) \\[2ex] S_i \; (\text{else}) \end{cases} \tag{7.12}$$

The 'max-generation' parameter is the trial number of an AF school search for food under given initial conditions and is one of the widely used criteria for terminating the AFVP simulation [Chen, et al. (2013)].

7.4.2 Multi-Objective Artificial Swarm Fish Algorithm

Furthermore, a multi-objective artificial swarm fish algorithm with the variable of population size using the non-dominated sorting method (MOAFNS) is discussed. Based on the AFVP, the MOAFNS is an artificial intelligence algorithm that addresses MO optimization problems by simulating the behaviors of AF and the interaction with other swarm members. Each AF searches its local optimal solutions for a few design objectives and passes information to its self-organized system, and the non-dominated sorting method is utilized to generate non-dominated fronts of an evolutionary MO optimization process. Finally, the global optimal solutions are achieved.

The MOAFNS workflow is illustrated in Figure 7.9. Similar to the artificial swarm fish algorithm [Chen, et al. (2012), Chen, et al. (2013)], the core MOAFNS algorithm includes four steps of operations: step (1) variable AF population P generation; step (2) behavior selection, in which one of the three types of behaviors (searching, swarming and following behaviors) will be selected for each loop; step (3) bulletin and update; and step (4) non-dominated sorting using the non-dominated sorting genetic algorithm II (NSGA-II) [Deb, et al. (2002)].

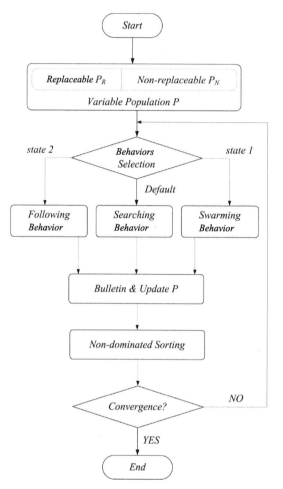

Fig. 7.9 Workflow of the multi-objective artificial swarm fish algorithm

Step (1) of the MOAFNS, as stated in Equation (7.13), provides a one-variable AF population P. The AF school population P is composed of two sub-schools: the replaceable school P_R and the non-replaceable school P_N, in which P_R at generation t is calculated based on its own size plus the off-size ΔP_R at generation $t-1$. The off-size replaceable population ΔP_R is proportional to the size of P_R, where λ is a reproduction rate set by users.

$$\begin{cases} P(t) & = P_R(t) + P_N(t) \\ P_R(t) & = P_R(t-1) + \Delta P_R(t-1) \\ \Delta P_R(t-1) = \lambda P(t-1) \end{cases} \quad (7.13)$$

In Step (2), according to the food density, the number of companions and the visual conditions, the AF school selects their behaviors, which include searching, following and swarming. The search behavior is considered default or initial behavior for each AF.

In Step (3), the algorithm compares each individual's current state with its historical states and the bulletin data are replaced and updated when the current state out-performs the last state.

In Step (4), the non-dominated sorting method has the following sub-steps: sub-step (1) fast non-dominated sorting; sub-step (2) crowding distance assignment; and sub-step (3) the crowded-comparison operator.

In this context, 'max-generation' is the convergence criterion of the swarm AF searching for food under a given initial condition and is one of the widely used termination conditions for MOAFNS optimization.

7.4.3 Case Study

As given by Equation (7.14), a standard test problem, ZDT6 [Zitzler, et al. (1999)] is solved by NSGA-II implemented in the MATLAB® toolboxes *SGALAB* [SGALAB (2009)], *Swarmfish* [Swarmfish (2011)] and *SECFLAB* [SECFLAB (2012)], in which $x_i \in [0,1]$ and $n = 10$ in this context.

$$
\begin{cases}
f_1(x) = 1 - \exp(-4x_1)\sin^6(6\pi x_1) \\[2ex]
f_2(x) = g(x)\left[1 - \left(\frac{f_1(x)}{g(x)}\right)^2\right] \\[2ex]
g(x) = 1 + 9\left[\frac{\sum\limits_{i=2}^{n} x_i}{n-1}\right]^{0.25}
\end{cases}
\tag{7.14}
$$

The parameters for this case are listed in Table 7.1, in which a max-generation of 200 is the termination condition of each round in the test, the total test number is 10, the population is 30, and the tournament selection operator, binary encoding/decoding method, and single-point crossover and mutation operators with $p_c = 0.8$ and $p_m = 0.01$, respectively, are used.

Figures 7.10 and 7.11 are the mAP $\pm$ mSTD diagrams for f_1 and f_2 over full simulation generations and indicate that a better solution for f_1 can be optimized without degrading a solution of f_2, which is not dominated by any other solution in the search space. As can be seen in Figure 7.10, the f_1's mAP $\pm$ mSTD curves increase quickly from generation = 1 to 8; they reach a stable status with a minor fluctuation at generation = 10 and remain there until the end of the simulation. Figure 7.11 shows that the f_2's mAP $\pm$ mSTD curves have a similar shape for generation = 1 to 8 and present a second jump from generation = 100 to 130, subsequently remaining stable until the end of the simulation.

Table 7.1 Parameters for MOAFNS Optimization

	Multi-objective algorithm	NSGA-II
	Max-generation	200
	Crossover probability (p_c)	0.8
	Mutation probability (p_m)	0.01
	Population	60
	Test number	10
	Selection operator	tournament
	Crossover operator	single point
	Mutation operator	single point
	Encoding/decoding method	binary
P_N	Non-replaceable population	50
P_R	Replaceable population	10
δ	Iteration step	0.5
υ	Visual	2.5
η	Crowd	0.618
	Attempt number	5

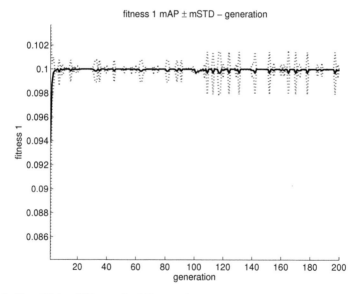

Fig. 7.10 f_1's mAP $\pm$ mSTD over the full generations

In Table 7.2, the recommended solutions are listed by the solution number in the column 'SOLUTION No.' with descending rank (RANK = 1 is the most recommended) using the values of β_1.

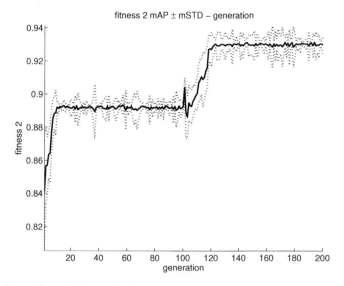

Fig. 7.11 f_2's mAP ± mSTD over the full generations

Table 7.2 Solution Recommendation for the ZDT6 Test Problem

RANK	SOLUTION No.	β_1
1	41	2.25e15
2	43	2.25e15
3	43	2.25e15
...	...	...
21	218	6.83
22	219	6.83
23	220	6.83
...	...	...
44	15	4.63
45	16	4.63
46	17	4.63
...	...	...
300	270	0.499

7.4.4 Conclusions

Using the index β_1, the FPR approach was utilized to provide a ranking list of Pareto-optimal solutions to facilitate decision making. The evolutionary trends were indicated by the indices mAP ± mSTD with variable uncertainty tolerances.

In further work, aimed at on-board industrial applications, this added-value approach has potential applications in MO optimization for robotic systems (e.g., exoskeletons, robotic space tethers, humanoid robots and industrial robotics), mobile devices and other systems with electrochemical power sources. In addition, other computational intelligence methods, such as swarm algorithms, will be applied to search for the recommended solutions of both an optimization process and a decision process.

7.5 Swarm Bat Algorithm

This section introduces a swarm bat algorithm with variable population (BAVP). Because the BAVP is inspired by the echolocation characteristics of swarm bats, it can be idealized to include the four following assumptions:

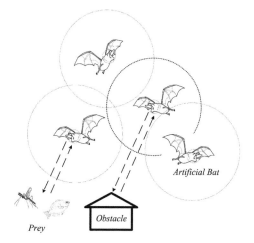

Fig. 7.12 The behaviors of swarm bats

- As shown in Figure 7.12, all artificial bats (ABs) utilize the same echolocation mechanism to measure distance, and each AB B_i can detect the difference between prey (food) and obstacles
- Each individual B_i can generate ultrasound to echolocate prey and obstacles with a velocity $v_{i,j}$ and position $x_{i,j}$ at time j, which are given by Equations (7.16) and (7.15), respectively, where x_* is the current global best position

$$x_{i,j+1} = v_{i,j} + x_{i,j} \qquad (7.15)$$

$$v_{i,j+1} = v_{i,j} + (x_{i,j} - x_*) f_{i,j} \qquad (7.16)$$

- Each individual B_i can adjust the ultrasound's frequency $f_{i,j}$ at time j within the range of $[f_{min}, f_{max}]$, corresponding to a wavelength λ in the range of $[\lambda_{min}, \lambda_{max}]$ and a loudness A in the range of $[A_{min}, A_{max}]$, as given by Equation (7.17), where β is a random vector of uniform distribution in the range of $[0,1]$

$$f_{i,j} = f_{min} + (f_{max} - f_{min}) \beta \qquad (7.17)$$

- As shown in Equation (7.23), the population P_j of ABs varies from time j to another time, which accelerates the optimization process, in which P_N is the non-replaceable population and P_{R_j} is the replaceable population at time j

$$P_j = P_N + P_{R_j} \tag{7.18}$$

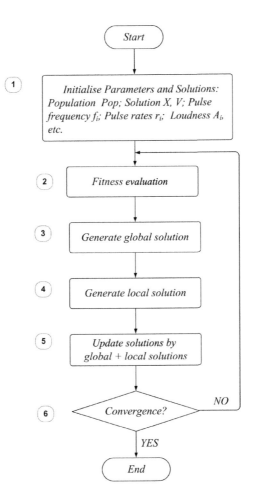

Fig. 7.13 The flow chart of BAVP

As shown in Figure 7.13, the following six steps are included in the BAVP flowchart: step (1) initialization; step (2) fitness evaluation; step (3) global solution generation; step (4) local solution generation; step (5) solution update using global and local solutions; and step (6) checking termination conditions for convergence. Specifically,

- Step (1) starts the program and initializes parameters and solutions; all artificial bats are moving randomly
- Step (2) evaluates the fitness for each solution
- Step (3) generates new global solutions x, updates velocities and adjusts frequencies using Equations (7.15) to (7.17)
- Step (4) generates new local solutions x^0 using Equation (7.19), where $\varepsilon \in [-1,1]$ is a random-walk factor. As defined by Equation (7.20), $A_{i,j}$ is the loudness of the bat B_i at time j, where $\alpha \in [0,1]$ is a reduction factor

$$x^0_{i,j+1} = x^0_{i,j} + \varepsilon A_{i,j} \qquad (7.19)$$

$$A_{i,j+1} = \alpha A_{i,j} \qquad (7.20)$$

- Step (5) compares the local and global solutions and updates solutions, as given by Equation (7.21)

$$x_{i,j} = \begin{cases} x_{i,j} \left(\text{if } x_{i,j} \geq x^0_{i,j} \right) \\ \\ x^0_{i,j} \left(\text{otherwise } x_{i,j} < x^0_{i,j} \right) \end{cases} \qquad (7.21)$$

- Step (6) continues running the calculation until the terminal conditions have been satisfied

7.6 Firefly Algorithm

The firefly algorithm (FA) was first proposed by Yang [Yang (2008)] in 2007. The FA is a metaheuristic algorithm inspired by the swarm behaviors of fireflies. The primary purpose of a firefly's flash is to act as a signal to attract its peers. Early FA algorithms were based on a fixed population, hindering their searching efficiency (we will discuss this later). This paper proposes a variable-population-based firefly algorithm (FAVP) to accelerate the computational speed. Based on the flashing characteristics of a firefly swarm, the working principle of the FAVP can be idealized through four behavioral rules:

- All individual fireflies (FF_i) are unisex and always move toward their neighbors with higher brightness. The brightness (also called the light intensity) I is given by Equation (7.22), in which r is the distance between two fireflies FF_i and FF_j, I_0 is the initial brightness, γ is the absorption coefficient for the decrease in the brightness and m is the multi-state factor, $m \geq 1$

$$I(r) = I_0 \exp\left(-\gamma r^m\right) \qquad (7.22)$$

- For any two fireflies FF_i and FF_j, their attractiveness β is proportional to the brightness, where, if FF_j is brighter than FF_i, FF_i will move toward FF_j. The brightness of FF_i and FF_j decreases if their distance increases; if no FF_i is brighter than any other firefly, all fireflies will move randomly and uniformly

- The brightness of a firefly is determined by the fitness function
- To speed up the computation and reduce the computation time cost, the population P of fireflies is allowed to vary from generation to generation. The variable population P is given by Equation (7.23), where P_N is the non-replaceable population and P_R is the replaceable population. The non-replaceable and replaceable population scheme provides flexible spatial storage for improving algorithm efficiency and reducing CPU time costs

$$P = P_N + P_R \tag{7.23}$$

The attractiveness function of FF_i to FF_j is defined by Equation (7.24), where β_0 is the attractiveness at the initial distance r_0 and the remaining parameters are the same as in Equation (7.22).

$$\beta_{ij}(r) = \beta_0 \exp\left(-\gamma r_{ij}^m\right) \tag{7.24}$$

The distance between any two fireflies FF_i and FF_j is measured by the Euclidean distance, as given by Equation (7.25) at positions x_i and x_j, where $x_{i,k}$ and $x_{j,k}$ are the k-th components of the spatial coordinates x_i and x_j of FF_i and FF_j, respectively, and d is the number of dimensions.

$$r_{ij} = \|x_i - x_j\| = \sqrt{\sum_{k=1}^{d} \left(x_{i,k} - x_{j,k}\right)^2} \tag{7.25}$$

The movement of FF_i to FF_j is determined by Equation (7.26), where ε is random movement in the case where equal brightness is generated by a uniform random variable in the range $[0,1]$. $\alpha \in [0,1]$ is a randomized factor determined through practice.

$$x_{i+1} = x_i + \beta_{ij}(r)(x_j - x_i) + \alpha\left(\varepsilon - \frac{1}{2}\right) \tag{7.26}$$

Figure 7.14 gives the flowchart (left) and pseudocode (right) of the FAVP algorithm. The algorithm starts by initializing the parameters and the population P based on Equation (7.23). The algorithm then compares the brightness I_i and I_j between any two fireflies. Based on the result of the comparison, the fireflies move and update the brightness and distance values simultaneously. The algorithm then ranks the solutions of the current iteration according to the fitness function. The iteration continues until terminal conditions are reached.

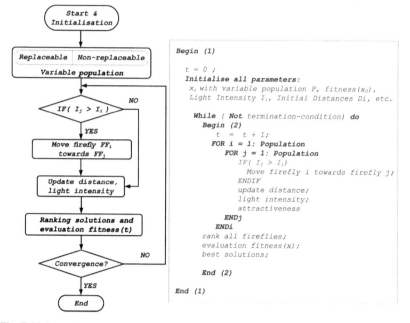

Fig. 7.14 Flowchart (left) and pseudocode (right) of the variable-population-based firefly algorithm

7.7 Artificial Dolphin Swarm Algorithm

7.7.1 Introduction

Dolphins have been regarded as one of the most intelligent animals on Earth. There are approximately 40 species of dolphins in 17 genera, which vary in size from 1.2 meters and 40 kilograms (e.g., Maui's dolphin) up to 9.5 meters and 10 tons (e.g., the orca and the killer whale). Dolphins can be found worldwide, mostly in the shallower seas of the continental shelves.

Dolphins are intelligent animals exhibiting a wide range of complex social behaviors. Compared to many other species, dolphins' behaviors have been widely studied, both in captivity and in the wild. Figure 7.15 presents the anatomy of a Bottlenose dolphin with a schematic diagram, including its eyes, tail and body shape. Dolphins always travel in social groups and communicate with each other through a complex system of squeaks and whistles. They have excellent hearing, which compensates for a poor sense of smell and for the uncertainties of visibility underwater. They have a high tolerance to carbon dioxide to help with lengthy dives and are two to three times more efficient than land mammals at using the oxygen in inhaled air. Finally, their rib cages are collapsible for deep diving, and they have layers of insulating fat to keep them warm.

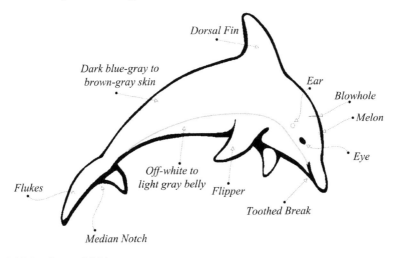

Fig. 7.15 Bottlenose dolphin anatomy

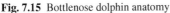

When dolphins are feeding, their target is often a bottom-dwelling fish, although they also eat shrimp and squid. Dolphins track their prey through echolocation using sounds with frequencies of up to 1000 Hz. These sounds travel underwater until they encounter objects and then bounce back to their dolphin senders, revealing the location, size and shape of their target. Based on the echolocation behaviors of swarm dolphins, we intend to propose a new metaheuristic method, namely, the Swarm Dolphin Algorithm (SDA). Dolphins' echolocation capability is fascinating, as they can track their prey and discriminate between different types of prey underwater.

The contributions can be highlighted in two aspects. First, a swarm intelligence method-SDA is devised as the search engine to optimize the validation cases; second, two new metrics, the index mmAP and the index mmSTD (as introduced in Section 10.5.2), are employed to characterize the dynamic behaviors of the evolutionary search process.

7.7.2 Dynamic Behaviors of Dolphins

In dolphins' fission-fusion societies, a primary group (PG) breaks up into a few smaller sub-groups (SGs) to explore, forage and later rejoin other SGs into a PG to both communicate and share resources while enhancing overall survival. Governed by a general social hierarchy, SGs may or not have leaders in each sub-group (SG), as it usually is the largest dolphin in body size or the most dominant. During daily activities, dolphins always exhibit cooperative behaviors, which can be briefly defined as two processes: foraging (search, detect and capture) and searching for partners [Schusterman, et al. (1986), Pryor (1998), Mann, et al. (2000), Haque, et al. (2009)].

7.7.2.1 Foraging

In proper formation, dolphins perform actions in three phases—**search**, **detect** and **capture**, which are the reactions to the threat level of the surroundings, such as that determined by the distribution of dolphin predators (killer whales and tiger sharks), commercial fishing nets and fishing boat activities, all representing threats [Pryor (1998)].

As shown in Figure 7.16, in the **search** phase, according to the threat level of the environmental risks, dolphins look for prey based on one of the three behaviors groups: *herds*, *scouts* or *groups*. Here, the environmental risk (ER) of a region is classified into three levels: high-risk regions, unexplored regions and safe regions. By monitoring the threat level of the environment, dolphins make the decision of how to react in one of the three ways:

(1) *herd*: When dolphins are in a high-risk region (ER = high risk), all individuals stay together to defend themselves.

(2) *scouts*: When dolphins are in an unexplored region (ER = unknown), dolphins send out two scouts (two dolphins) to explore unknown areas, and the remaining dolphins follow the scouts from a safe distance.

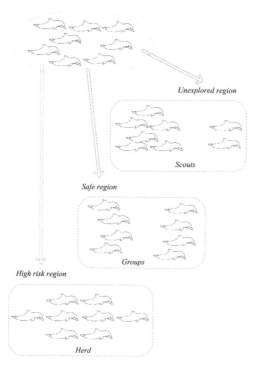

Fig. 7.16 Dolphins' search phase consisting of three processes

(3) *groups*: When dolphins are in a safe region (ER = safe), all members form small groups randomly.

In the **detect** phase, the individual dolphin who located the prey (a swarm of fish) notifies the position of the prey to its peers; they will then select one of two methods for capturing the prey in the subsequent phase: the *wall* method (W-method) or the *carousel* method (C-method).

In the **capture** phase, there are two methods for capturing prey:

(1) The W-method.

As shown in Figure 7.17, the W-method, in which the dolphins drive the fish towards a natural barrier (e.g., the shore) and capture them, is employed by bottlenose dolphins quite often. As shown in Figure 7.18, four types of formations, (a) front, (b) double front, (c) line and (d) tight group, are employed by bottlenose dolphins during foraging, where formations (c) and (d) require a leader (black dot) and the arrows denote the headings of the dolphins. Dolphins have a well-defined social hierarchy, where the role of the leader dolphin goes to the largest male, who plays two important roles: (1) it determines the threat level of an environment, and (2) it is the first scout to enter an unexplored area.

(2) The C-method.

This carousel-like movement by the dolphins is initiated by either 'curving' in from one side of the fish or by simultaneously surrounding them from both sides. As shown in Figure 7.19, there are two sub-steps in the C-method: (a) when the Dolphins find a sizeable amount of fish school, they create a large circle to trap the fish school inside it; (b) they start to tighten the encirclement by forming increasingly smaller circles to constrict the movement of the fish, and then, the encirclements become small enough for the dolphins to dive into the fish school and feed on their prey.

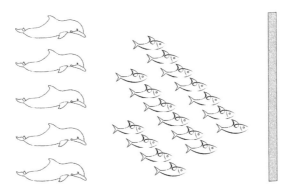

Fig. 7.17 The W-method for capturing prey

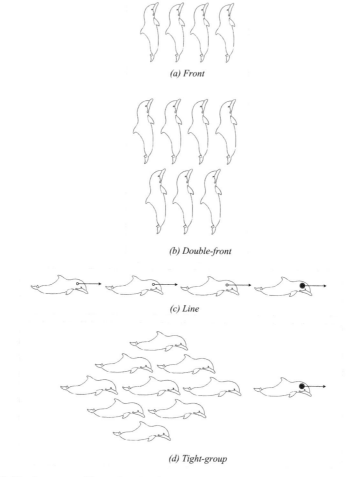

(a) Front

(b) Double-front

(c) Line

(d) Tight-group

Fig. 7.18 The four types of formations for the W-method

7.7.2.2 Partner Searching

Dolphins have many partners over a lifetime and mate year-round. When searching for a partner, dolphins are performing sexual selection, which is selection for any trait that increases mating success by increasing the attractiveness of an organism to potential mates. Traits that evolved through sexual selection are particularly prominent among males of several animal species. Although sexually favored, traits such as large body size and bright colors often stimulate the senses of sight, hearing, taste and touch to be utilized during navigation, predation, feeding, breeding and communication, which compromises the survival of the dolphins.

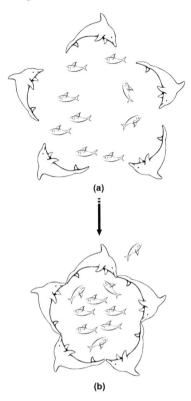

Fig. 7.19 The C-method for prey capturing

7.7.3 *k-Nearest Neighbor Classification*

The *k*-nearest neighbor (*k*-NN) algorithm is a method for classifying features based on the closest training samples in a spatial domain [Altman (1992)], wherein a query object is classified by votes of its neighbors and assigned to a class of its *k*-nearest neighbors. *k* is a small user-defined positive integer; for example, if *k*=1, then the query object is assigned to the class of its nearest neighbor (Figure 7.20).

Given a query object P_0 with a feature vector $\{x_{01}, x_{02}, ..., x_{0N}\}$, where N is the feature size and the training object is P_{ij} $(i = 1, 2, ..., M; j = 1, 2, ..., N)$, in which M is the size of the training objects with the same size of P_0, the *k*-NN algorithm can be summarized as follows:

(1) Definition of distance metric: The Euclidean distance is usually used as the distance metric.

(2) Identification of the *k*-nearest neighbors, in which *k* is usually odd for a two-class problem and not a multiple of the number of classes M. As given in Figure 7.20, if *k*=3 or *k*=5, the query object P_0 can be assigned to either classes $\triangle$ or •.

(3) Assignment of P_0 to class i with the maximum number of samples *k*.

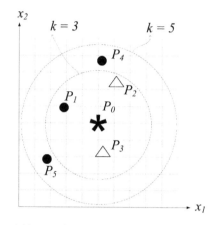

Fig. 7.20 The k-nearest neighbors and k-NN classification

7.7.4 Swarm Dolphin Algorithm

The SDA is an artificial intelligence algorithm inspired by swarm intelligence, characterizing the social behavior of dolphin herds. The SDA initially simulates the behavior of an artificial dolphin individual (ad_i) and then constructs an artificial dolphin (AD). Each ad_i searches its own local optimal solution and passes specific information to its peers until it finally arrives at global optimal solutions. The SDA is based on the following assumptions:

- Prey and obstacles can be identified by each ad_i.
- Only one species of dolphin is in a dolphin herd AD, for example, bottlenose dolphins only.

The SDA workflow is illustrated in Figure 7.21 and involves four steps: search, detect, capture and bulletin. Upon starting the SDA program, all the parameters, such as the population P and behavior status, are initialized and then processed in the four steps as follows:

1. *Search.* Based on the food density and the number of peers and visual conditions, one of the three reactions of the AD (herd, scout and group) is then selected by the k-NN classifier, in which scout is the default or initial behavior of each ad_i.
2. *Detect.* The ad_i finds prey, relays the position of the prey to the remainder of the herd, and then decides on a method for the capture action in the next step.
3. *Capture.* Given the location of prey, the AD converges to the location, traps and captures the prey either through the C-method or the W-method and feeds on it.
4. *Bulletin.* This step compares the current solution x_i of each ad_i with the historical state data; the bulletin data are replaced and updated only when the current solution is better than that in the last generation.

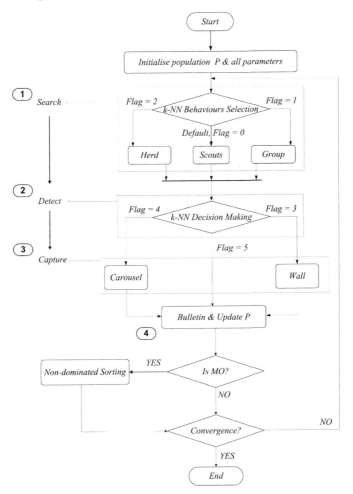

Fig. 7.21 Workflow chart of swarm dolphin algorithm

After the four steps, the program checks if it is an MO application. If 'YES', then the non-dominated sorting genetic algorithm II (NSGA-II) is utilized to perform the MO optimizations; if 'NO', then the SDA converges to meet the given termination condition, which is the 'max-generation' of *AD* searching for food under a given initial condition [SwarmDolphin (2014)].

8

Evolving Artificial Neural Networks
in a Closed Loop

8.1 Introduction

The intelligent information processing system of human beings is different from conventional computers and mathematics. An example of such a system is the eye-brain system, which consists of many concurrent neurons. These neurons process information and switch at an individual speed much slower than a digital computer. Such a system is, however, much more effective and efficient than a computer in pattern recognition, decision making and learning.

Simulating this system in inference and decision making, the *artificial neural network* (ANN) incorporates the learning capability with a concurrent information-processing architecture. Behaving like a 'black-box', such a network is usually a physical cellular system with a known architecture but with unknown or time-varying parameters that reflect the strength of inter-connections. When trained, it can acquire, store, judge and utilize experimental data and knowledge that may be inaccurate and incomplete. Its successful applications have been widely reported, including those to feedback control system design [Psaltis, et al. (1988), Ichikawa and Sawa (1992), Rogers and Li (1993)] and to very closely-related modeling and identification of complicated, irregular, irrational, stochastic, nonlinear or time-varying systems [Chen, et al. (1990), Pham and Liu (1993), Li, et al. (1995a)].

In this chapter, a novel ANN is developed and a GA is applied to the structure and design of the network. We use an application-specific example to illustrate how to evolve a challenging ANN. The following section develops the structure of the ANN used as a neurocontroller from the structure of a widely used *proportional plus integral plus derivative* (PID) controller for direct embedment within a feedback loop. An interesting feature of such a controller is that the embedded nonlinearity and input bias of the ANN can help eliminate steady-state errors of the closed-loop system of no energy storage (termed a 'Type 0' system), and hence no integral term would be required of the PID controller, simplifying it to a proportional plus derivative (PD) controller.

Difficulties associated with traditional design methods for ANNs, such as the *backpropagation* (BP)-based method, are highlighted in Section 3, which have motivated a GA-based evolutionary design method. In Section 4, design examples of direct neurocontrollers using the GA are demonstrated and whose performance confirms the strength of the structural ANN strategies. Techniques are also developed to optimize the architectures in the same process of parameter training. Finally, conclusions and further work are highlighted in Section 5.

8.2 Directly Evolving a Neural Network in a Closed Loop

8.2.1 Existing Architectures for Neural Control

The majority of artificial neural networks in use have an architecture of *multilayer perceptrons*. In such a network, the perceptrons, or *neurons*, of each layer are not connected to one another but *completely connected* to the perceptrons in the adjacent layers. This architecture forms a *partitioned complete digraph* [Rogers and Li (1993), Li and Rogers (1995)]. In the design process for practical applications, the network must be trained to gain an optimal connection-strength represented by *weights* in the connecting path. The weighted input to a neuron is further processed by an *activation function*, together with a linear shift operation determined by another parameter, the *threshold* level, which also needs to be trained before the network is applied to solve problems.

There are two different ways of applying a neural network to solve decision-making or control engineering problems. One is to use the ANN to adjust the parameters of a conventional controller, and therefore the ANN is a modeler and not in effect a controller [Rogers and Li (1993)]. The other is the use of the ANN as a direct controller [Psaltis, et al. (1988), Ichikawa and Sawa (1992), Rogers and Li (1993)], which is termed a *neurocontroller*. The latter forms the underlying theme of this chapter.

The majority of existing neurocontrollers utilize a single error as input to the ANN to form a *proportional* type control action and thus may not be able to cope with overshoots and oscillations in the controlled behavior. Further, in the design process, they are mainly trained relatively independently to use the *plant* input-output data and lack the systematic integration with the plant to be controlled. To improve the controllability and performance, this chapter develops a direct neurocontrol architecture from the standing point of classical control systems and embeds the neurocontroller directly within the feedback loop.

8.2.2 Architecture of a Neural Network Controller

To begin with, the development of a feedback neurocontroller, recall the input-output relationship of a digital PD controller. This is given by:

$$u(k) = K \left\{ e(k) + \frac{T_D}{T_0} \left[e(k) - e(k-1) \right] \right\}$$

$$= \left(K + \frac{KT_D}{T_0} \right) e(k) + \left(-\frac{KT_D}{T_0} \right) e(k-1) \tag{8.1}$$

where k is the time index, K the proportional (and overall) gain, T_D the time-constant of the differentiator, T_0 the sampling period, $u(k)$ the output of the controller and $e(k)$ the input to the controller, being discrete error signal between the desired output and actual output of the plant. The inclusion of the derivative term overcomes difficulties encountered in the transient and stability, such as overshoots and oscillations. In this chapter, an attempt is made to design a neurocontroller inspired by the PD controller, which utilizes both the error and the change of error signals. The construction of the *nonlinear* ANN does not, however, need to follow Equation (8.1) exactly and thus a second order 'differential' term, $e(k-2)$, can be added as an additional input. This makes simultaneous use of available information better and predicts actions more thoroughly. It is also employed to compensate for the differentiation action lost by nonlinear activation. Further, the use of this multiple excitation set facilitates direct application of neural networks to embedded control. Following these discussions, the structure of a direct feedback neural control system is developed as shown in Figure 8.1, where $r(k)$ represents the discrete command signal and $y(k)$ the discrete output of the plant.

With the moving error signals as inputs that provide differentiation actions, this structure provides a means to combat stable and unstable oscillations and improve the transient response. Specifying this structure prior to training is important, since 'proportional' type of neurocontrollers cannot be trained to gain such a structure or to lead to compensated transient behavior. Note that, incorporating an integral term similar to that in a *proportional plus derivative plus integral* controller would have the effect of reducing the steady-state errors. In a closed-loop control system, however, this tends to destabilize the system. Integration is thus not used in this neurocontroller. Further, the nonlinear activation functions of the ANN can be trained to tackle the steady-state problems, if the training criterion, or the performance index, is designed to penalize the steady-state errors.

Note also that there are no restrictions on the connecting architectures to be taken and any multilayer or Hopfield networks [Rogers and Li (1993)] with three inputs and one output can be used here. An example of a multilayer architecture is shown in Figure 8.2, where the fixed inputs with value 1 represent the threshold levels to be trained via their connecting weights. With the structure imposed by the closed-loop system, as shown in Figure 8.1, the feedback neurocontroller behaves similar to a recurrent neural network and can be viewed as a modified form of such type of network.

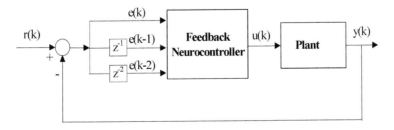

Fig. 8.1 The structure of a direct feedback neural control system

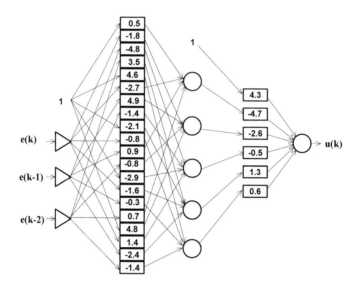

Fig. 8.2 An example of a multilayer feedback neurocontroller (whose weights are evolved by a genetic algorithm discussed in Section 8.4.1)

8.2.3 The Problem of Neurocontroller Design

Same as all other applications of neural networks, training to gain the optimal weights that yield the best performance is the key in the design of a neural control system, once the architecture of the network is known or determined. A set of weights thus uniquely represents a *candidate* neurocontroller in a design process. This forms a uniform *parameter vector* representation of the neural network given by:

$$P_i = \left\{ p_1, ..., p_n \right\} \in \mathbf{R}^n \qquad (8.2)$$

where i stands for the ith design candidate, n the maximum number of parameters required by the neurocontroller, $p_j \in \mathbf{R}$ the jth parameter of the ith candidate with $j \in \{1, ..., n\}$, and R^n the n-dimensional real Euclidean space. For the architecture shown in Figures 8.1 and 8.2, $n = (3 + 1) \times 5 + (5 + 1) \times 1 = 26$, since there is one hidden layer with 5 completely connected neurons. Therefore, all the candidate designs form a solution space, as given by:

$$S = \left\{ P_i, \forall i \mid p_j \in \mathbf{R} \text{ and } j \in \{1, ..., n\} \right\} \subseteq \mathbf{R}^n. \qquad (8.3)$$

Here, clearly, n may be time-varying if the architecture is also to be trained or an adaptive architecture is to be used (see Section 4 for examples).

In the design process of an ANN, the performance of the network being trained is usually measured against the error norm. A commonly used performance index, $f(Pi): \mathbf{R}^n \rightarrow \mathbf{R}^+$, is the inverse of a scaled root mean-square (rms) error given by:

$$f(\mathbf{P}_i) = \alpha \sqrt{\frac{N}{\sum_{k=1}^{N} e^2(k)}} = \alpha \frac{\sqrt{N}}{\|e(k)\|_{L_2}} \tag{8.4}$$

where N is the number of input-output data sets used in training. Here α is a scaling factor, which affects the absolute value of the performance index in all the candidate designs, but does not affect the relative values of individual designs. In a neural control system design, however, the performance index needs to reflect the following *design criteria* in the presence of practical system constraints:

- An excellent transient response in terms of rise-time, overshoots and settling-time
- An excellent steady-state response in terms of steady-state error
- Acceptable stability margins
- Robustness in terms of disturbance rejection
- Robustness in terms of parameter sensitivity

To reflect the neurocontroller design criteria, Equation (8.4) needs to be slightly modified to penalize the errors occurred in the steady-state [Li, et al. (1995a), Ng and Li (1994), Li, et al. (1995b)]. The simplest example of such indices is the time-weighted L_1 norm, as given by:

$$f(\mathbf{P}_i) = \frac{\alpha}{\sum_{k=0}^{N} k^m |e(k)|} \tag{8.5}$$

where m represents the degree of depressing the errors. Note that, with the inclusion of the scaling factor α, there are no fundamental differences between the indices based on different types of norms [Li, et al. (1995b)], since homogeneous norms are equivalent (i.e., bound by one another with a scaling factor not greater than N). Thus, the L_1 based index given by Equation (8.5) is used throughout this chapter for a fast evaluation.

Based on the above discussions, the *design problem* of a neural control system can be uniformly defined as the problem of finding, through training, a neurocontroller given by:

$$\mathbf{P}_o = \left\{ \mathbf{P}_o \in \mathbf{S} \;\middle|\; f(\mathbf{P}_o) = \sup_{\forall \mathbf{P}_i \in \mathbf{S}} f(\mathbf{P}_i) \right\}. \tag{8.6}$$

8.3 Globally Optimized Design Through a Genetic Algorithm

8.3.1 Difficulties with Conventional Neural Network Training Methods

In the design of a neurocontroller, the most popular training method for multilayer neural networks is the *backpropagation* algorithm [Psaltis, et al. (1988), Ichikawa and Sawa (1992), Rogers and Li (1993)]. For a neural control system, there are two types of training and application mechanisms [Psaltis, et al. (1988)], namely, *general learning* and *specialized learning*. In general learning, the network is usually trained off-line. For certain training patterns given to the input of the plant, the plant output is considered as the input of the neural network. Similarly, the training patterns at the input of the plant are considered as the output of the neural network. By training the neural network with these data, it learns the *inverse dynamics* of the plant. Although it is common for a neural network to control the plant using its inverse model, there are several disadvantages of this method [Psaltis, et al. (1988), Ichikawa and Sawa (1992), Rogers and Li (1993)]. For instance, special control tactics cannot be learned and varying parameters, noise and disturbances that occur very often in real plants will cause difficulties. Specialized learning as an alternative method is a way of training the neurocontroller on-line or from the plant model directly [Psaltis, et al. (1988), Ichikawa and Sawa (1992), Rogers and Li (1993)]. Because of the desired control signal, which implies that the desired output of the neural network is unknown, the plant is compared with the command signal and the error has to be back-propagated through both the plant and the neural network.

Since both the general and specialized learning methods are based on the BP algorithm, the weights connected to the output layer can be adjusted directly from the output error signals and the performance index. The errors are back-propagated and the weights back-adjusted layer by layer, in the reverse order, from the output layer to the input layer. This process is repeated for every *epoch*, until all training data have been used up. Complete training typically requires many epochs in order to gain an optimal connection-strength and the associated threshold. Further, the guidance of the parameter optimization process is sequentially based on the gradient of the error. This training method thus suffers from the drawbacks highlighted as follows:

- Gradient guidance passively adjusts parameters from the performance index, which must be differentiable or 'well-behaved' and thus may not allow modified error terms that suit real engineering problems
- It is difficult to train a direct feedback neurocontroller that meets constraint conditions in practical applications
- The trained parameters may be local optima, although parallel training [Li, et al. (1995)] may overcome this problem to a certain extent
- Different *control parameters* in ANN training (e.g., the *learning rate* and *momentum rate*) may result in different minimum rms errors for the same (large) number of epochs
- The architecture of the network usually needs to be fixed prior to parameter training and thus optimal architectures or topologies cannot be revealed for different types of data or applications
- Using BP is difficult to obtain an ANN that matches multiple data sets that are extracted from different operating conditions of the same plant
- It is usually difficult to incorporate knowledge and expertise that the designer may already have on the network design
- It is almost impossible to minimize the input to an ANN

8.3.2 Training with a Genetic Algorithm

Emulating Darwin's evolutionary principle of *survival-of-the-fittest* in natural selection and genetics, *genetic algorithm*-based search techniques have been found very effective and powerful in searching poorly understood, irregular and complex spaces for optimization and machine learning [Goldberg (1989)]. Such an algorithm can simultaneously evaluate performance at multiple points while intelligently searching through the solution space and can thus approach the global optima for almost any type of objectives without the need for differentiation. This characteristic, together with a *non-deterministic polynomial* (NP) search-time required by this method, is achieved by trading off precision slightly for improved tractability, robustness and ease of design [Li, et al. (1995b), Li, et al. (1995c)]. Recent development in this area and the general area of *evolutionary computing* show the strength of artificial evolution in overcoming all the seven drawbacks of BP listed in the previous subsection [Li, et al. (1995a), Harp and Samad (1992), Yoon, et al. (1994), Sharman, et al. (1995)]. For example, it is seen that training by an adaptive genetic algorithm can be more efficient, in terms of the convergence speed, than the backpropagation method [Yoon, et al. (1994)] and that GA-trained networks can effectively lead to low rms errors in estimation tasks [Li, et al. (1995a)]. Further, the experimental studies on an artificial problem [Keane (1995)] show that a GA, particularly when it is fine-tuned by *simulated annealing* [Tan, et al. (1995), Li, et al. (1995c)], provides a much higher convergence rate and robustness than conventional optimization means, such as the well-developed linear approximation and heuristic search algorithms [Keane (1995)]. This method has also been successfully applied to assisting the optimization of parameters that are estimated by traditional methods [Tan, et al. (1995), Sharman and McClurkin (1989)]. Thus, in this chapter, the GA (instead of BP)-based training methods are developed for the neurocontroller design. This is constructed in a way similar to the specialized learning method.

In the design of a neural network represented by Equation (8.2), a candidate parameter set of all weights and thresholds can be encoded by, for example, an integer string. Such a string is termed a *chromosome* in the GA context and a digit of the string is termed a *gene*. Initially, many such chromosomes are randomly generated to form a *population* of candidate designs. In this initial population, existing or known good designs can be conveniently included, which usually lead to a faster convergence [Ng and Li (1994), Li, et al. (1995b), Tan et al. (1995), Li, et al. (1995c)]. The GA uses three basic operators termed *selection, crossover* and *mutation* to evolve, by the NP approach, a globally optimized network parameter set. A fourth operator, *inversion*, can be derived from crossover and mutation and is thus not commonly used. The size of the population and the probability rates for crossover and mutation are termed *control parameters* of the GA. For details of the operators and procedures of a GA, refer to Goldberg [Goldberg (1989)]. Note that, however, the task of selecting crossover and mutation rates in a GA is much easier than determining the learning and momentum rates in a BP-based algorithm. In the optimization process of ANN parameters, the GA search is guided by analysis and exploration of the *fitness* of every individual in the evolving population. The relative fitness of an individual in the population is the criterion that determines the probability of its reproduction and survival. This is usually evaluated by a performance index that reflects the design specifications, such as that given by Equation (8.5) in this chapter.

Since the GA is a fitness evaluation-based search, as opposed to calculus based optimization, method, the analytical requirement of the fitness function is thus much more relaxed than an objective or cost function used in traditional optimization techniques. This is very useful in systems and control engineering applications, since the errors that need to be minimized for high-performance specifications may be weighted by time, for example, as shown in by Equation (8.5) which also leads to stability indirectly [Ng and Li (1994), Li, et al. (1995b) Li, et al. (1995c)]. The fitness function may also include criteria for robust control [Li, et al. (1995b)]. It is worth noting that, however, the advantage of a GA in incorporating practical actuator constraints need not to be employed here, as an ANN can easily tackle this problem by its activation functions in the output layer.

Based upon the above discussions, it can also be concluded that a neurocontroller, as well as other decision-making and control systems, can always be designed by the GA under the following conditions:

- The solution space for design is known (which is unnecessary if the genetic programming [Sharman, et al. (1995)], instead of genetic algorithm, technique is used) and can be represented by a finite quantization
- The system is analyzable, i.e., the performance of candidate designs can be evaluated
- There exists a performance index that has values with more information than a simple *True-or-False* answer, as otherwise the reproduction will be completely polarized and the GA will become a multi-point random search.

8.4 Neural Network Control for Linear and Nonlinear System Control

In this chapter, the weights and thresholds of the neural network are coded by simple decimal strings [Li, et al. (1995a), Li, et al. (1995b), Tan et al. (1995), Li, et al. (1995c)]. This yields a direct mapping of decimal numerals and can smooth, in some degree, the *Hamming Cliff* diversion [Ng and Li (1994), Li, et al. (1995b)] that is usually encountered in GAs based on binary coding. Real logarithmic coding [Flexible Intelligence Group (1995)] may also be developed, however, if a high accuracy coding for reducing the parameter quantization errors is desired. In the selection and reproduction process throughout the designs reported in this chapter, the simple '*roulette-wheel*' selection mechanism [Goldberg (1989)] is used to show the powerfulness of the simple GA, although more advanced selection mechanisms can also be employed (see also Section 5) [Ng and Li (1994), Li, et al. (1995b), Li, et al. (1995c)].

8.4.1 GA-based Design for Neural Control of a Linear Plant

In order to illustrate the GA-based design methodology in detail, consider a typical second order servo-system for velocity control, which is to follow a step command as rapidly as possible without oscillations or steady-state errors. The control input is applied to the field winding of the DC motor, whose open-loop behavior of the system before a step-down gear-box is described by:

$$\frac{d^2\omega}{dt} + \left(\frac{JR + LB}{LJ}\right)\frac{d\omega}{dt} + \left(\frac{RB}{LJ}\right)\omega = \left(\frac{K_T}{LJ}\right)v_{in} \tag{8.7}$$

where $v_{in} \in [-5V, 5V]$ is the input control voltage which implies an indirect constraint on the neurocontroller parameters; K_T the torque constant; R the resistance of the motor winding; L the inductance; B the friction coefficient of the shaft; and J the moment of inertia of the load and the machine. For simplicity, the power drive is modeled as a linear amplifier, which enables a gain of 1 for the overall open-loop system. This is thus a *Type 0 system*, which will result in steady-state errors if no integral action is taken when using a linear controller. The transfer function of this system in the Z-transform with a zero-order hold is given by:

$$G(z^{-1}) = \frac{Y(z^{-1})}{U(z^{-1})} = \frac{b_1 z^{-1} + b_2 z^{-2}}{1 + a_1 z^{-1} + a_2 z^{-2}} \tag{8.8}$$

In discrete-time, the parameters of the transfer function of the system incorporating a 9:1 step-down gear-box are:

$a_1 = -1.7826, a_2 = 0.8187$

$b_1 = 0.01867, b_2 = 0.01746$

To demonstrate the neurocontroller design method, a simple two-layered neural network with a 3-5-1 structure is used, which stands for 3 input neurons, 5 neurons in one hidden layer and 1 output neuron. For every neuron in the hidden layer, a sigmoid function is chosen as the activation function. For the output neuron, a piece-wise saturated *ramp* activation function is used to limit the practical actuation control signal.

For use with the GA, every weight has been coded by two digits with values in the range (−4.9, 5.0). The thresholds are also treated as weights, with an input value of 1. The simulation time steps, N, is set to 80. The fitness function used is that given by Equation (8.5) with $m = 1$. The control parameters of the GA were chosen as:

Population size = 80, mutation rate = 0.5%

crossover rate = 60%

according to the size and complexity of this design problem. On an Intel 66 MHz processor, it took less then 10 minutes to train the neural network for 200 generations, using Turbo-Pascal programming. Figure 8.3 depicts the average and the highest fitness of the neural networks in every generation. Note that both fitness values shown are normalized to the maximum value found in the 200 generations and the average, as opposed to the highest, fitness is used to determine the convergence and the stop-criterion of the algorithm. It shows the network with best parameter values found in these generations.

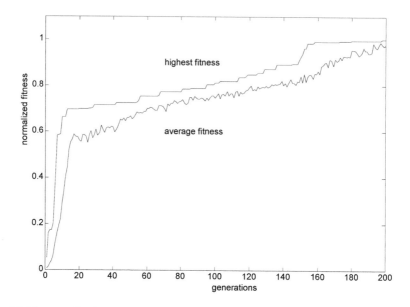

Fig. 8.3 The normalized average and highest fitness values in 200 generations

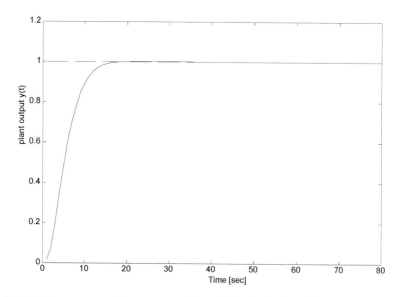

Fig. 8.4 Unit step response of the neurocontroller embedded in the closed-loop system

The resulting response of the ANN-controlled servo-system that needs to follow a unit step command is shown in Figure 8.4. It can be seen that there are virtually no oscillations or steady-state errors and despite that a relatively very short rise-

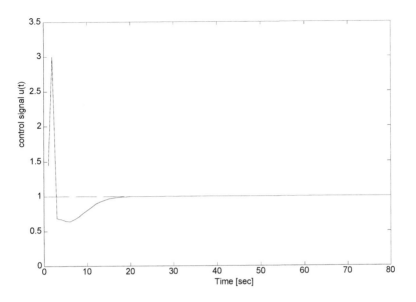

Fig. 8.5 The confined control signal provided by the neurocontroller

time has been obtained. If a classical or modern linear controller is used, such a short rise-time and a zero overshoot cannot be achieved at the same time. Further, without an integrator incorporated, a linear controller will not result in a zero steady-state error for this Type 0 servo-system. This confirms the correct strategy of designing the neurocontroller structure shown in Figure 8.1 and of selecting the fitness function given by Equation (8.5). The corresponding control signal provided by the neurocontroller is depicted in Figure 8.5. Clearly, this signal is well confined within the range [–5V, 5V]. Further, due to nonlinear activation, the neurocontroller provides, from a zero steady-state error signal, a constant control effort to maintain the required steady-state performance of the controlled system, without explicitly incorporating a destabilizing integral action.

Training to Cope with Transport Delay

The developed neurocontroller structure shown in Figure 8.1 should provide the ability to deal with pure time delays usually encountered in a practical system due to delays in actuation and sensing. In order to verify this point, a transport-delay of 4 s is artificially inserted into the system by discrete means. The training using the same GA yielded a very slight oscillation as one may have expected from the destabilizing time-delay. This oscillation is, however, easily suppressed by modifying the fitness function slightly by setting $m = 2$. In the evolutionary design process, this fitness converged in similar rates as depicted by Figure 8.3. The resulting network parameters are given by Figure 8.6. The step response of this closed-loop system is shown in Figure 8.7, which indicates a response as good as that of Figure 8.4.

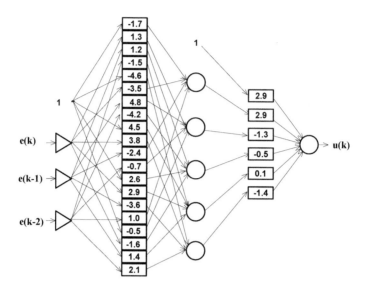

Fig. 8.6 The GA-trained neurocontroller for dealing with pure time delay

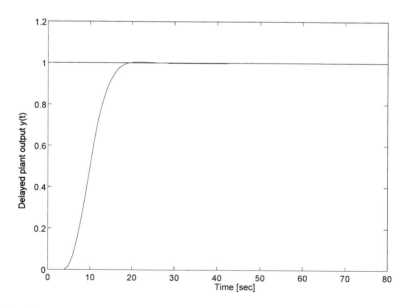

Fig. 8.7 Step response of the closed-loop system with a 4 s time delay

Evolving the Architecture

Another advantage of the genetic algorithm in designing neurocontrollers lies in its ability in searching for an optimized topology and architecture of a given problem. Here, the GA is to be used to search the architecture at the same time of optimizing the weights. To utilize the GA used earlier in a convenient way, the *pruning method* [Sanker, et al. (1993)] is adopted for adaptive selection of the architecture, which will result in an optimized architecture with smaller or equal size of the parent.

It is known that, in a training process, an ANN can blindly 'grow' its connection weights for every *available link* that is allowed by the architecture. This means that almost none of the weights can be trained to have zero values. Thus, in order to design an efficient architecture, artificial elimination of the connections is needed. This is easily achieved by the GA starting the evolution from a parent architecture for pruning. In order to include a wider possibility of architectures, the number of the hidden layers is thus extended to four as shown in Figure 8.8, where every unit area describes a neuron whose connection link and weight need to be determined by training. In the design, the parent neurocontroller is now described by an enlarged parameter set.

In detail, the three input neurons and the output neuron are coded as before. Coding of the hidden neurons is augmented by one additional digit (gene). For the odd values of this gene, the neuron and all of its local connections will be interpreted as 'exist' and for the even values they will not. When decoding, the existence of a hidden neuron is signaled by a flag that represents either 'true' or 'false' for its existence. Figure 8.9 shows the structure of such a chromosome, where w_{ij} is the jth digit of weight i. The weights are numbered by columns.

By using such chromosomes, a GA can search for the weights and the architecture at the same time. The population size for this GA is enlarged to 150 in view of the increased number of parameters and thus the enlarged solution space. The fitness function is as before with $m = 1$. The convergence curves are shown in Figure 8.10, for running the same GA. For 200 generations, the GA has found a 3-3-4-1 structure, shown by the shaded areas in Figure 8.8, to be the most appropriate for the given problem.

The step response of the servo-system controlled by this neurocontroller is shown in Figure 8.11. It can be seen that, while maintaining similarly negligible overshoots and steady-state errors, this architecturally optimised network offers a better performance in terms of rise-time. Such a performance is not obtainable by low order classical or modern linear controllers. It may be achieved by a relatively very high order linear controller, but this would degrade the performance in terms

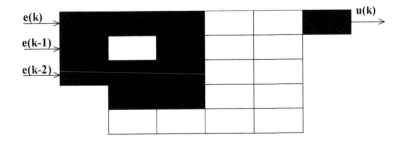

Fig. 8.8 An augmented 3-5-5-5-5-1 parent neurocontroller for pruning (where the clear areas represent the hidden neurons that will have been pruned by the GA)

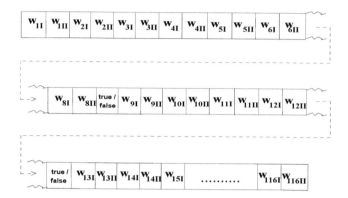

Fig. 8.9 The structure of a chromosome for a 3-5-5-5-1 parent network

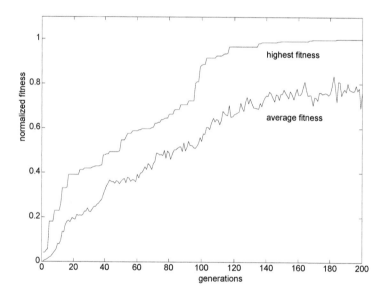

Fig. 8.10 The convergence curves in simultaneous evolution of the architecture and weights

of sensitivity, oscillations and stability and would require a high and rapid control action not generated by a practical actuator. Also to compare, the corresponding control signal provided by this neurocontroller is shown in Figure 8.12. Note that, during the transient, the control signal provided by this nonlinear controller goes negative in order to suppress the overshoot, but is still confined within the actuation limit specified.

8.4.2 GA-Based Design for Neural Control of a Nonlinear Plant

This subsection is to show that the multilayer neurocontroller structured by Figure 8.13 can be trained by the genetic algorithm also to control an asymmetric

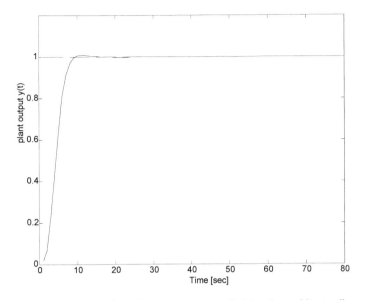

Fig. 8.11 Performance of the closed-loop system controlled by the architecturally optimised network

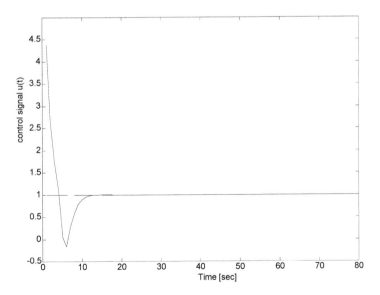

Fig. 8.12 Control signal provided by the neurocontroller with GA optimised architecture

nonlinear system. Here, a laboratory-scale, second-order liquid-level regulation system shown in Figure 8.13 is used to demonstrate the design. In this simplified example, only is the input to Tank 1 used as the input flow, u (cm^3/sec), which is mapped from an actuator voltage in the implementations that follow. It is used to control the liquid-level of Tank 2, h_2 (cm), through the liquid-level of Tank 1, h_1 (cm).

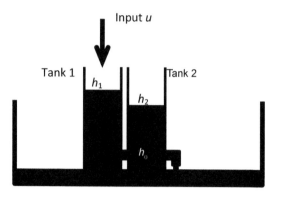

Fig. 8.13 Laboratory-scale liquid-tank demonstration system

A nonlinear equation of this system is given by [Tan, et al. (1995)]:

$$\begin{cases} A\dot{h}_1 = u - a_1 c_1 \sqrt{2g(h_1 - h_2)} \\ A\dot{h}_2 = a_1 c_1 \sqrt{2g(h_1 - h_2)} - a_2 c_2 \sqrt{2g(h_2 - h_0)} \end{cases} \tag{8.9}$$

where $A = 100$ cm², is the cross-section area of both tanks; $a_1 = 0.396$ cm² and $a_2 = 0.385$ cm², the orifice areas of Tank 1 and Tank 2, respectively; $c_1 = c_2 = 0.58$, the discharge constants; $h_0 = 3$ cm, the height of the orifices and of the coupling path; and $g = 981$ cm/sec², the gravitational constant. At rest, there was no input flow for a 'long' time. The initial conditions of h_1 and h_2 are thus the same as h_0. The objective of this control system is to drive the liquid-level at Tank 2 towards the desired level of 10 cm as fast as possible with minimum overshoots and steady-state errors. In this design, the same neurocontroller of Figure 8.2 is used with activation voltage of the output neuron limited within the range [0, +4.5V], which maps to an input liquid flow $u \in$ [0, 2000 cm³/min]. This network is trained by the same GA with fitness given by Equation (8.5) with $m = 1$ and $N = 400$. For every control signal provided by the controller, the output of the plant was calculated using the classical fourth-order *Runge-Kutta* method. The control parameters of the genetic algorithm used are the same as those in the previous subsection. Figure 8.14 shows the neurocontroller found by the genetic algorithm in 200 generations of evolution.

The response of this controlled system is shown in Figure 8.15, which exhibits a satisfactory performance in both transient and steady-state behavior although a small overshoot is observed. Figure 8.16 shows the control signal produced by the neural controller. It can be seen that the nonlinear activation mechanism provides a constant control action during steady-state that results in a zero steady-state error without the need to use a harmfully destabilizing integrator. This is in direct contrast to the use of a linear controller. Further, the plant model used for direct neural control needs not to be linearized first. It is worth noting that, in the work reported in this chapter, another neurocontroller has been designed using a linearized model of the nonlinear-coupled liquid-level regulation system. It resulted in an inferior performance to that obtained by directly using the original nonlinear model given by Equation (8.9).

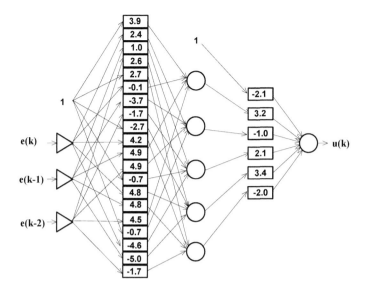

Fig. 8.14 The GA trained neurocontroller for control of a nonlinear coupled liquid-tank system

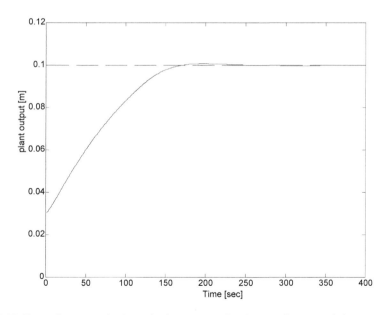

Fig. 8.15 The performance of a GA-trained neurocontroller for a nonlinear coupled system

Evolving the Architecture

Here the parent network shown in Figure 8.8 and the same method used in the previous subsection are employed to select an optimized network by the same GA. It was found that the architecture in the form of 3-4-1 connection yielded the best performance.

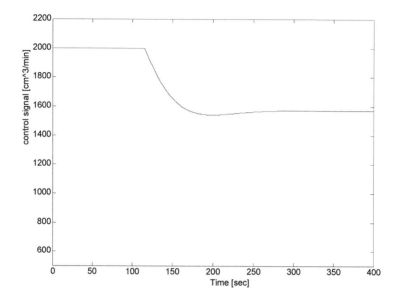

Fig. 8.16 The control signal provided by the GA trained neurocontroller

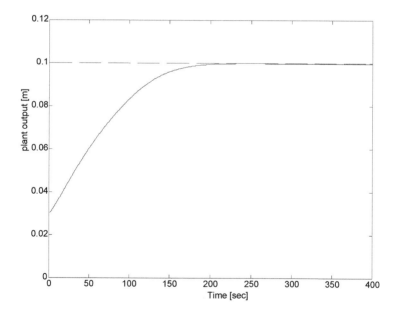

Fig. 8.17 Step response of the closed-loop system with an architecturally optimized neurocontroller

The closed-loop step response for the same control task is shown in Figure 8.17. Compared with Figure 8.15, the performance offered by the neurocontroller without architectural optimization, this response is slightly better, in terms of overshoot. It is worth noting that the manually specified 3-5-1 architecture was already very close to this optimized architecture and thus the performance improvement is not as significant as that obtained in Subsection 4.1.

8.5 Conclusions

This chapter has developed a novel mechanism to devise a neurocontroller for direct feedback embedment. It is inspired by the PD controller characteristics. Problems existing in neural-network training by mathematically-guided methods are presented. The parameter vector representation has been used to unify the problem of neural network design, which naturally leads to the development of a genetic algorithm based design technique that overcomes difficulties encountered in the traditional methods. The conditions of valid use of GAs have also been presented. Using this GA-based design method, neurocontrollers for control of linear and nonlinear plants have been devised as examples, whose performances confirm the correct strategy of the structural development and of the GA-based design method.

9

Evolving Fuzzy Decision-Making Systems

9.1 Introduction

With rapid development in modern computing technology, fuzzy decision-making systems and fuzzy logic control (FLC) have received increasing attention across the systems and control community. The FLC scheme incorporates the uncertainty and abstract nature inherent in human decision making and captures approximate and qualitative boundaries of a system variable by membership degrees to a *fuzzy set*. Such a system implements control actions in near human terms, such as **If-Then** linguistic rules. It has been demonstrated that these systems are reliable, robust and relatively straightforward to implement.

A common fuzzy controller makes decisions from the input error signal and its changing trend. It is thus structurally similar to the classical PD controller. Extending from this system, there exist other three major types of fuzzy control architectures—the proportional (P), proportional plus integral (PI) and PID architectures. These utilize one, two and three input variables and thus one-dimensional 1-D, 2-D and 3-D rule-bases, respectively.

The size of a rule-base increases with the number of fuzzy sets used for every input variable in a polynomial manner. With an underlying PID architecture, for example, the increase will be of the third order. This hinders both off-line design and on-line implementation (or adaptation). Thus a reduced rule-base will be desirable in order to reduce the computational time and to improve the performance/cost ratio. By fitting into a symmetrical lookup table, [Abdelnour (1991)] has achieved rule-base reductions, which are accurate if the system to be controlled exhibits a symmetrical input-output (I/O) behavior. In practice, however, asymmetrical systems are widely encountered.

To avoid using a 3-D rule-base for PID architectures, [Lee (1993)] has proposed a method that 'gain-schedules' a fuzzy PD controller gradually to become a PI one when the response approaches the steady-state. Similar to this method, [Brehm and Rattan (1994)] have developed a hybrid fuzzy PID controller switching between a fuzzy PD and a PI controller. Clearly, the schedule or switching strategy will not only be difficult to design, but also needs to vary with system input and/or operating levels. To avoid this, the authors have achieved PID rule-base reductions using a PD component with a 2-D lookup table and a PI component of a 1-D table [Li and Ng (1995)]. Such reduction may, however, lead to loss of control information if the design is not sophisticatedly optimized [Ng and Li (1994)].

In this chapter, a novel approach to rule-base reduction is developed. This scheme provides a high-fidelity reduction and is elaborated in Section 2. Design parameters and solution space of such fuzzy controllers are outlined in Section 3. Section 4 then demonstrates two design and application examples for a nonlinear plant. Both the simulated and implemented results show that the proposed method yields high performances. Finally, Section 5 concludes the chapter.

9.2 Formulation of a Fuzzy Decision-Making System

A fuzzy decision-making or FLC system consists of a fuzzy relationship or algorithm which relates significant and observable variables to the control actions [Mamdani (1974)]. The time-sequenced crisp variables are first converted to fuzzy variables and then serve as conditions to the rule-base. Commonly used such variables are $e(k)$ and $\Delta e(k)$ in a PD manner, where $e(k)$ is the error between the reference and output and $\Delta e(k)$ the change-in-error. A follow-up defuzzification operation will be needed to reconvert the fuzzy control actions back to crisp values that drive actuators.

Similar to classical PD control, the PD type fuzzy control also suffers from steady-state errors [Li and Ng (1995)]. This is overcome by PI type fuzzy control.

9.2.1 PI-Type Fuzzy Decision Making

Consider a classical PI controller whose control law is given by:

$$u_{\mathrm{PI}}(k) = K_{\mathrm{P}}\, e(k) + K_{\mathrm{I}} \sum_{i=0}^{k} e(i)\Delta t \qquad (9.1)$$

where K_{p} is the gain for proportional action $u_{\mathrm{p}}(k)$, K_{I} the gain for integral action $u_{\mathrm{I}}(k)$, $e(k)$ the error signal and Δt the sampling period. For PI type FLC, however, scaling difficulties in fuzzifying the integral of the error signal have led to the controller being realized first by a PD type FLC [Chen and Tong (1994), Lee (1993)]. From (9.1), the underlying PD control is described by

$$\Delta u_{\mathrm{PI}}(k) = K_{\mathrm{P}}\, \Delta e(k) + K_{\mathrm{I}}\, e(k)\Delta t$$
$$= \left(K_{\mathrm{I}}\Delta t\right)e(k) + \left(K_{\mathrm{P}}\right)\Delta e(k) \qquad (9.2\mathrm{a})$$

The defuzzified output of this FLC is then integrated to form the eventual PI action:

$$u_{\mathrm{PI}}(k) = u_{\mathrm{PI}}(k-1) + \Delta u_{\mathrm{PI}}(k) \qquad (9.2\mathrm{b})$$

In an environment encountered in real world, however, all signals are coupled with noises. Implementations by way of (9.2a) and (9.2b) are reported to exhibit poor transient responses [Chen and Tong (1994), Lee (1993)]. This is mainly due to double numerical errors derived from the unnecessary differentiation and accumulated in the unnecessary integration. Differentiation of a real-world signal will highlight the noises and integrating the derivative back to the original signal cannot be of high fidelity. In addition, a winding-up problem can also be encountered in the implementations. Note that, however, the numerical errors have a less effect on the steady-state performance, since the integral action tends to correct the steady-state errors.

In order to eliminate numerical errors resulting from unnecessary computations, a *Direct Implementation Fuzzy PI* (DI-FPI) controller is developed here. To proceed, note that (9.1) can be rewritten as:

$$u_{PI}(k) = K_P \, e(k) + \phi \sum_{i=0}^{k} K_P e(i) \Delta t$$

$$= u_P(k) + \left(\phi \Delta t\right) \sum_{i=0}^{k} u_P(i) \tag{9.3a}$$

where the newly introduced parameter is

$$\phi = \frac{K_I}{K_P} \tag{9.3b}$$

and the underlying proportional control law is

$$u_P(k) = K_P \, e(k) \tag{9.3c}$$

Thus $u_{PI}(k)$ can be obtained from $u_P(k)$, which involves only one input variable. Hence, a 1-D rule-base suffices. In this FPI approach, the controller can directly switch to P control alone, by letting $\phi \to 0$, if necessary.

Design of such a fuzzy controller, however, becomes more sensitive and difficult, as the fuzzy logic decision is now made only upon information on $e(k)$ and not upon the relationship between $e(k)$ and $\Delta e(k)$. This problem is overcome by design optimization, which is feasible by using an evolutionary computing technique, such as a genetic algorithm [Ng and Li (1994)].

9.2.2 PID-Type Decision Making

In PID-type fuzzy control, the commonly adopted 'velocity algorithm' [Abdelnour, et al. (1991), Huang and Lin (1993)] is given as follows, using a similar strategy as in (9.2a) and (9.2b):

$$u_{PID}(k) = K_P \, e(k) + K_I \sum_{i=0}^{k} e(i) \Delta t + K_D \frac{\Delta e(k)}{\Delta t}$$

$$= u_{PID}(k-1) + \Delta u_{PID}(k) \tag{9.4a}$$

where

$$\Delta u_{\text{PID}}(k) = \left\{ K_{\text{I}} e(k) + K_{\text{P}} \frac{\Delta e(k)}{\Delta t} + K_{\text{D}} \frac{\Delta^2 e(k)}{\Delta t^2} \right\} \Delta t \qquad (9.4b)$$

A 3-D rule-base, defined by $e(k)$, $\Delta e(k)/\Delta t$ and, $\Delta^2 e(k)/\Delta t^2$ is used in existing designs and implementations. Similar to fuzzy PI control, this implementation may also experience poor transient due to the double numerical errors resulting from unnecessary computations. The transient may, however, be improved in some degree by the derivative action.

Similarly, to overcome these problems, this chapter presents a 2-D *Direct Implementation Fuzzy PID* (DI-FPID) control scheme is developed in this chapter. Here the underlying control law is first expressed by

$$\left\{ \begin{aligned} & u_{\text{PID}}(k) = u_{\text{PD}}(k) + u_{\text{PI}}(k) && (9.5a) \\[2mm] & u_{\text{PD}}(k) = K_{\text{P}}' e(k) + K_{\text{D}} \frac{\Delta e(k)}{\Delta t} && (9.5b) \\[2mm] & u_{\text{PI}}(k) = K_{\text{P}}'' e(k) + K_{\text{I}} \sum_{i=0}^{k} e(i)\Delta t && (9.5c) \end{aligned} \right.$$

where

$$K_{\text{P}} = K_{\text{P}}' + K_{\text{P}}'' \qquad (9.5d)$$

To achieve a 1-D reduction in the 3-D rule-base, first re-arrange (5c) to become

$$u_{\text{PI}}(k) = \sum_{i=0}^{k} \left\{ K_{\text{I}} e(i) + K_{\text{P}}'' \frac{\Delta e(i)}{\Delta t} \right\} \Delta t \qquad (9.6a)$$

and let

$$K_{\text{P}}' = \frac{K_{\text{I}}}{\varphi} \qquad (9.6b)$$

$$K_{\text{P}}'' = \varphi K_{\text{D}} \qquad (9.6c)$$

Then

$$u_{\text{PI}}(k) = (\varphi \Delta t) \sum_{i=0}^{k} u_{\text{PD}}(i) \qquad (9.6d)$$

Thus, from Equation (9.5a), the underlying control law is rewritten as

$$u_{\mathrm{PID}}(k) = u_{\mathrm{PD}}(k) + \varphi \sum_{i=0}^{k} u_{\mathrm{PD}}(i)\Delta t \qquad (9.6\mathrm{e})$$

where $u_{\mathrm{PD}}(k)$ is given by (9.5b). Hence realizing (9.6e) involves two input variables, requiring a 2-D rule-base. Note that only part of the proportional action here is subject to double numerical errors and there is no second-order derivatives.

There are three gain parameters to be defined in this DI-FPID control scheme, namely, K_{P}', K_{D}' and φ. These are related to the three traditional parameters in (9.4a) and can be obtained uniquely from (9.5d), (9.6b) and (9.6c), if the values of K_{P}, K_{I} and K_{D} need to be pre-specified. Similarly, this FPID control can also switch to FPD control by defining $\varphi = 0$.

9.3 Decision-Making Parameters

9.3.1 Membership Functions

To reflect fuzzy nature of classification, bell-type of exponential membership functions are adopted in this chapter, as given by [Ng and Li (1994)]:

$$\mu_i(x) = \exp\left(-\frac{|x - \alpha_i|^{\beta_i}}{\sigma_i}\right) \qquad \forall x \in [-Big, +Big] \qquad (9.7\mathrm{a})$$

where $i \in \{Zero, \pm Small, \pm Medium, \pm Big\}$ and

$$\mu_{+Big}(x) = 1, \qquad \forall x > \alpha_{+Big} \qquad (9.7\mathrm{b})$$

$$\mu_{-Big}(x) = 1, \qquad \forall x < \alpha_{-Big} \qquad (9.7\mathrm{c})$$

Here $\alpha_i \in [-Big, Big]$ is termed the *position parameter* which describes the center point of the membership function along the universe of discourse, $\beta_i \in [1.5, 5.5]$ is the *shape parameter* which resembles gradual bell shapes, including those approximating the triangles and trapezoids shapes, and $\sigma_i \in [0.05, 5.0]$ is the *scale parameter* which modifies the base-length of the membership functions and determines the amount of overlapping. Note also that in (9.7), σ_i is not included in the power of β_i to allow for a more gradual change in base-length.

In simplified designs, trapezoid and triangle membership functions are given by:

$$\mu_i(x) = \begin{cases} 1 & \text{if } |x - \alpha_i| \le a_i \\[2mm] \dfrac{b_i - |x - \alpha_i|}{b_i - a_i} & \text{if } a_i < |x - \alpha_i| < b_i \\[2mm] 0 & \text{if } |x - \alpha_i| \ge b_i \end{cases} \qquad (9.8)$$

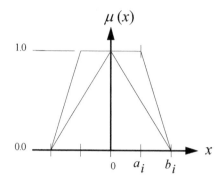

Fig. 9.1 Symmetrical trapezoid and triangle membership functions with $a_i = 0$

		X_e						
		-3	-2	-1	0	+1	+2	+3
		-B	-M	-S	ZO	+S	+M	+B
+3 +B		ZO	-S	-M	-B	-B	-B	-B
+2 +M		+S	ZO	-S	-M	-B	-B	-B
+1 +S		+M	+S	ZO	-S	-M	-B	-B
0 ZO		+B	+M	+S	ZO	-S	-M	-B
-1 -S		+B	+B	+M	+S	ZO	-S	-M
-2 -M		+B	+B	+B	+M	+S	ZO	-S
-3 -B		+B	+B	+B	+B	+M	+S	ZO

$X_{\Delta e}$ (row label at left)

Fig. 9.2 Typical lookup table for a nominal FLC system

Here, as shown in Figure 9.1, a_i is the half top length from the center point α_i; b_i the half base length. A triangle membership function is obtained when $a_i = 0$.

9.3.2 Fuzzy Rule Base

The rules of a 2-D fuzzy controller can initially be designed in a phase plane similar to sliding mode control [Ng (1995)]. Such a phase plane diagram is partitioned according to the number of fuzzy sets used for the control actions, as shown in Figure 9.2. In most reported designs involving asymmetric dynamics of a nonlinear system, the table has usually been pre-defined skew-symmetrically, since an asymmetrical rule-base is difficult to determine manually. In this chapter, however, the more accurate asymmetric table is constructed, using the powerful evolutionary search method [Li and Ng (1995)].

9.3.3 Scaling Factors

In the condition interface, mapping the range of the values of the controller I/O variables to a pre-defined universe of discourse requires scaling factors. The choice of scaling factors has a significant effect on the controller performance [Linkens and Abbod (1992)], in terms of the controller sensitivity, steady-state errors, stability, transients and number of rules to be activated. In the 2-D DI-FPID control system given by (9.5b) and (9.6e), two input scaling factors are to be defined for $e(k)$ and $\Delta e(k)$ through artificial evolution.

The action interface converts the fuzzy control decisions into a unique control action in the so-called combination and defuzzification process. Combining of control actions is essential as more than one rule may be activated for any given set of inputs. Currently, the most commonly used method is the center of gravity method. The resulting single action (combined from the actions of activated rules) is then transformed from a fuzzy output value to a crisp and executable value scaled by an output factor. The output scaling factor has an effect on the overall gain of the system [Linkens and Abbod (1992)] and is also to be optimized by evolution.

9.4 Design Example for a Nonlinear System to Control

The design task of an FLC lies in the optimal choice of a parameter set, including selection among triangular, trapezoidal and bell-shaped memberships, membership parameters (α_i, β_i, b_i), a rule-base, input scaling factors, an output scaling factor and an integral constant (ϕ or φ). With these integral constants, a genetic algorithm can automatically recommend a P controller from a PD one or a PD controller from a PID one.

All parametric variables form a complete parameter set and the solution space to the FLC design problem. Due to difficulties in optimizing FLC parameters by conventional numerical techniques, a systematic design approach based on evolutionary computation is used in this chapter. Here, all parameters are coded in a base-7 string [Ng and Li (1994), Li and Ng (1995)], as seven fuzzy sets are used for all input and output variables. Therefore, in the rule-base coding, one digit is used to represent one level of control action. On contrast, the use of binary coding will require three bits to represent one fuzzy action.

Without loss of generality, a twin-tank liquid level regulation system is used to demonstrate an FLC design and implementation problem. This system can be modeled by a state-space equation given by

$$A \frac{dx_1}{dt} = u - a_1 c_1 \sqrt{2g(x_1 - x_2)}$$

$$A \frac{dx_2}{dt} = a_1 c_1 \sqrt{2g(x_1 - x_2)} - a_2 c_2 \sqrt{2g(x_2 - x_0)} + d$$

(9.9)

where $A = 100$ cm^2 is the cross-section of both tanks; $a_1 = 0.386$ cm^2 and $a_2 = 0.976$ cm^2, the orifice areas of Tank 1 and Tank 2, respectively; $c_1 = c_2 = 0.58$, the discharge constants; x_1 and x_2 (cm), the liquid level of Tank 1 and Tank 2, respectively; x_0 (cm), the height of the orifices in both tanks; and $g = 981$ cm/s^2 the gravitational constant.

In this system, the flow to Tank 1 is used as the system input, u (cm^3/s), which maps from an actuator voltage controlled by the FLCs. The liquid level in Tank 2 is regulated by this control action. An input to Tank 2, $d = 500$ cm^3/min, is applied at an interval of 400 s as a constantly biased disturbance for robustness test of fuzzy logic control. The objective of this control system is to drive, through input to Tank 1, the liquid level at Tank 2 towards the desired level of 10 cm as fast as possible with minimum overshoots and steady-state errors. Subsequently a step-down command of 5 cm is given at 800 s with the same objective as the step-up operation.

Here, FPI and FPID controller designs are automated by a base-7 genetic algorithm. The fourth-order Runge Kutta method is used in the simulations and performance evaluations that guide the evolution. For the 'best' evolved DI-FPI and DI-FPID controllers, simulated control performances are depicted in Figures 9.3 and 9.4, respectively. For comparison, performances of 'best' evolved traditional FPI and FPID controllers are also depicted in these figures. As can be seen, the DI fuzzy controllers offer superior performances compared with their traditional counterparts, in terms of overshoots and disturbance rejection.

In numerical simulations, the computing environment is almost noise-free with short step-length and long word-length. These conditions, however, do not hold in real world implementation. Captured responses in real-time implementations of the traditional and DI fuzzy controllers are depicted in Figures 9.5 and 9.6. Despite

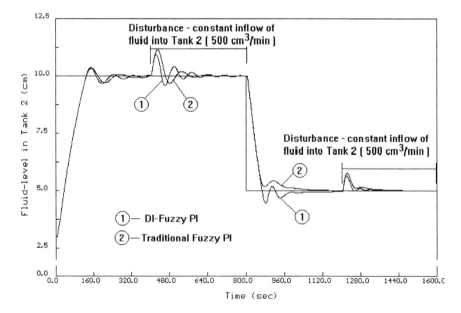

Fig. 9.3 Simulated performances of the best evolved traditional and DI-FPI controllers

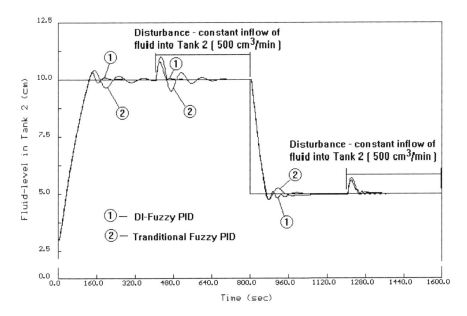

Fig. 9.4 Simulated performances of the best evolved traditional and DI-FPID controllers

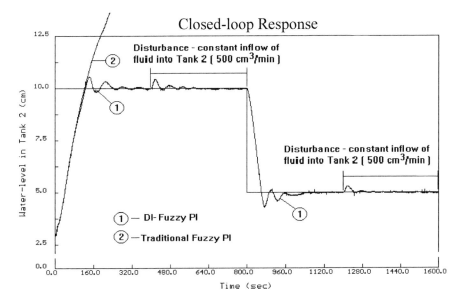

Fig. 9.5 Implemented performances of the best evolved traditional and DI-FPI controllers

discrepancies between the plant model of Equation (9.9) and the physical system, the DI fuzzy controllers are shown to provide significantly close performances, confirming the robustness of the evolved controllers. In contrast the traditional fuzzy controllers,

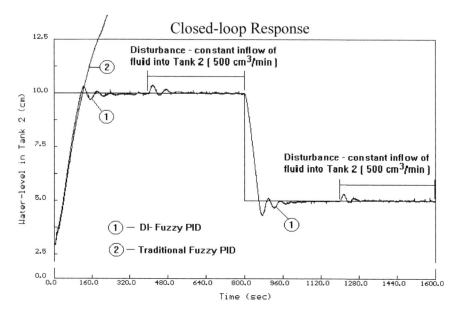

Fig. 9.6 Implemented performances of the best evolved traditional and DI-FPID controllers

although globally optimized by evolution, have failed completely. A windup integral has occurred in both FPI and FPID controllers due to saturated actuation. Although on-line manual adjustments of scaling factors have been attempted, no sign of useful improvements have been observed.

9.5 Conclusion

This chapter reports the development of direct-implementation oriented fuzzy decision-making system in the form of a fuzzy PI and PID controller, and their automated designs via genetic algorithm based evolutionary computing techniques. Using this architecture, the genetic algorithm can automatically recommend a P controller from a PD one and a PD controller from a PID one by optimizing the integral constants.

Such fuzzy control architectures allow a 1-D reduction of the rule-base. It has been achieved for both PI and PID type FLCs. This reduces computational efforts not only in the off-line designs but also in the implementations of real-time control actions. Further, the DI fuzzy control approach has excelled both in simulated and in implemented results. Currently, this work is being extended to multiple-input and multiple-output nonlinear systems.

10

Performance Assessment and Metric Indices

10.1 Introduction

It is essential to assess the performance of evolutionary algorithms (EAs) quantitatively using a set of metric indices or metrics. An EA's performance includes both the quality of the outcome and the computational resources required to generate the outcome.

Due to the EA's effectiveness and robustness in searching for the optimal solutions in multidisciplinary applications, especially approaches for the MO optimization problems, various metric indices have been proposed for performance assessment of the quality of an EA's outcome. Defining a set of appropriate metrics is the key to validating and assessing EAs. However, when addressing MO optimization problems, there are several reasons why the qualitative assessment of results becomes difficult:

- Multiple results. Usually, the MO will produce several solutions (instead of only one) to generate as many elements as possible of the Pareto-optimal set.
- Stochasticity. The stochastic nature of EAs makes it necessary to perform several runs to assess their performance. Thus, our results have to be validated using statistical analysis tools, e.g., MEAN $\pm$ STD.
- Need for multiple metrics. We may be interested in measuring different aspects, e.g., an EA's robustness, trend, diversity and progressive convergence to a set of solutions close to the global Pareto front of a problem.

In this chapter, a few metric indices are introduced to build fitness functions for the EA performance assessment. Specifically, Section 10.2 introduces some traditional metric indices; Sections 10.3 to 10.6 introduce a few newly developed approaches for metric indices; Section 10.7 discusses how to define a fitness function; and finally, in Section 10.8, some test functions, that are able to validate optimization algorithms and to be used to compare the performance of various algorithms, are introduced.

10.2 Metric Indices

For MO optimizations, the main performance assessments on a set of Pareto-optimal solutions can be summarized in the three main aspects:

- the *accuracy* of the solutions in the set, i.e., the closeness of the solutions to the theoretical Pareto front
- the *number* of Pareto-optimal solutions in the set
- the *distribution* and spread of the solutions

Accordingly, to achieve the 'good' properties of a solution to an MO problem as discussed in Section 6.3, there are three issues that need to be considered in designing a good metric in this domain, therein having the following goals:

- A1: minimize the distance of the Pareto front produced by our algorithm with respect to the theoretical global Pareto front (assuming that we know its location)
- A2: maximize the number of elements of the Pareto-optimal set found
- A3: maximize the spread of solutions found so that we can obtain a distribution of vectors that is as smooth and uniform as possible.

This section introduces some general metric indices that assess the performance of EAs, as discussed below:

1. Error Ratio (ER)

   The metric index ER was proposed by Van Veldhuizen [?] and defines the percentage of solutions (from the non-dominated vectors found so far) that are not members of the true Pareto-optimal set:

   $$ER = \frac{\sum_{i=1}^{n} e_i}{n} \qquad (10.1)$$

   where n is the number of vectors in the current set of non-dominated vectors available, $e_i = 0$ when vector i is a member of the Pareto-optimal set and $e_i = 1$ otherwise. It should be clear that $ER = 0$ indicates an ideal behavior since it would mean that all the vectors generated by the MOEA belong to the Pareto-optimal set of the problem. This index addresses issue $A1$ from the list previously provided.

2. Generational Distance (GD)

   The concept of generational distance was introduced as a way of estimating how far away the elements in the set of non-dominated vectors are found from those in the Pareto-optimal set and is defined as [?]

   $$GD = \frac{\left(\sum_{i=1}^{n} d_i^p\right)^{\frac{1}{p}}}{n} \qquad (10.2)$$

   where n is the number of vectors in the current set of non-dominated vectors available and d_i is the Euclidean distance (measured in objective space) between each of these and the nearest member of the Pareto-optimal set. For any p value, regions of the Pareto front can use the same equation; then, a weighted sum can be obtained across the entire Pareto front, usually $p = 2$.

It should be clear that a value of $GD = 0$ indicates that all the elements generated are in the Pareto-optimal set. Therefore, any other value will indicate how 'far away' we are from the global Pareto front of our problem. This metric addresses issue $A2$ from the list previously provided.

3. Distributed Spacing
   This metric addresses issue $A3$ from the list previously provided, which defines a similar measure expressing how well an MOEA has distributed Pareto-optimal solutions over a non-dominated region (the Pareto-optimal set). This Pareto non-compliant metric is defined as [Srinivas and Deb (1994)]

$$\iota = \left(\sum_{i=1}^{q+1} \left(\frac{n_i - \bar{n}_i}{\sigma} \right)^p \right)^{\frac{1}{p}} \tag{10.3}$$

   where q is the number of desired optimal points, the $(q+1)$th subregion is the dominated region, n_i is the actual number of individuals in the ith subregion (niche) of the non-dominated region, $\bar{n}_i$ is the expected number of individuals in the ith subregion of the non-dominated region, $p = 2$, and σ_i^2 is the variance of individuals serving the ith subregion of the non-dominated region. They show that if the distribution of points is ideal with $\bar{n}_i$ number of points in the ith subregion, the performance measure $\iota = 0$. Thus, a low performance measure characterizes an algorithm with a good distribution capacity.

4. Spacing (S)
   As given by Equation (10.4) [SchottMaster (1995)], the spacing (S) metric numerically describes the spread of the vectors in F_n, where $\bar{d}$ is the mean of all d_i; d_i is given in Equation (10.5) and n is the number of non-dominated vectors found so far. This Pareto non-compliant metric measures the distance variance of neighboring vectors in F_n.

$$S = \sqrt{\frac{1}{n-1} \sum_{i=1}^{n} (\bar{d} - d_i)^2} \tag{10.4}$$

$$d_i = min_j \left(\left| f_1^i(x) - f_1^j(x) \right| + \left| f_2^i(x) - f_2^j(x) \right| \right) \tag{10.5}$$

This metric can be modified to measure the distribution of vectors within the Pareto front. In that case, both metrics (S and ι) measure only the uniformity of the vector distribution and thus complement the generational distance and maximum Pareto-front error metrics. This metric addresses issue $A3$.

Many other metrics are listed in Table 10.1; however, some recent theoretical results seem to indicate that they may not be as reliable as we believe and further research in this direction is necessary.

Table 10.1 Summary of Metric Indices [Coello Coello, et al. (2007)]

Indices	Author(s)	Year
Distributed Spacing	Srinivas and Deb [Srinivas and Deb (1994)]	1994
Spacing (SP)	Schott [SchottMaster (1995)]	1995
Maximum Pareto-Front Error	Schott [Schott (1995)]	1995
Progress Measure	Bäck [Back (1996)]	1996
Generational Distance (GD)	Van Veldhuizen and Lamont [Van Veldhuizen and Lamont (1998)]	1998
R_2 and R_{2R} Indicators	Hansen and Jaszkiewicz [Hansen and Jaszkiewicz (1998)]	1998
R_3 and R_{3R} Indicators	Hansen and Jaszkiewicz [Hansen and Jaszkiewicz (1998)]	1998
Hypervolume (HV) or the S metric	Zitzler and Thiele [Zitzler and Thiele (1998)]	1998
Error Ratio (ER)	Van Veldhuizen [Van Veldhuizen (1999)]	1999
Hyperarea and Hyperarea Ratio	Zitzler and Thiele [Zitzler and Thiele (1998)]	1998
Overall non-dominated Vector Generation (ONVG)	Van Veldhuizen [Van Veldhuizen (1999)]	1999
Overall non-dominated Vector Generation Ratio (ONVGR)	Van Veldhuizen [Van Veldhuizen (1999)]	1999
Maximum Pareto-Front Error (ME)	Van Veldhuizen [Van Veldhuizen (1999)]	1999
ε-indicator	Zitzler and Thiele [Zitzler and Zitzler (1999)]	1999
R_1 and R_{1R} Indicators	Hansen and Jaszkiewicz [Hansen and Jaszkiewicz (1998)]	1998
Two-Set Coverage (CS)	Zitzler et al. [Zitzler, et al. (2000)]	2000
Generational non-dominated Vector Generation (GNVG)	Van Veldhuizen and Lamont [Van Veldhuizen and Lamont (2000)]	2000
Generational non-dominated Vector Generation Ratio (GNVGR)	Van Veldhuizen and Lamont [Van Veldhuizen and Lamont (2000)]	2000
Non-dominated Vector Addition (NVA)	Van Veldhuizen and Lamont [Van Veldhuizen and Lamont (2000)]	2000

10.3 Measure of Fitness of Fitting—Coefficients of Determination

To investigate the functional relationship between two or more variables of a regression model, the quantity of the coefficient of determination (R^2) is utilized to measure the proportion of total variability in the responses, where larger values of R^2 indicate the better variability of the data, as given by Equation (10.6) [Montgomery and Runger (2003), Gujarati (2004)],

$$R^2(y_i, \hat{y}_i) = 1 - \frac{SS_{err}}{SS_{tot}} = 1 - \frac{\sum\limits_{i=1}^{k} (y_i - \hat{y}_i)^2}{\sum\limits_{i=1}^{k} (y_i - \bar{y}_i)^2} \qquad (10.6)$$

where
  ∘ R^2 is the coefficient of determination for the two given data sequences y_i and $\hat{y}_i$, as shown in Figure 10.1; y_i are the dashed data and $\hat{y}_i$ are the dotted data

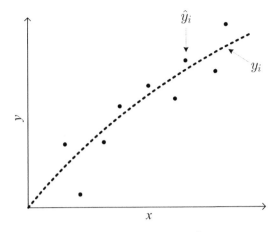

Fig. 10.1 The definition of the coefficient of determination (R^2) [Montgomery and Runger (2003), Gujarati (2004)]

○ y_i are the observed data, which are generated by the micro-GA over the evaluation process
○ $\hat{y}_i$ are the specimen data, which are obtained from the existing THz spectroscopic database for the THz-TDS experiment
○ $\bar{y}_i$ are the mean values of the observed data y_i
○ k are the sample sizes of both of the given data sequences y_i and $\hat{y}_i$
○ SS_{tot} is the total corrected sum of squares, as given by Equation (10.7) [Montgomery and Runger (2003), Gujarati (2004)]:

$$SS_{tot} = \sum_{i=1}^{k} (y_i - \bar{y}_i)^2 \tag{10.7}$$

○ SS_{err} is the residual sum of squares, as given by Equation (10.8) [Montgomery and Runger (2003), Gujarati (2004)]:

$$SS_{err} = \sum_{i=1}^{k} (y_i - \hat{y}_i)^2 \tag{10.8}$$

Meanwhile, R^2 also measures the proportion of the variance in the dependent variable explained by the independent variable [Allen (2007)]; thus, Equation (10.6) can be re-written as Equation (10.9) in terms of the variances, which compare the explained variance σ_e^2 to the total variance σ_y^2.

$$R^2 (y_i, \hat{y}_i) = 1 - \frac{\sigma_e^2}{\sigma_y^2} \tag{10.9}$$

where

○ σ_e^2 is the explained variance, which is the variance of the model's predictions, as given by Equation (10.10):

$$\sigma_e^2 = \frac{\sum_{i=1}^{k}(y_i - \hat{y}_i)^2}{k} = \frac{SS_{err}}{k} \tag{10.10}$$

○ σ_y^2 is the total variance, as given by Equation (10.11):

$$\sigma_y^2 = \frac{\sum_{i=1}^{k}(y_i - \bar{y}_i)^2}{k} = \frac{SS_{tot}}{k} \tag{10.11}$$

In later chapters, R^2 will be utilized to define the fitness functions of the optimal design to measure the goodness of fit.

10.4 Measure of Error Heterogeneity—Relative Gini Index

$$\beta_4 = G = 1 - \sum_{i=1}^{k} p_i^2 \tag{10.12}$$

Table 10.2 presents the general representation of the frequency distribution of a qualitative variable among k levels. The Gini index of heterogeneity is defined by Equation (10.12) [Fantini and Figini (2009)], in which
○ ε_i is the residual, as given by Equation (10.13):

$$\varepsilon_i = y_i - \hat{y}_i \tag{10.13}$$

○ p_i and n_i are the relative and absolute frequencies for level ε_i, which are transformed by Equation (10.14):

$$p_i = \frac{n_i}{k} \tag{10.14}$$

○ $G = 0$ in the case of full homogeneity, which is one of the objectives of the optimization approach
○ $G = 1 - \dfrac{1}{k}$ in the case of maximum heterogeneity

Table 10.2 Univariate Frequency Distribution [Fantini and Figini (2009)]

Residual Levels	Relative Frequencies	Absolute Frequencies
ε_1	p_1	n_1
ε_2	p_2	n_2
$\vdots$	$\vdots$	$\vdots$
ε_k	p_k	n_k

According to Equation (10.12), to obtain the range of [0,1], a normalized index of the Gini index G' can be rescaled by its maximum value, which expresses the relative index of heterogeneity, as stated in Equation (10.15).

$$\beta_{4r} = G' = \frac{G}{(k-1)/k} \tag{10.15}$$

10.5 Measure of Trend—Trend Indices

Table 10.3 Evolutionary Trend Indices for a Single Objective

Indices	Variable	Definition
Mean average precision	mAP	Equation (10.16)
Mean standard deviation	$mSTD$	Equation (10.17)
Mean variance	$mVAR$	Equation (10.18)
Moving mean average precision	$mmAP$	Equation (10.19)
Moving mean standard deviation	$mmSTD$	Equation (10.20)
Moving mean variance	$mmVAR$	Equation (10.21)

Table 10.4 Evolutionary Trend Indices for MO Problems

Indices	Variable	Definition
Pareto Reliability Index	β_1	Equation (10.22)
Pareto Risk Index	β_2	Equation (10.27)
Pareto Sensitive Index	$\beta_{3,1}$ and $\beta_{3,2}$	Equations (10.31) and (10.32)
Pareto Gini Index	β_4	Equation (10.12)
Pareto Relative Gini Index	β_{4r}	Equation (10.15)

10.5.1 The Mean Variables

In this section, the factors mAP [Manning, et al. (2008)] and mSTD, which are defined by Equations (10.16) and (10.17), are introduced as the trend indices for the optimization process.

As shown in Figure 10.2, the solid curve presents the mAP scores for each vector f_j as given by Equation (10.16), and the dashed curves are the mAP $\pm$ mSTD for each vector f_j as given by Equation (10.17), in which p is the population of the data set, $AVG(\cdot)$ is the average function and $VAR(\cdot)$ is the variance function.

$$mAP(f_j) = \frac{1}{p} \sum_{j=1}^{p} (AVG(f_j)) \tag{10.16}$$

$$mSTD(f_j) = \frac{1}{p} \sum_{j=1}^{p} (STD(f_j)) \tag{10.17}$$

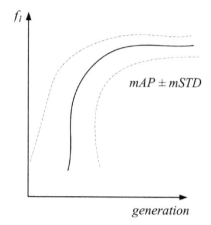

Fig. 10.2 The diagram of mAP $\pm$ mSTD over the full generations

$$\text{mVAR}(f_j) = \frac{1}{p}\sum_{j=1}^{p}(VAR(f_j)) \qquad (10.18)$$

10.5.2 The Moving Mean Variables

To assess the optimization performance, this section introduces three trend indices: mmAP, mmSTD and mmVAR, as given by Equations (10.19), (10.20) and (10.21) respectively.

As stated in Equation (10.19), the index of mmAP is a moving average score of the mean value of the vector f_j, where $i = 1, 2, \cdots, p$; p is the population of the dataset and $MEAN(\cdot)$ is the average function. The index of mmSTD is a moving average score of the STD value of the vector f_j, as given by Equation (10.20), where $STD(\cdot)$ is the standard deviation function. Both indices are used to mitigate the short-term fluctuations by capturing the longer-term trend across the evolutionary process.

$$mmAP(f_j) = \frac{1}{p}\sum_{i=1}^{p}\left(\frac{1}{i}\sum_{j=1}^{i}MEAN(f_j)\right) \qquad (10.19)$$

$$mmSTD(f_j) = \frac{1}{p}\sum_{i=1}^{p}\left(\frac{1}{i}\sum_{j=1}^{i}STD(f_j)\right) \qquad (10.20)$$

$$mmVAR(f_j) = \frac{1}{p}\sum_{i=1}^{p}\left(\frac{1}{i}\sum_{j=1}^{i}VAR(f_j)\right) \qquad (10.21)$$

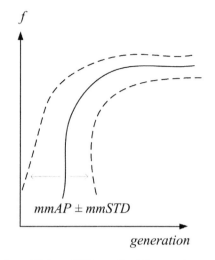

Fig. 10.3 The diagram of mmAP ± mmSTD over the full generations

As shown in Figure 10.3, the solid line gives the mmAP scores for each vector f_j as given by Equation (10.19). The dashed lines are the mmAP ± mmSTD for each vector f_j as given by Equation (10.20), which demonstrates the evolutionary path of the optimization process (generation vs. fitness f) with the upper and lower boundaries.

10.5.3 Pareto Reliability Index

The Pareto reliability index (PRI1, β_1) is defined by Equation (10.22), where μ_f is the mean and σ_f is the standard deviation of the normalized objectives, as given by Equations (10.28) and (10.24) respectively. W_i are the weighted normalized objectives in Equation (10.25) and w_i are the weight factors as given by Equation (10.26), which balance the weights of all the normalized objectives, $w_i \in [0,1]$.

$$\beta_1 = \frac{\mu_f}{\sigma_f} \tag{10.22}$$

$$\mu_f = \frac{\sum_i^N W_i}{N} \tag{10.23}$$

$$\sigma_f^2 = \frac{\sum_i^N \left(W_i - \mu_f \right)^2}{N-1} \tag{10.24}$$

$$W_i = w_i \cdot f_{i_h} \tag{10.25}$$

$$\sum_i^N w_i = 1 \tag{10.26}$$

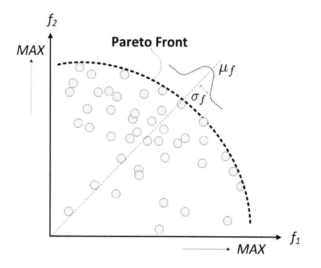

Fig. 10.4 Pareto Reliability Index

As shown in Figure 10.4, without loss of generality, a case of two objective functions, f_1 and f_2, is utilized to present the definition of β_1. The figure shows a geometrical illustration of the β_1 index in a dual-objective case, which indicates the distance of the mean margin of a multi-criterion range. The idea behind β_1 is the distance from the location measure μ_f to the limit states σ_f, which provides a good measure of the reliability of the Pareto solutions, that is, a larger value of β_1 leads to a better solution.

10.5.4 Pareto Risk Index

The Pareto Risk Index (PRI2, β_2) is defined by Equation (10.22), where μ_f is the mean and σ_f is the standard deviation of the normalized objectives, as given by Equations (10.28) and (10.24) respectively. W_i are the weighted normalized objectives given in Equation (10.25), and w_i are the weight factors, which balance the weights of all the normalized objectives, $w_i \in [0,1]$.

$$\beta_2 = \frac{\sigma_f}{\mu_f} \tag{10.27}$$

$$\mu_f = \frac{\sum_i^J W_i}{J} \tag{10.28}$$

$$\sigma_f^2 = \frac{\sum_i^J \left(W_i - \mu_f\right)^2}{J-1} \tag{10.29}$$

$$W_i = w_i \cdot f_{i_h}, \sum_i^J w_i = 1 \tag{10.30}$$

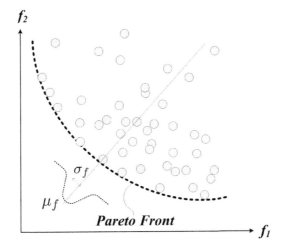

Fig. 10.5 Pareto Risk Index

As shown in Figure 10.5, without loss of generality, a case of two objective functions, f_1 and f_2, is utilized to present the definition of β_2, which shows a geometrical illustration of the β_2 index in a dual-objective case, indicating the distance of the mean margin of a multi-criterion range. The distance from the location measure μ_f to the limit states σ_f provides a good measure of the reliability of the Pareto solutions, that is, a smaller value of β_2 leads to a better solution.

10.5.5 Pareto Sensitivity Indices

To assess the relative importance of each objective to the overall optimal performance of β_1 and β_2, the Pareto sensitivity indices are proposed through Equations (10.31) and (10.32), which measure the small changes of β_1 and β_2 with respect to the fitness function f_i and examine the dynamic behaviors over the optimization process.

$$\beta_{3,1,i} = \frac{\partial \beta_1}{\partial f_i} \tag{10.31}$$

$$\beta_{3,2,i} = \frac{\partial \beta_2}{\partial f_i} \tag{10.32}$$

10.6 Fast Approach to Pareto-Optimal Solution Recommendation

In this section, a fast approach for Pareto-optimal solution recommendation (FPR) using the Pareto risk index is proposed. This approach provides users with a recommendation list of optimal ranking and optimal trend indications with different risk tolerances.

10.6.1 Normalization

The normalization process involves a mapping of the variables from their original value range to a normalized value range, e.g., $[0, 1]$, through the two operations *scale* and *shift* [Chen (2010a)]. As defined by Equation (6.1), the vector of objective functions F_i with the values $[f_1(\mathbf{x}), f_2(\mathbf{x}), \cdots, f_J(\mathbf{x})]$ is the raw data source of the normalization block, in which $f_i \in [f_{min}, f_{max}]$.

First, as given by Equation (16.3), the *scale* operation calculates the scale factor according to the input range $[f_{min}, f_{max}]$ of the raw data f_i, and then, all the input data are scaled to the range of $[c_l, c_u]$. Specifically, the fitness values are mapped from the practical value range $[f_{min}, f_{max}]$ to the normalized value range $[c_l, c_u]$, which is $[0, 1]$ in this context.

$$f_{i_c} = (c_u - c_l) \times \frac{f_i - f_{min}}{f_{max} - f_{min}} \tag{10.33}$$

Then, in the *shift* operation, the scaled data f_{i_c} are shifted to the new range of $[c_l, c_u]$, as given by Equation (16.4), where f_{i_h} represents the normalized fitness data.

$$f_{i_h} = c_l + f_{i_c} \tag{10.34}$$

10.6.2 FPR Steps

As shown in Figure 10.6, the flow chart of the FPR can be divided into 6 steps:

- initializing parameters and starting the optimization process
- performing optimization using multi-objective algorithms
- generating Pareto-optimal solutions
- performing PRI assessment, which has the following four sub-steps

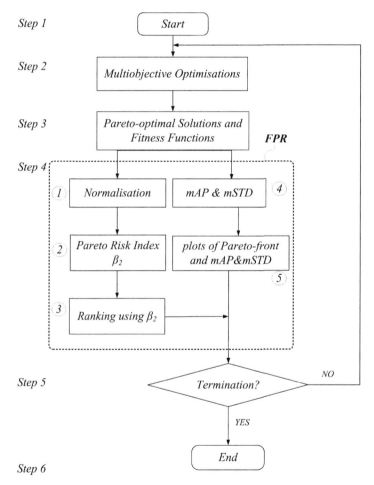

Fig. 10.6 Workflow of the FPR

- normalization for the multi-objective fitness functions of the Pareto-optimal solutions
- calculation of the PRI index β_2
- calculation of the evolutionary trend indices, mean average precision (mAP) and mean standard deviation (mSTD)
- visualization of Pareto front and evolutionary trend indices

- checking optimization termination conditions
- ending the program and post-calculation process

10.7 Fitness Functions

A fitness function is a particular type of objective function that is used to summarize as a single figure of merit how close a given design solution is to achieving the set goals. Usually, a practical problem needs to be transformed from a practical problem to a fitness function for evolutionary calculation. A fitness value is assigned to each solution depending on how close it actually is to solving the problem.

Using a single-objective problem as an example [?], if the objective function, either as a cost function $J \in [0, \infty)$, or inversely, as a fitness function $f \in (0, 1]$, where,

$$f = \frac{J}{1+J} \tag{10.35}$$

Equation (10.35) is differentiable under practical constraints in the multidimensional space and the design problem may be solved analytically. Finding the parameter sets that result in a zero first-order derivative and satisfy the second-order derivative conditions would reveal all local optima. Then comparing the values of the performance index of all the local optima, together with those of all boundary parameter sets, would lead to the global optimum, whose corresponding 'parameter' set will thus represent the best design. However, in practice, the optimization usually involves multiple objectives and the matters involving derivatives are a lot more complex [Wikipedia (2011)].

The fitness function usually actively affects the performance of the optimization and calculation. An evolutionary algorithm will either converge on an inappropriate solution or will have difficulty in converging at all if given an inappropriate fitness function. Two main types of fitness functions exist:

- Type 1, where the fitness function has analytical expressions, which do not change during the evolutionary iteration
- Type 2, where the fitness function is co-evolving with the evolutionary process

In this case, the definition of the fitness function is not straightforward in many cases and often is performed iteratively if the fittest solutions produced by CI algorithms are not what is desired.

Moreover, the fitness function should not only correlate closely to the designer's goal but also be computationally very efficient for the fitness approximation, especially in those cases that require large amounts of CPU time, including where (1) the computation time of a single solution is extremely high, (2) the precise model for the fitness computation is missing or is too complex to define, or (3) the fitness function is uncertain or noisy.

10.8 Test Functions

One of the most fundamental issues when proposing an algorithm is to have a standard methodology to test and validate its performance. Test functions, such as convergence speed, precision, robustness and other performance indicators, have been proposed as benchmarks for evaluating the characteristics of optimization algorithms.

Basically, the test functions can be divide into 2 types: Type-1 (T1), which are test functions for single-objective optimization problems, and Type-2 (T2), which are test functions for multi-objective optimization problems. Tables 10.5 and 10.6 list a few test functions widely known in the literature for T1 and T2, respectively. More test functions can be found in the given references [Whitley, et al. (1996), Jamil and Yang (2013), Montaz Ali (2005), Test Functions Index (2017), Virtual Library of Simulation Experiments (2017)].

Test functions are commonly used to evaluate the effectiveness of different search algorithms. However, the results of evaluation are as dependent on the test problems as they are on the algorithms that are subject to comparison.

Table 10.5 T1: Test Functions for Single-objective Optimization Problems [Whitley, et al. (1996), Jamil and Yang (2013), Montaz Ali (2005), Test Functions Index (2017), Virtual Library of Simulation Experiments (2017)]

Function	Formula	Search Domain	Plot	m-file	Minimum
Ackley function	$f(x_i) =$ $-a\exp\left(-b\sqrt{\dfrac{1}{d}\sum_{i=1}^{d}x_i^2}\right)$ $-\exp\left(\dfrac{1}{d}\sum_{i=1}^{d}\cos(cx_i)\right)$ $+a+\exp(1)$ $\quad$ (10.36)	$-5 \le \{x_i\} \le 5$	Figure 10.7	TF1-Ackley	$f(x_i^*) = 0$, at $x_i^* = [0,\cdots,0]$
Sphere function	$f(x_i) = \sum_{i=1}^{n} x_i^2$ $\quad$ (10.37)	$-\infty \le \{x_i\} \le \infty, 1 \le i \le n$	Figure 10.8	$TF1\_Sphere$	$f(x_i^*) = 0$, at $x_i^* = [0,\cdots,0]$
Rosenbrock function	$f(x_i) =$ $\sum_{i=1}^{n-1}\left[100\left(x_{i+1}-x_i^2\right)^2\right]$ $+\sum_{i=1}^{n-1}\left[(x_i-1)^2\right]$ $\quad$ (10.38)	$-5 \le \{x_i\} \le 10, 1 \le i \le n$	Figure 10.9	$TF1\_Rosenbrock$	$f(x_i^*) = 0$, at $x_i^* = \underbrace{[1,\ldots,1]}_{(n)\text{ times}}$
Beale function	$f(x,y) =$ $(1.5 - x + xy)^2$ $+ (2.25 - x + xy^2)^2$ $+ (2.625 - x + xy^3)^2$ $\quad$ (10.39)	$-4.5 \le \{x,y\} \le 4.5$	Figure 10.10	$TF1\_Beale$	$f(3, 0.5) = 0$

Function	Formula	Search Domain	Plot	m-file	Minimum
Goldstein Price function	$f(x,y) =$ $\left(1 + (x+y+1)^2 \right.$ $\left(\begin{array}{c} 19 - 14x + 3x^2 \\ -14y + 6xy + 3y^2 \end{array} \right)$ $\left. 30 + (2x - 3y)^2 \right.$ $\left(\begin{array}{c} 18 - 32x + 12x^2 \\ +48y - 36xy + 27y^2 \end{array} \right)$ $\quad(10.40)$	$-2 \leq \{x,y\} \leq 2$	Figure 10.11	TF1_GPrice	$f(0,-1) = 3$
Booth function	$f(x,y) =$ $(x + 2y - 7)^2 + (2x + y - 5)^2$ $\quad(10.41)$	$-10 \leq \{x,y\} \leq 10$	Figure 10.12	$TF1\_Booth$	$f(0,3) = 0$
Bukin function No. 6	$f(x,y) =$ $100\sqrt{|y - 0.01x^2|}$ $+ 0.01|x + 10|$ $\quad(10.42)$	$-15 \leq x \leq -5,\ -3 \leq y \leq 3$	Figure 10.13	$TF1\_Bukin6$	$f(-10,1) = 0$
Matyas function	$f(x,y) =$ $0.26(x^2 + y^2) - 0.48xy$ $\quad(10.43)$	$-10 \leq \{x,y\} \leq 10$	Figure 10.14	$TF1\_Matyas$	$f(0,0) = 0$
Lévi function No. 13	$f(x,y) =$ $\sin^2(3\pi x)$ $+ (x - 1)^2 (1 + \sin^2(3\pi y))$ $+ (y - 1)^2 (1 + \sin^2(2\pi y))$ $\quad(10.44)$	$-10 \leq \{x,y\} \leq 10$	Figure 10.15	$TF1\_Levi13$	$f(1,1) = 0$

Continued on next page

Table 10.5 T1: Test Functions for Single-objective Optimization Problems

Function	Formula	Search Domain	Plot	m-file	Minimum
Three-hump camel function	$f(x,y) =$ $2x^2 - 1.05x^4 + \dfrac{x^6}{6} + xy + y^2$ $\qquad$ (10.45)	$-5 \leq \{x,y\} \leq 5$	Figure 10.16	$TF1\_threecamel$	$f(0,0) = 0$
Easom function	$f(x,y) =$ $-\cos(x)\cos(y)$ $\exp\left(-\left((x-\pi)^2 + (y-\pi)^2\right)\right)$ $\qquad$ (10.46)	$-100 \leq \{x,y\} \leq 100$	Figure 10.17	$TF1\_Easom$	$f(\pi,\pi) = -1$
Cross-in-tray function	$f(x,y) = -0.0001$ $\left(\left\| \sin(x)\sin(y) \right.\right.$ $\left.\left. \exp\left(\left\|100 - \dfrac{\sqrt{x^2+y^2}}{\pi}\right\|\right) \right\| + 1\right)^{0.1}$ $\qquad$ (10.47)	$-10 \leq \{x,y\} \leq 10$	Figure 10.18	$TF1\_Crossintray$	$f(x_i^*) = -2.06261$ $at \ x_i^* =$ $\{1.34941, -1.34941\}$ $\{1.34941, 1.34941\}$ $\{-1.34941, 1.34941\}$ $\{-1.34941, -1.34941\}$ $\qquad$ (10.48)
Eggholder func-tion	$f(x,y) =$ $-(y+47)\sin\left(\sqrt{\left\|y+\dfrac{x}{2}+47\right\|}\right)$ $-x\sin\left(\sqrt{\|x-(y+47)\|}\right)$ $\qquad$ (10.49)	$-512 \leq \{x,y\} \leq 512$	Figure 10.19	$TF1\_Eggholder$	$f(512, 404.2319) = -959.6407$

Function	Formula	Search Domain	Plot	m-file	Minimum
Hölder table function	$f(x,y) =$ $-\left\lvert \sin(x)\cos(y) \exp\left(\left\lvert 1 - \dfrac{\sqrt{x^2+y^2}}{\pi}\right\rvert\right) \right\rvert$ (10.50)	$-10 \le \{x,y\} \le 10$	Figure 10.20	$TF1\_Holdertable$	$f(x_i^*) = -19.2085$ $at \quad x_i^* =$ $\{8.05502, 9.66459\}$ $\{-8.05502, 9.66459\}$ $\{8.05502, -9.66459\}$ $\{-8.05502, -9.66459\}$ (10.51)
McCormick function	$f(x,y) =$ $\sin(x+y) + (x-y)^2 -$ $1.5x + 2.5y + 1$ (10.52)	$-1.5 \le x \le 4, -3 \le y \le 4$	Figure 10.21	$TF1\_McCormick$	$f(-0.54719, -1.54719) =$ -1.9133
Schaffer function No. 2	$f(x,y) =$ $0.5 + \dfrac{\sin^2\left(x^2 - y^2\right) - 0.5}{\left(1 + 0.001\left(x^2 + y^2\right)\right)^2}$ (10.53)	$-100 \le \{x,y\} \le 100$	Figure 10.22	$TF1\_Schaffer2$	$f(0,0) = 0$
Schaffer function No. 4	$f(x,y) =$ $0.5 + \dfrac{\cos^2\left(\sin\left(\lvert x^2 - y^2\rvert\right)\right) - 0.5}{\left(1 + 0.001\left(x^2 + y^2\right)\right)^2}$ (10.54)	$-100 \le \{x,y\} \le 100$	Figure 10.23	$TF1\_Schaffer4$	$f(0, 1.25313) = 0.292579$

Continued on next page

Table 10.5 T1: Test Functions for Single-objective Optimization Problems

Function	Formula	Search Domain	Plot	m-file	Minimum
Styblinski-Tang function	$f(x,y) = \dfrac{\sum_{i=1}^{n} x_i^4 - 16x_i^2 + 5x_i}{2}$ (10.55)	$-5 \leq x_i \leq 5, 1 \leq i \leq n$	Figure 10.24	$TF1\_StyblinskiT$	$-39.16617n \leq$ $f\left(\underbrace{-2.903534,\ldots,-2.903534}_{(n)\text{times}}\right)$ $\leq -39.16616n$ (10.56)
Himmelblau function	$f(x,y) =$ $(x^2 + y - 11)^2 + (x + y^2 - 7)^2$ (10.57)	$-5 \leq \{x, y\} \leq 5$	Figure 10.25	$TF1\_Himmelblau$	$f(3,2) = 0$ $f(-3.779310, -3.283186) = 0$ $f(-2.805118, 3.131312) = 0$ $f(3.584428, -1.848126) = 0$ (10.58)

Table 10.6 T2: Test Functions for Multi-objective Optimization Problems [Whitley, et al. (1996), Jamil and Yang (2013), Montaz Ali (2005), Test Functions Index (2017), Virtual Library of Simulation Experiments (2017)]

Function	Formula	Search Domain	m-file
Binh and Korn function	$f_1(x,y) = 4x^2 + 4y^2$ $f_2(x,y) = (x-5)^2 + (y-5)^2$ s.t. $g_1(x,y) = (x-5)^2 + y^2 \leq 25$ $g_2(x,y) = (x-8)^2 + (y+3)^2 \geq 7.7$ $\qquad\qquad\qquad\qquad (10.59)$	$0 \leq \{x\} \leq 5, 0 \leq \{y\} \leq 3$	TF2_BinhKorn
Chakong and Haimes function	$f_1(x,y) = 2 + (x-2)^2 + (y-1)^2$ $f_2(x,y) = 9x - (y-1)^2$ s.t. $g_1(x,y) = x^2 + y^2 \leq 225$ $g_2(x,y) = x - 3y + 10 \leq 0$ $\qquad\qquad\qquad\qquad (10.60)$	$-20 \leq \{x,y\} \leq 20$	TF2_ChakongHaimes
Fonseca and Fleming function	$f_1(x) = 1 - \exp\left(-\sum_{i=1}^{n}\left(x_i - \frac{1}{\sqrt{n}}\right)^2\right)$ $f_2(x) = 1 - \exp\left(-\sum_{i=1}^{n}\left(x_i + \frac{1}{\sqrt{n}}\right)^2\right)$ $\qquad\qquad\qquad\qquad (10.61)$	$-4 \leq x_i \leq 4, 1 \leq i \leq n$	TF2_FonsecaFleming

Continued on next page

Table 10.6 T2: Test Functions for Multi-objective Optimization Problems

Function	Formula	Search Domain	m-file
Binh function No. 4	$f_1(x,y) = x^2 - y$ $f_2(x,y) = -0.5x - y - 1$ s.t. $g_1(x,y) = 6.5 - \frac{x}{6} - y \geq 0$ $g_2(x,y) = 7.5 - 0.5x - y \geq 0$ $g_3(x,y) = 30 - 5x - y \geq 0$ (10.62)	$-7 \leq \{x,y\} \leq 4$	TF2_Binh4
Zitzler-Deb-Thiele's function No. 1	$f_1(x) = x_1$ $f_2(x) = g(x)h(f_1(x), g(x))$ $g(x) = 1 + \frac{9}{29}\sum_{i=2}^{30} x_i$ $h(f_1(x), g(x)) = 1 - \sqrt{\frac{f_1(x)}{g(x)}}$ (10.63)	$0 \leq x_i \leq 1, 1 \leq i \leq 30$	TF2_ZDT1
Zitzler-Deb-Thiele's function No. 2	$f_1(x) = x_1$ $f_2(x) = g(x)h(f_1(x), g(x))$ $g(x) = 1 + \frac{9}{29}\sum_{i=2}^{30} x_i$ $h(f_1(x), g(x)) = 1 - \left(\frac{f_1(x)}{g(x)}\right)^2$ (10.64)	$0 \leq x_i \leq 1, 1 \leq i \leq 30$	TF2_ZDT2

Function	Formula	Search Domain	m-file
Zitzler-Deb-Thiele's function No. 4	$f_1(x) = x_1$ $f_2(x) = g(x)h(f_1(x), g(x))$ $g(x) = 91 + \sum_{i=2}^{10}(x_i^2 - 10\cos(4\pi x_i))$ $h(f_1(x), g(x)) = 1 - \sqrt{\dfrac{f_1(x)}{g(x)}}$ (10.65)	$0 \le x_1 \le 1, -5 \le x_i \le 5, 2 \le i \le 10$	TF2_ZDT4
Zitzler-Deb-Thiele's function No. 6	$f_1(x) = 1 - \exp(-4x_1)\sin^6(6\pi x_1)$ $f_2(x) = g(x)h(f_1(x), g(x))$ $g(x) = 1 + 9\left[\dfrac{\sum_{i=2}^{10}x_i}{9}\right]^{0.25}$ $h(f_1(x), g(x)) = 1 - \left(\dfrac{f_1(x)}{g(x)}\right)^2$ (10.66)	$0 \le x_i \le 1, 1 \le i \le 10$	TF2_ZDT6

1. Ackley function

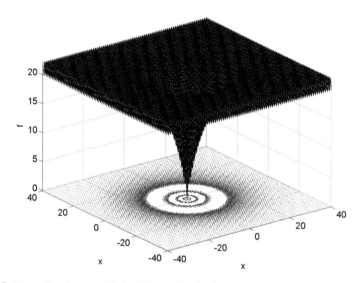

Fig. 10.7 Ackley function, $a = 20, b = 0.2, c = 2\pi, d = 2$

2. Sphere function

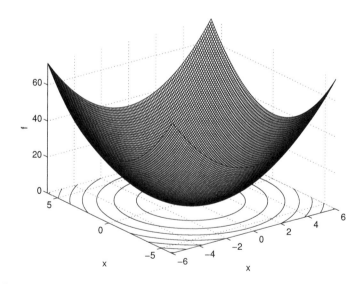

Fig. 10.8 Sphere function

3. Rosenbrock function

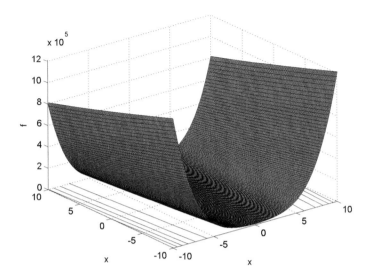

Fig. 10.9 Rosenbrock function

4. Beale function

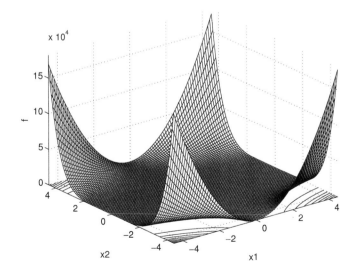

Fig. 10.10 Beale function

5. Goldstein Price function

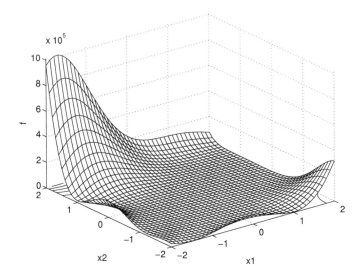

Fig. 10.11 Goldstein Price function

6. Booth function

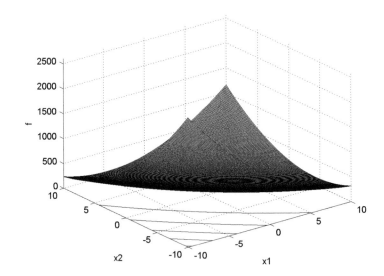

Fig. 10.12 Booth function

7. Bukin function No. 6

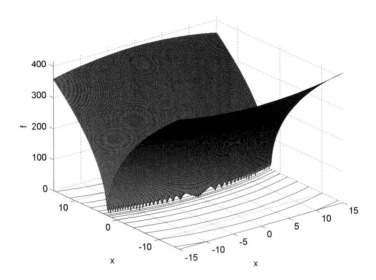

Fig. 10.13 Bukin function No. 6

8. Matyas function

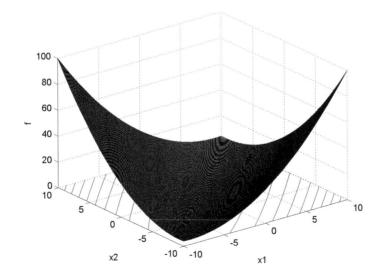

Fig. 10.14 Matyas function

9. Lévi function No. 13

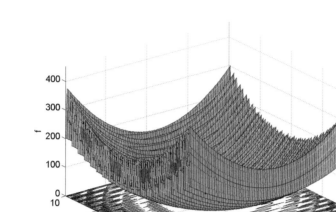

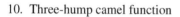

Fig. 10.15 Lévi function

10. Three-hump camel function

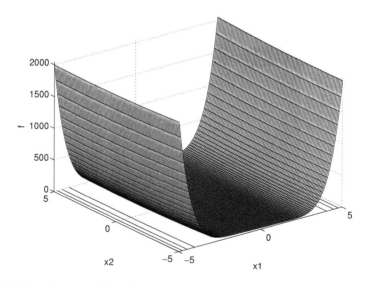

Fig. 10.16 Three-hump camel function

11. Easom function

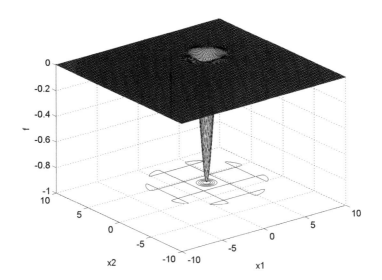

Fig. 10.17 Easom function

12. Cross-in-tray function

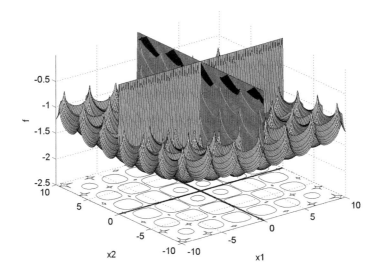

Fig. 10.18 Cross-in-tray function

13. Eggholder function

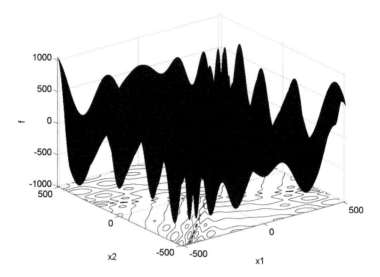

Fig. 10.19 Eggholder function

14. Hölder table function

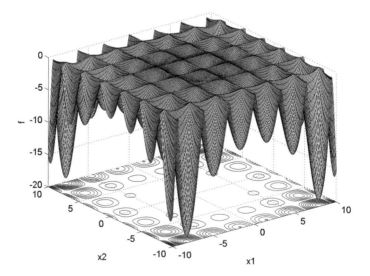

Fig. 10.20 Hölder table function

15. McCormick function

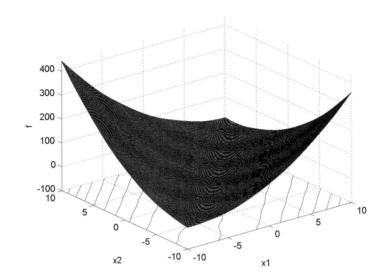

Fig. 10.21 McCormick function

16. Schaffer function No. 2

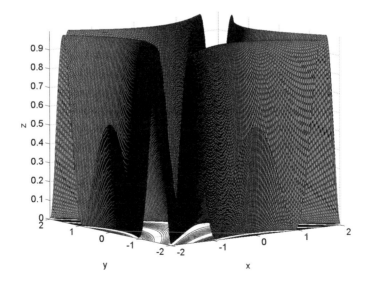

Fig. 10.22 Schaffer function No. 2

17. Schaffer function No. 4

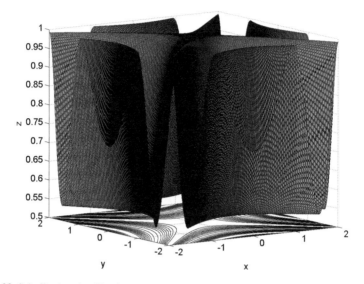

Fig. 10.23 Schaffer function No. 4

18. Styblinski-Tang function

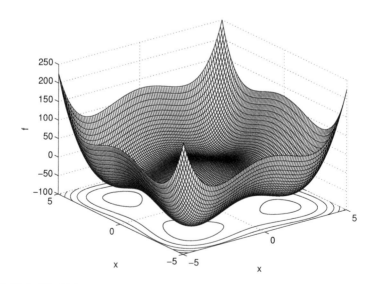

Fig. 10.24 Styblinski-Tang function

19. Himmelblau function

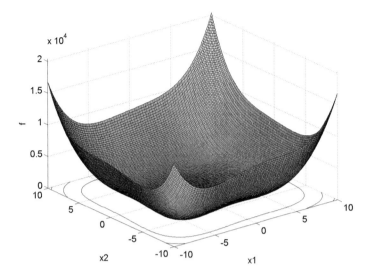

Fig. 10.25 Himmelblau function

Part III
CIAD for Science and Technology

Part III introduces the conceptual framework of CIAD and its applications in science and technology such as control systems and battery capacity prediction.

11

Adaptive Bathtub-Shaped Curve

11.1 Introduction

The bathtub-shaped failure rate (BFR) function is a well-known concept for representing the failure behavior of various products. A BFR function can be divided into three portions—'Infant Mortality', 'Steady State' and 'Wearout Failures', as demonstrated in Figure 11.1, which denotes the three phases that a product passes through during its full life cycle [Dhillon (1979), Hjorth (1980), AL-Hussaini and ABD-EL-Hakim (1989), Klutke, et al. (2003), Dhillon (2005), Crowe and Feinberg (2000), Jiang and Murthy (1998), Wang, et al. (2002)]. The adaptive bathtub failure rate function (ABF) was proposed by Chen et al. [Chen, et al. (2011)] to provide a parametric representation of the failure behaviors of universal engineering applications. The ABF method can be utilized as the design objective for determination and optimization of the 'useful period' lifetime parameter.

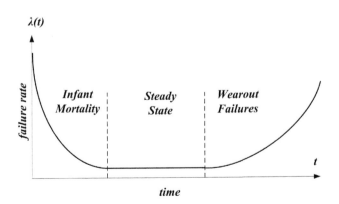

Fig. 11.1 The Bathtub curve [Dhillon (2005)]

The artificial fish swarm algorithm (AFSA) was first proposed in 2002 [Li, et al. (2002)], inspired by the social behaviors of fish schools during searching, swarming and following. Schooling fish can quickly respond to changes in the direction and speed of their neighbors; information on their behaviors is passed to others, thereby helping the fish move from one configuration to another almost as a single unit. By borrowing this intelligence characterizing social behavior, the AFSA is made parallel, independent of the initial values and is able to achieve a global optimum.

BFR modeling is one of the most important studies concerning CMOS devices. As CMOS device feature sizes continue to scale down, the accompanying increases in both electric field strength and current density act to reduce device lifetime. To understand and meet target device lifetime requirements, it is important to estimate all aspects of device reliability, measure the performance and provide a necessary prediction for device lifetime under real-world conditions.

In this chapter, an AFSA method with a variable population size (AFSAVP), as introduced in Section 7.4, is utilized and applied as the intelligence approach for a case study of a CMOS device acceleration test. An empirical analysis is devised to estimate the parameters for the ABF behaviors of the studied CMOS device.

11.2 Parameterization Method via Radial Basis Functions

As illustrated in Figure 11.2, without loss of generality, the radial basis function (RBF) can be defined as μ_1 by Equation (11.1), in which $b(t)$ is a segment function, as given by Equation (11.10). In Equation (11.10), the parameters of $b(t)$ are defined as follows: α is the support factor, β is the boundary factor, γ is the gain factor, and ζ is the shape factor. Referring to the above definition of μ_1, μ_2 is mirrored as the reflection of μ_1 over the x-axis by Equation (11.3).

Figure 11.3 demonstrates three of the further transformations of μ_2, in which μ_3 is translated by $\mu_2 + 1$, μ_4 is $\mu_3 \times \eta$ and μ_5 is $\mu_4 + \delta$, as expressed by Equations (11.4), (11.5) and (11.6). Specifically, the original RBF μ_1 is translated by reflection, shifting upward one unit and scaling and shifting upward δ units, as stated in the functions μ_2 to μ_5, respectively.

$$\mu_1(x) = e^{b(x)} \tag{11.1}$$

$$b(x) = \begin{cases} -\dfrac{(|x - \alpha| - \beta)^\zeta}{\gamma} & \text{if } |x - \alpha| \geq \beta \\ 0 & \text{if } |x - \alpha| < \beta \end{cases} \tag{11.2}$$

$$\mu_2(x) = -e^{b(x)} \tag{11.3}$$

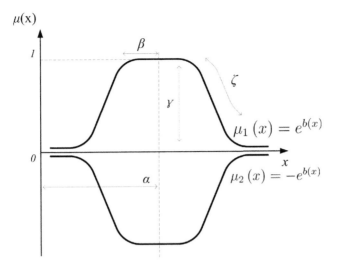

Fig. 11.2 Radial basis function μ_1 and its reflection μ_2 over the x-axis

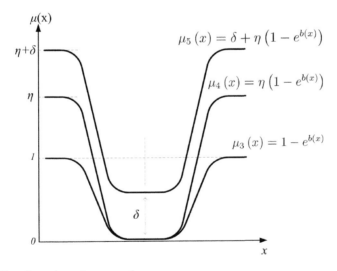

Fig. 11.3 Transformations of μ_3, μ_4 and μ_5

$$\mu_3 (x) = 1 - e^{b(x)} \tag{11.4}$$

$$\mu_4 (x) = \eta \left(1 - e^{b(x)} \right) \tag{11.5}$$

$$\mu_5 (x) = \delta + \eta \left(1 - e^{b(x)} \right) \tag{11.6}$$

11.3 Adaptive Bathtub-Shaped Failure Rate Function

By borrowing the form of the transformed RBF μ_5, this chapter attempts to propose an adaptive model concerned with the behavior of the failure rate in the full range of the bathtub curve, named the 'adaptive bathtub-shaped failure rate function' (ABF). There are two types of ABF functions—the symmetric ABF (SABF) in Figure 11.4 and the asymmetric ABF (AABF) in Figure 11.5, as defined by Equations (16.1) and (11.8), respectively.

$$\lambda_1(t) = \delta + \eta\left(1 - e^{b(t)}\right), t \geq 0 \tag{11.7}$$

$$\lambda_2(t) = \begin{cases} \delta_1 + \eta_1\left(1 - e^{b_1(t)}\right) & \text{if } t \geq \alpha \\ \delta_2 + \eta_2\left(1 - e^{b_2(t)}\right) & \text{if } 0 \leq t < \alpha \end{cases} \tag{11.8}$$

As shown in Figure 11.6, the original form of the RBF is defined by μ_1. μ_2 and μ_3 are two transformations of μ_1 by mirror reflections with respect to the x-axis and subsequent shifting upward by one unit; μ_2 and μ_3 are called the transformed radial basis functions (TRBFs). By scaling η and shifting upward by δ units, we can obtain a TRBF stated as $\lambda(t)$ for the ABF definition.

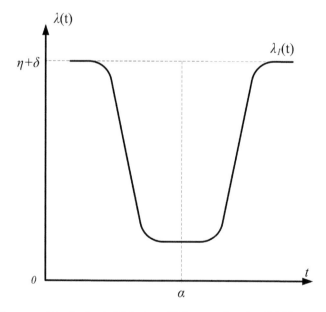

Fig. 11.4 The symmetric adaptive bathtub-shaped failure rate function, SABF

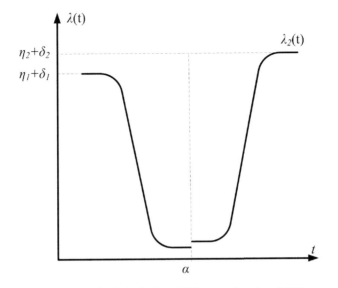

Fig. 11.5 The asymmetric adaptive bathtub-shaped failure rate function, AABF

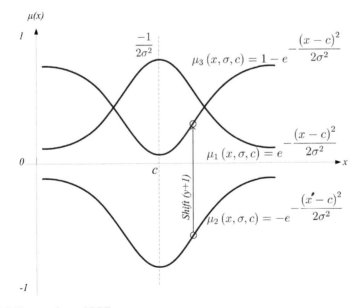

Fig. 11.6 The transformed RBF

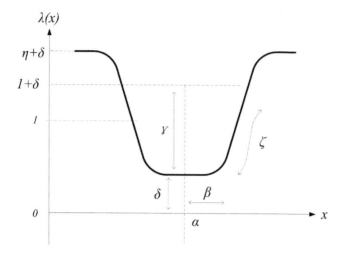

Fig. 11.7 The adaptive bathtub-shaped failure rate function [Chen, et al. (2011)]

By using the TRBF $\lambda\,(t)$ as given by Equation (16.1), the ABF can be introduced. The ABF concerns the behavior of the failure rate in the full range of the bathtub curve [Chen, et al. (2011)]. As illustrated in Figures 11.6 and 11.7, without loss of generality, the ABF can be defined as $\lambda\,(t)$ by Equation (16.1), in which $b(x)$ is a segment function for the RBF, as given by Equation (11.10). As shown in Equation (11.10), the parameters of $b(x)$ are defined as follows: α is the support factor, β is the boundary factor, γ is a gain, ζ is the shape factor, η is a gain, and δ is the shift factor.

$$\lambda\,(t) = \delta + \eta\left(1 - e^{b(t)}\right), t \geq 0 \tag{11.9}$$

$$b(x) = \begin{cases} -\dfrac{(|x - \alpha| - \beta)^{\zeta}}{\gamma} & \text{if } |x - \alpha| \geq \beta \\ 0 & \text{if } |x - \alpha| < \beta \end{cases} \tag{11.10}$$

$$R(t) = e^{-\int_0^t \left(\delta + \eta\left(1 - e^{b(t)}\right)\right)dt}, t \geq 0 \tag{11.11}$$

Accordingly, Equation (11.11) is the reliability function, which can be utilized to obtain the adaptive reliability functions for engineering applications and optimizations, given the conditions $t \geq 0$, $\frac{(|t-\alpha|-\beta)^{\zeta}}{\gamma} > 0$, $\zeta \geq 0$.

11.4 Fitness Function Definition

The coefficient of determination (R^2) is taken as a quantity for quantitative analysis and measures the proportion of total components for the parameter determination via AFSAVP. R^2 is defined by Equation (10.6), in which SS_t and SS_e are the total corrected sum of squares and the residual sum of squares, respectively; $\vec{y}_i$ are the spectral data calculated by AFSAVP over the evaluation process; $\vec{y}_i$ are the actual spectral data obtained experimentally; $\vec{\bar{y}}_i$ is the mean value of the calculated data $\vec{y}_i$; and k is the number of observations. In this context, R^2 is employed as the fitness function F to evaluate AFSAVP through quantitative analysis, where a larger value of R^2 indicates a better approach within the simulation data ranges.

Over the full simulation time, AFSAVP compares the calculated characteristics (failure rate) with the experimental data by maximizing the fitness function F, as shown by Equation (11.12), where λ_{fish} is the failure rate estimated by the ABF method and λ_0 are the obtained experimental data.

$$Maximize : F = R^2\left(\lambda_{fish}, \lambda_0\right) \tag{11.12}$$

11.5 Simulations and Discussion

Numerical results are obtained using the simulation toolbox for $MATLAB^{\circledR}$, known henceforth as *SwarmsFish* [Swarmfish (2011)]. Unless otherwise stated, all the results are generated using the parameters listed in Table 11.1 for the simulation process, in which max generations = 50 is the termination condition. The fish visual factor is 2.5, the attempt number for the searching behavior is 5, the iteration step is 0.1, and the population is 30, in which the non-replaceable population P_N and replaceable population P_R are 20 and 10, respectively.

Table 11.1 Parameters for the ASFAVP Numerical Simulation

	Max generations	50
	Visual	2.5
	Attempt number	5
	Step	0.1
	Population	30
	Crowd	0.618
P_N	Non-replaceable population	20
P_R	Replaceable population	10
	Sample size	60

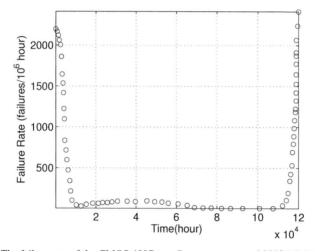

Fig. 11.8 The failure rate of the CMOS 4007 test @ a temperature of $200^\circ C$ [Moltofta (1983)]

Table 11.2 Results for the CMOS 4007 Case

α	0.218658
β	3.468728
γ	9.664110
ζ	4.429524
δ	3.137874
η	14.036155

As shown in Figure 11.8, the experimental failure rate data λ_0 came from the CMOS 4007 test at a temperature of $200^\circ C$ [Moltofta (1983)] and are to be normalized to the ranges of [-5,5] on the x-axis and [0,10] on the y-axis for the fitness function. The specifications of the computer used for the simulations are as follows: 2.1 GHz Intel dual-core processor, Windows XP Professional v5.01 build 2600 Service Pack 3, 2.0 GB of 800 MHz dual-channel DDR2 SDRAM, MATLAB R2008a, and *SwarmsFish*1001. The optimized results for the CMOS4007 case are listed in Table 11.2.

As shown in Figure 11.9, the fitness values of max-fitness (solid line), mean-fitness (dash-dot line) and min-fitness (dotted line) increase with slightly random oscillations when the generation size starts increasing. From generations 15 to 50, the fitness values reach a steady state, with a subsequent plateau ending. For a small volume data in this computation, the simulation process converges quickly when generation = 15, that is, in this case, the fitness values remain constant from generation = 15 to 50.

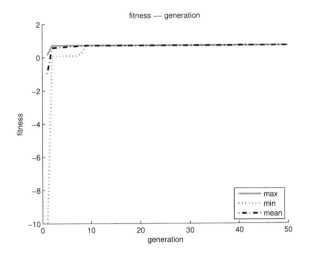

Fig. 11.9 The AFSAVP fitness over the full simulation

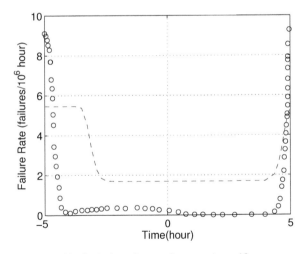

Fig. 11.10 The normalized bathtub-shaped curve @ generation = 10

Figures 11.10 and 11.11 are the normalized bathtub-shaped curve @ generation = 10 and 15 to 50, respectively, in which the 'o' line denotes the CMOS 4007 tested data and the '−' dashed line denotes the AFSAVP generated data. There are two evolutional stages that are captured at generation sizes of 10 and from generations 15 to 50, as shown in Figures 11.10 to 11.11. When the generation size increases, the data obtained by the AFSAVP produce a better fit to the experimental data.

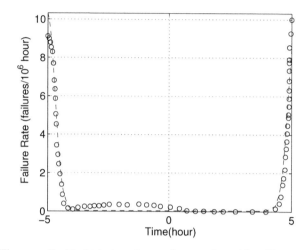

Fig. 11.11 The normalized bathtub-shaped curve @ generation = 15 to 50

Figures 11.9 to 11.11 demonstrate the AFSAVP's high efficiency in convergence, where the CMOS 4007 failure rate data can be parameterized by the ABF properly via optimizing parameter settings.

11.6 Conclusions and Future Work

The AFSAVP has been proposed and subsequently utilized for the quantitative representation of CMOS 4007 failure rate data by the ABF method. The results indicate that the ABF can represent the failure behavior of bathtub-shaped curves accordingly, if provided proper experimental data. The quick convergence of the fitness curve over full simulation demonstrates the efficiency of this method. The method also provides valuable guidelines for planning a solution that monitors critical maintenance processes and assets.

By following the proposed design framework, this AFSAVP technique has potential applications in product reliability analysis and modeling, electrical device parameter determination, prognostic health management data mining and real-time condition-based maintenance systems by establishing proper fitness functions for specific applications.

Notations

$\lambda(t)$ The bathtub-shaped failure rate function
$b(x)$ The segment function for RBF
α The support factor
β The boundary factor
γ The gain factor
ζ The shape factor
η A gain factor
δ The shift factor
S_k The kth fish state
F The fitness function

12

Terahertz Spectroscopic Analysis

12.1 Introduction

The terahertz (THz) region of the electromagnetic spectrum lies between the microwave and mid-infrared regions and is usually defined by the frequency range of $0.1 - 10$ THz, that is, 10^{12} cycles per second, as shown in Figure 12.1.

In the last 30 years, terahertz spectroscopy has been extensively used in applications in the security, pharmaceutical and semiconductor industries. Many materials have adequate spectral fingerprints in the terahertz range. These spectral fingerprints can be used for material identification. Additionally, it has been reported that terahertz spectroscopy can be used to investigate the collective vibration and rotation modes of large biological molecules, such as proteins, DNA and amino acids. Such information is very useful for studies of the structures and dynamics of biological molecules. For security applications, terahertz spectroscopy is an emerging tool to identify illegal drugs and explosives. Numerous terahertz spectroscopic studies of illegal drugs and explosives, including RDX, HMX, TNT, PETN and DNT, have been reported recently [Son (2009), Kanamorit, et al. (2005), Davies, et al. (2008), Burnett, et al. (2006), Cook, et al.(2005)]. It has been shown that most such chemical materials can be distinguished by monitoring their terahertz spectral fingerprints. Most of the samples used for corresponding experiments are purified. However, the samples in actual applications may be a mixture of several chemical components, which makes the spectral analysis difficult. Several computing techniques have been reported to identify the components in a mixture and determine the concentration of each component. Watanabe and his co-workers utilized a nonlinear least squares algorithm with non-negative constraints to determine the presence or absence of a specific chemical in a target; they also estimated the concentration [Watanabe, et al. (2004)]. Both Zhang et al. [Zhang, et al. (2006)] and Ma et al. [Ma, et al. (2009)] analyzed several chemical mixtures using a linear regression method. The results showed agreement with experimentally recorded terahertz spectra. Liu et al. [Liu, et al. (2009)] experimentally demonstrated that the absorption coefficient of the components in a mixture is linearly proportional to their concentrations. Neural networks were also used to analyze the terahertz spectra of explosive and bio-agent mixtures [Oliveira, et al. (2004), Zhong, et al. (2006)].

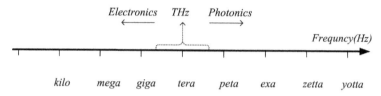

Fig. 12.1 THz range in the electromagnetic spectrum

In this chapter, an alternative technique based on an MO micro-GA for the uncertainty analysis of chemical mixtures is presented, where two of the objectives are expressed by the two fitness functions J_1 and J_2. The components and their corresponding concentrations in the mixture samples can be identified within a simulation time of the order of ten seconds. Mixtures with 10 possible components are analyzed in the study cases. This technique is supposed to work even if there are substantially more possible explosive components involved. In addition, some analytical results are compared with other computing techniques, therein obtaining a good agreement between each other. The results show that the micro-GA and its derivatives have potential applications in the fields of security, medicine and in the food industry for quickly identifying mixtures.

12.2 THz-TDS Experimental Setup Sketch

Time-domain spectroscopy is a well-established technique used for analyzing a wide range of materials in the UV-visible [Samuels, et al. (2004)] and infrared regions [Chen, et al. (2004), Ferguson and Zhang (2002)] since the development of short and ultra-short pulsed laser sources. Recently, the application of this technique has been extended to the THz region, where it is known as Terahertz Time-Domain Spectroscopy (THz-TDS) [Hangyo, et al. (2005), Schmuttenmaer (2004), Walther, et al. (2000), Ding and Shi (2006)]. THz-TDS provides a much better signal-to-noise ratio as compared to conventional far-infrared spectroscopy [Hargreaves and Lewis (2007)]. This technique can provide amplitude as well as phase information, in contrast to far-infrared spectroscopy, which can only provide intensity information [Katzenellenbogen and Grischkowsky (1991), Schneider, et al. (2006), Nahata et al. (1996), Mangeney and Crozat (2008)].

Figure 12.2 shows a typical THz-TDS experimental configuration. A femtosecond pulsed laser source is used to excite the THz emitter, which can be either a nonlinear crystal or semiconductor photo-conductive antenna (PCA). A parabolic mirror is used to focus the THz beam on to the sample cell and another parabolic mirror is used to collect the transmitted THz beam from the sample cell and focus it onto the THz detector. The probe beam analyzes the THz pulse via either electro-optic sampling in the nonlinear crystal or the photo-conductive effect in the PCA. Consequently, the time-domain THz pulse envelope and its cor-

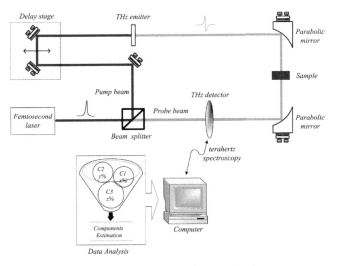

Fig. 12.2 Experimental setup of THz-TDS based on photo-conductive antennas

responding frequency-domain spectrum are virtualized, using a programmed data acquisition system and Fast Fourier Transform (FFT) computing algorithm. In our work, the THz absorption spectra of ten typical explosives were used as the database. The spectra were obtained from Zhang's [Zhang, et al. (2006)] and Liu's [Liu and Zhang (2007)] papers, using terahertz time-domain spectroscopy in transmission mode. According to their work, the samples for the THz-TDS measurements were prepared as pellets while some samples were prepared in pure form, and some others by mixing polyethylene (PE) powders at different weight ratios. The sample powders were compressed into pellets and measured using THz-TDS under the same experimental condition. Although the absorption and dispersions of PE mixture samples are not exactly the same as those of the pure samples with the same explosive substances, this imperfection does not interfere with the demonstration of the feasibility of the mixture determination technique described in this chapter. The mixture samples were used, as it was difficult and expensive to obtain pure samples. The spectral information of all the explosive samples can be obtained in pure form and under constant experimental conditions to provide more useful information for practical applications.

12.3 Statement of Mixture Component Determination

As shown in Figure 12.3, a 'statement' $W_{M \times N}$ of the mixture component determination is defined, where $M \times N$ are the dimensions of the statement W. In this context, the statement or table W for the statistical analysis is utilized to determine the explosive mixture components using the micro-GA process, in which

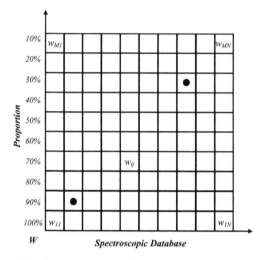

Fig. 12.3 Statement of the mixture component determination

- ○ M is the row number of the statement W, which states the proportional resolution. As shown in Figure 12.3, in this case, $i = 10$ per cent to 100 per cent with an increment or resolution of 10 per cent
- ○ N is the column number of the statement W, which numbers the THz spectroscopic database and indicates the types of specimens. As shown in the figure, j = 1 to 10 with an increment or resolution of 1, which means that there are 10 types of specimens in the THz spectroscopic database

w_{ij} is the element of W. There are two statuses of w_{ij} $(i = 1, 2, \ldots, M,\ j = 1, 2, \ldots, N)$, namely, ● or empty (1 or 0), as stated in Equation (12.1).

$$w_{ij} = \begin{cases} 0 \text{ empty} \\ 1 \ \bullet \end{cases} \tag{12.1}$$

- ○ When $w_{ij} = 1$, the state is an active state, which indicates that the current mixture consists of a combination of the specimen type j with the weight i. The mixture is now active and will be subject to the micro-GA fitness assessment
- ○ When $w_{ij} = 0$, the state is a non-active state. This combination will not be subject to the micro-GA process

As shown in Figure 12.3, a 10×10 statement shows that there are 10 types of specimens in the THz database ($N = 10$) and 10 choices of proportions as percentages ($M = 10$). For example, $w_{2,2} = 1$ and $w_{8,8} = 1$ indicate that the current mixture is a combination of specimen type No. 2 at 20 per cent and specimen type No. 8 at 80 per cent.

Within the same THz frequency range, the selected specimens' absorptions have been summed with respective proportion, as given in Figure 12.4. The absorption data, which were adopted from [Liu and Zhang (2007)], are for comparison with the micro-GA with the fitness function discussed below.

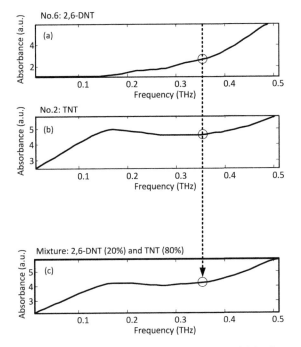

Fig. 12.4 Comparison of (a) 2,6-DNT spectra, (b) TNT spectra and (c) mixture spectra, where 2,6-DNT is No. 6 and TNT is No. 2 in the THz database (in Table 12.3). Mixture sample (c) is composed of 20 per cent 2,6-DNT (a) and 80 per cent TNT (b)

12.4 Fitness Function Definition

To evaluate the performance of the micro-GA comparison process with the exist-ing THz spectral feature, the R^2 metric, which is a statistical measurement of the agreement between the micro-GA observed data sequence y_i and the specimen data sequence $\hat{y}_i$, has been borrowed for the fitness function definition.

Applying the basic idea of the R^2 metric, which measures the goodness of fit of the regression for the THz spectroscopic data set, the investigation of the explosive mixture component determination has been performed using the micro-GA:

- The micro-GA optimizes and generates the THz spectroscopic combination as the observed data
- The micro-GA compares the observed THz spectroscopic feature (absorption) with the specimen data in the THz database using the first fitness function, as shown in Equation (12.2), which maximizes the R^2, that is, to minimize the spec-troscopic difference between the specimen and experimental data

$$Maximize : J_1 = R^2 \left(\sum_{i=1}^{M} \sum_{j=1}^{N} F\left(w_{ij}\right), F\left(w_0\right) \right) \qquad (12.2)$$

where

- J_1 is the fitness function of the first design objective for the mixture determination
- w_0 is the given sample element, whose components are to be compared with the specimens in the THz database and then to be determined
- w_{ij} are the micro-GA-generated elements of the statement over the evaluation process, in which $i = 1, 2, \ldots, M$, $j = 1, 2, \ldots, N$, as discussed in Section 12.3; and
- $F(*)$ is the lookup function, which can return the weighted spectroscopic data (absorption) from the THz spectroscopic database, using the input statements w_{ij} and w_0, as given by Equations (12.3) and (12.4), where $D(j)$ is the type-j THz spectroscopic data sequence, $P_{er}(i)$ is the percentage index for the current proportion, and D_0 is the given sample THz spectroscopic data sequence.

$$F(w_{ij}) = w_{ij} D(j) P_{er}(i) \tag{12.3}$$

$$F(w_0) = D_0 \tag{12.4}$$

- $R^2(*, *)$ is the function for the coefficient of determination with two input arguments

The second fitness function J_2 is defined as Equation (12.5) with the implementation of the relative Gini index for the THz spectroscopic residual analysis, which measures the agreement between the micro-GA observed data y_i and the specimen data $\hat{y}_i$.

$$Minimize\ :J_2 = G'(F(w_{ij}), F(w_0)) \tag{12.5}$$

where

- $\circ\ G'(*)$ is the relative Gini index function, whose input arguments are the lookup function $F(*)$ and the THz specimen database, which is listed in the mixture component determination statement, as in Figure 12.3.

12.5 Uncertainty Studies

The functional block diagram of the explosive component determination is presented in Figure 12.5, which shows the micro-GA-driven alignment process. U is the input state variable generated by the micro-GA with stochastic statements and Y is output state variable with the objectives J_1 and J_2.

As shown in Figure 12.6, the uncertainty analysis with a feedback loop is proposed corresponding to the explosive component determination process, as discussed in Section 12.3.

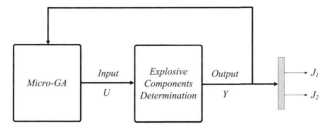

Fig. 12.5 Functional block diagram for the explosive component determination

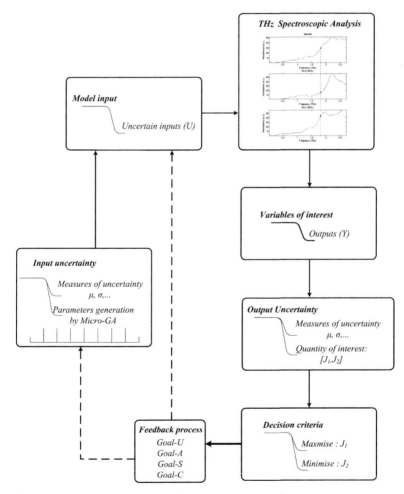

Fig. 12.6 Uncertainty analysis framework for THz spectroscopic data

Table 12.1 The Goals of the Quantitative Uncertainty Assessment [Rocquigny, et al. (2008)]

Type	Goal
Goal-U (Understand)	To understand the importance of uncertain and then to establish the measurements and modeling
Goal-A (Accredit)	To give credit to a method of measurement and to reach an acceptable quality level for its use. This may involve calibrating sensors and estimating the parameters of the model inputs
Goal-S (Select)	To compare the relative performance and optimize the choice of objective policy, operation or design of the system
Goal-C (Comply)	To demonstrate compliance of the system with an explicit criterion

The goal of explosive component determination via terahertz spectroscopic analysis is primarily to demonstrate the compliance of uncertainty criteria embodied by the target values in the guidelines (Goal-U, A, S and C), as given in Table 19.3 [Rocquigny, et al. (2008)]. According to Figure 12.6 and Table 12.1, a metrological chain is compared to investigate which complies best with the proper criterion guidelines (Goal-A, S and C). Through the establishment of industrial emission practices, it also appears that understanding the importance of the various sources of uncertainty will become even more important in improving the metrological options in the long term (Goal-U).

Primarily, two objectives with uncertainty are considered, as expressed in Figure 12.6: J_1 and J_2. Based on the international standard in metrological uncertainty (GUM) [International Organization for Standardization (2008)], the target values are specified by the following uncertainty decision criterion: a quantity of interest representing the relative uncertainty is compared to a maximal/minimal percentage of relative uncertainty.

The uncertainty of this THz spectroscopic study comes mainly from the following three sources: (1) the errors in the experimental data adopted from Zhang's paper [Zhang, et al. (2006)], which depend on the imaging digitalization algorithms; (2) the errors in the interpolation for the THz frequency and absorption ranges; and (3) the natural errors in the micro-GA evolution process. Under the uncertainty involved in this context, there are upper and lower error boundaries that limit the micro-GA-generated solutions for the component determination procedure, as will be discussed in Section 12.6.

12.6 Empirical Studies and Discussion

There are various experimental configurations for THz-TDS. Briefly, the basic configuration is similar to time-domain spectroscopy in the UV-visible [Weichert, et al. (2001)] and infrared regions [Wu and Zhang (1997)]. The mechanisms of both the generation and detection of THz pulses can be photo-conductive antennas, electro-optic (EO) sampling or the combination of the two methods [Katzenellenbogen and Grischkowsky (1991), Mangeney and Crozat (2008), Han, et al. (2000), Han, et al. (2001)]. Here, we take THz-TDS based on photo-conductive antennas as the examples for explanation. Additional details about the EO sampling method can be found in references [Han, et al. (2000), Han, et al. (2001)]. As shown in Figure 12.2, to generate and detect THz pulses, a ultra-short pulsed laser is used. The ultra-short-pulse laser is split into two beams by a beam splitter. One beam is focused on to the DC-biased photo-conductive antenna to generate THz pulses, called the pump beam; the other beam travels through the mechanical delay stage and is focused on to another photo-conductive antenna, called the probe beam. The THz pulses pass through the sample and are collected and focused on to the PCA detector. The THz pulses create an electrical field on the surface of the PCA detector, which accelerates and decelerates the electron-hole carriers generated by the excitation of the probe beam. Consequently, a photo-excited current signal that is proportional to the amplitude of the THz pulses is generated. A lock-in amplifier is used to amplify the detector signal. The temporal profile of the THz pulses can be obtained by scanning the THz pulses with optical probe pulses via the movement of the delay stage. The corresponding THz spectra can be obtained by converting the signal from the temporal domain to the frequency domain via a Fast Fourier Transform (FFT) algorithm. The spectra of the signal and that of the reference are recorded with and without subjecting the sample to THz-TDS. The transmission spectrum of the sample can be obtained by comparing the THz spectra of the sample and that of the reference.

The numerical results are obtained using a specially-devised simulation toolkit in *MATLAB*®, known henceforth as *SGALAB* [SGALAB (2009)]. Unless stated otherwise, all the results are generated using the following parameters for the GA parameters in Table 12.2. There are two test mixture samples for statistical analysis, which are adopted from Zhang's work [Zhang, et al. (2006)], as listed in Table 12.3 and the chemical structures of benzoic acid, o-toluic acid, m-toluic acid and p-toluic acid are presented in Figure 12.7. The Test-1 mixture is called MixBO, whose weight ratio is 80 per cent benzoic acid and 20 per cent o-toluic acid, and Test-2 mixture is called MixBOP, whose weight ratio is 50 per cent benzoic acid, 30 per cent o-toluic acid and 20 per cent p-toluic acid.

Table 12.2 Parameters for Micro-GA

	Max generations	1000
	Number of experiments	100
	Crossover probability	0.8
	Max internal generation	4
	Internal population	4
	External population	100
	Selection operator	tournament
	Crossover operator	single point
	Mutation operator	no
	Encoding method	binary
M	Proportion increment step	10
N	Specimen types	10
	Tournament size	2
	Pareto ranking	Goldberg ranking [Goldberg (1989)]

Table 12.3 Numbering THz Database Components and Test Contents

No.	Components	Test-1:MixBO [Zhang, et al. (2006)]	Test-2:MixBOP [Zhang, et al. (2006)]
1	1,3-DNB		
2	TNT		
3	RDX		
4	PENTN		
5	HMX		
6	2,6-DNT		
7	benzoic acid	80%	50%
8	o-toluic acid	20%	30%
9	m-toluic acid		
10	p-toluic acid		20%

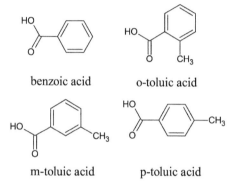

benzoic acid   o-toluic acid

m-toluic acid   p-toluic acid

Fig. 12.7 Chemical structure of benzoic acid, o-toluic acid, m-toluic acid and p-toluic acid

Table 12.4 Measures Used in Uncertainty Analysis

Measures	Test-1: J_1	Test-1: J_2	Test-2: J_1	Test-2: J_2
Max	0.9212	0.5001	0.9752	0.4574
Min	0.0121	0.0051	0.0118	0.0050
Mean	0.8693	0.0266	0.8035	0.0267
Median	0.4944	0.2604	0.5023	0.2624
STD	0.3704	0.4640	0.3704	0.5140
Variance	0.1372	0.2153	0.1372	0.2642
Skewness	5.1995	−5.0674	5.2042	−5.0255
Kurtosis	28.0345	25.4532	28.0724	25.7343

The characteristics of the THz spectroscopic analysis for the uncertainty measures of the explosive mixture components are given in Table 12.4, in which the measures of the maximum (*Max*), the minimum (*Min*), the average (*Mean*, *Median* and *Mode*), the dispersion (*STD*, *Variance*), the asymmetry (*Skewness*) and the flatness (*Kurtosis*) [Rocquigny, et al. (2008), International Organization for Standardization (2008), Choi, et al. (2007)] can be obtained by estimating the corresponding values of the objectives J_1 and J_2, and the uncertainty measures have been matched to the figures for the system outputs, as discussed below.

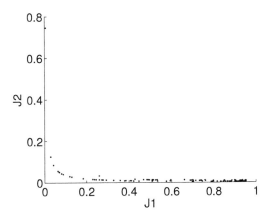

Fig. 12.8 Pareto-optimal set of test-1: J_1 vs. J_2

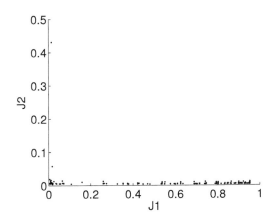

Fig. 12.9 Pareto-optimal set of test-2: J_1 vs. J_2

The evolutionary processes of the fitness functions are presented in Figures 12.8 and 12.9, in which Figure 12.8 and 12.9 are the Pareto fronts of J_1 vs. J_2 for Test-1 and Test-2, respectively, and the positions with higher scatter data density indicate the Pareto optimal of the 'trade-off' solutions generated by the micro-GA between maximizing J_1 and minimizing J_2. Specifically, the micro-GA evaluation process for the THz spectroscopic statistical analysis with 1000 generations is shown in Figures 12.8 and 12.9, in which the max generation is taken as the termination criterion for the evaluation. In addition, the following can be observed:

▷ Figures 12.8 and 12.9 show that the optimized solutions approach the area with maximum J_1 and minimum J_2

▷ Figures 12.8 and 12.9 present a slightly different scatter Pareto dataset, which indicates that the micro-GA has driven the statistical fitness values to a stable status so far

Figures 12.10 and 12.11 are statements and data comparison processes for sample mixtures Test-1 and Test-2, which are the samples of MixBO and MixBOP mixed with benzoic acid, o-toluic acid and p-toluic acid, as given in Table 12.3. Figures 12.10(a), 12.10(c) and 12.10(d) are the statement evolution for the MixBO sample during the micro-GA process when the generation number is 10, 100 and 1000.

▷ Figure 12.10(a) is the statement for generation = 10, which gives a micro-GA-generated mixture determination solution. Correspondingly, 12.10(b) is the THz spectroscopic data comparison, in which the solid line is for the sample mixture THz spectroscopic absorption and the dashed line is the micro-GA-generated THz spectroscopic absorption; they present an obvious difference.

▷ Figure 12.10(c) is the statement for generation = 100, which presents a better solution and is closer to the given experimental THz spectra of MixBO. Meanwhile, 12.10(d) shows that the absorption curves present a more similar shape than that of 12.10(b).

▷ Figure 12.10(e) shows that the statement matches the experimental THz spectra of the Test-1 sample, as given in Table 12.3.

▷ Figure 12.10(f) shows that the micro-GA provides ±10 per cent error boundaries for generation = 1000, which can account for the error between the experimental data and the micro-GA optimized data.

▷ Figures 12.10(b), (d) and (f) show that the micro-GA-driven THz spectra (∗) are approaching the experimental spectroscopic curve through the multiple objectives J_1 and J_2, indicating an effective determination procedure with acceptable error limits (±10 per cent in this case).

Similar phenomena can be observed in the micro-GA-driven statistical analysis for the sample mixture Test-2, as given in Table 12.3. In this case, the experimental spectra of MixBOP are loaded in the micro-GA optimization process, as expressed in Figure 12.11,

▷ Figures 12.11(a)-(b), (c)-(d) and (e)-(f) present the statements and THz spectroscopic data comparisons for generation = 10, 100 and 1000.

▷ Figures 12.11(a)-(b) → (c)-(d) → (e)-(f) express the incremental comparison and matching process and then determine the mixture components within given error boundaries.

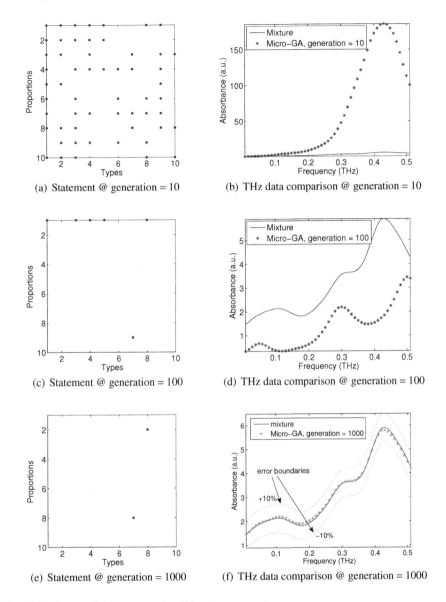

(a) Statement @ generation = 10

(b) THz data comparison @ generation = 10

(c) Statement @ generation = 100

(d) THz data comparison @ generation = 100

(e) Statement @ generation = 1000

(f) THz data comparison @ generation = 1000

Fig. 12.10 Test-1: *MixBO*, statement and THz data comparison

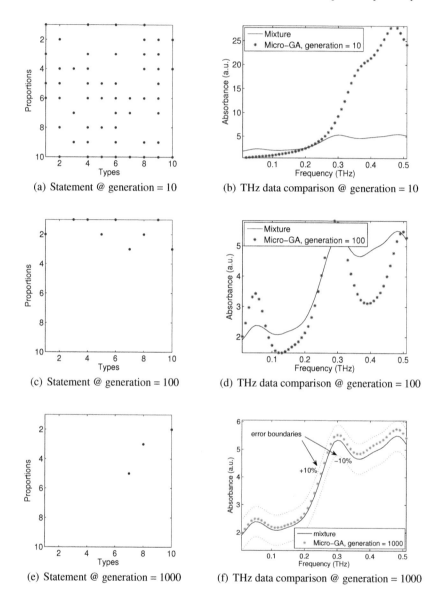

(a) Statement @ generation = 10

(b) THz data comparison @ generation = 10

(c) Statement @ generation = 100

(d) THz data comparison @ generation = 100

(e) Statement @ generation = 1000

(f) THz data comparison @ generation = 1000

Fig. 12.11 Test-2: *MixBOP*, statement and THz data comparison

12.7 Conclusions and Future Work

In this chapter, the MO micro-GA method was utilized to determine explosive mixture components via THz spectroscopic statistical analysis, supported by two experimental test cases. Taking the coefficient of determination (R^2) and the relative Gini index (G') as the multiple objectives, it was demonstrated that the micro-GA can be used as an efficient alternative to conventional methods for mixture component determination. In principle, it can also be applied to other mixture determination applications, such as those concerning illegal drugs, poisons and micro devices in bio-terrorism and agricultural products, with a THz spectroscopy database.

The simulation results with the component statement suggested good prospects for developing an intelligent THz spectroscopic data-based mixture determination system. Further investigations are required to establish both a real-time measurement and interpretation methods for more complex test samples under more complex environments, such as by using additional types of THz spectroscopic data.

Notations

W	The component statement
w_{ij}	The element of W
M	The row number of the statement W
N	The column number of the statement W
R^2	The coefficient of determination
y_i	The observed data
$\hat{y}_i$	The specimen data
$\bar{y}_i$	The mean value of the observed data y_i
SS_{tot}	The total corrected sum of squares
SS_{err}	The error (residual) sum of squares
w_0	The given sample element
n_i	The sampling absolute frequency of level i
p_i	The sampling relative frequency of level i
k	The level size
$F(*)$	The lookup function
$D(j)$	The sum of the type-j THz spectroscopic data sequence
$P_{er}(i)$	The percentage index for the current proportion
D_0	The sum of the given sample THz spectroscopic data sequence
J	The fitness function
ε_i	The residual

13

Evolving a Sliding Robust Fuzzy System

13.1 Introduction

Sliding mode control (SMC) achieves robustness by adding a nonlinear control signal across the sliding surface. A control strategy that satisfies the sliding condition forces the system to the origin of the state space, whereby achieving 0 error and 0 change of error without the need for an integrator like action for a Type 0 system. However, a sliding mode control system has a particularly high control gain and decision chattering due to the robust nonlinear compensation. For practical engineering applications, SMC suffer from the effects of actuator chattering due to the switching and imperfect implementations [Slotine and Li (1991)].

Conversely, FLC has excelled in dealing with this by its fuzzification tolerance. In addition, FLC is relatively easy to implement and can be developed as part of other control methodologies. The feature of a smooth control action of FLC can thus be used to overcome the disadvantages of the SMC systems. This is achieved by merging the FLC with the variable structure of the SMC to form a *fuzzy-sliding mode controller* (FSMC). In this hybrid control system, the strength of the sliding mode control lies in its ability to account for modeling imprecision and external disturbances while the FLC provides better damping and reduced chattering.

13.2 Application of Fuzzy Logic to Sliding Mode Control

In the literature, two common areas where the possible applications of fuzzy logic system are identified are

- the use of fuzzy logic control scheme to replace the non-linear element in the sliding mode control law [Ting, et al. (1994)]
- the use of fuzzy logic control to adjust the switching gains of the nonlinear element [Ishigame, et al. (1991), Kung and Lin (1992), Meyer, et al. (1993), Ting, et al. (1994)].

Due to the close-relationship between the sliding mode and fuzzy control system as both of them are designed based on the phase-plane partitioning of control actions, the third area where the hybrid system can be developed is proposed. It proceeds by using the sliding mode control system to smoothen the fuzzy control. Such hybrid system consists of a fuzzy PD controller and a sliding mode I controller. The sliding mode control is used to improve the drawback of the steady-state response of a fuzzy PD controller and also provides a switching capability with the whole control structure.

13.2.1 Fuzzy Switching Element for the SMC System

Consider SMC for a single-input single-output nonlinear system again. The control action u is to be discontinuous across the sliding hyperplane in order to satisfy the robustness condition. Undesirable chattering will occur due to the use of the discontinuous term. To improve such inadequacy, the discontinuous term is replaced by a FLC term to form a FSMC control action given by [Ting, et al. (1994)]:

$$u = \hat{u}_{eq} - u_f(s) \tag{13.1}$$

where $u_f(s)$ provides the switching power determined by fuzzy control means. It is assumed that in this system, the sliding mode s is measurable by a state observer and to be used to form the input space A of the fuzzy controller. Similarly, the control $u_f(s)$ will also form the output space F of the fuzzy controller. The major switching control rules can be written in the form of:

$$R_j:\text{If} s(t) \text{ is } A_i \text{ Then } u_f \text{ is } \widetilde{F}_i, \ i = -m, \ ...., \ +m , j = 1, \ ...., r \tag{13.2}$$

where $2m+1$ represents the maximum number of linguistic variables, e.g., NB, PB, ZO, etc., and r, the number of rules. Here $\sim$ is used to mean $\widetilde{F}_i$ is a fuzzy variable that belongs to F_i.

For this scheme, [Ting, et al. (1994)] studied the case where all membership functions are of identical triangular shapes that are uniformly spread across the universe of discourse as shown in Figure 13.1. Their controller is, in effect, a saturation controller, since $s(t)$ can only belong to A_i and A_{i+1}, where $\mu_{A_i} + \mu_{A_{i+1}} = 1$. To obtain a true FSMC, therefore, let us study the membership with minimum restriction. For simplicity, the commonly used symmetrical membership functions are used to demonstrate the robust stability of the system, using such fuzzy scheme. In addition, the following conditions of membership functions are made:

- A_i, $i = -m$, ..., $+m$ is spread across the universe of discourse $A = [-\xi, \xi]$ and similarly for F_i, $i = -m$, ...,$+m$ is spread across $F = [-K, K]$.
- For boundary membership functions:

$$\mu_{A_{+m}}(s) = 1 \qquad \forall s(t) > \xi$$

$$\mu_{A_{-m}}(s) = 1 \quad \forall s(t) < -\xi$$

$$\mu_{F_{+m}}(\widetilde{F}) = 1 \quad \forall \widetilde{F} > K$$

$$\mu_{F_{-m}}(\widetilde{F}) = 1 \quad \forall \widetilde{F} < -K$$

The graphical representation of A_i and F_i defined on the universe of discourse A and F are shown in Figure 13.2.

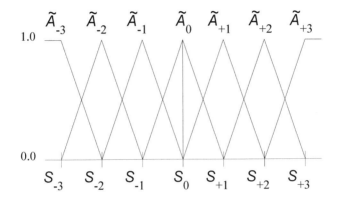

Fig. 13.1 Membership functions

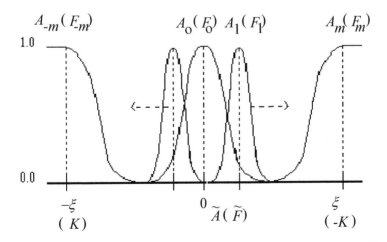

Fig. 13.2 Membership functions of sets A_i and F_i for input and output fuzzy variables $\widetilde{A}_i$ and $\widetilde{F}_i$, respectively

Based upon above membership conditions and by rule (5.2), we can infer that if $s \geq \xi$, i.e.,

$$
\begin{cases}
\mu_{A_i}(s) = 0, & \forall i \neq m \\
\mu_{A_m}(s) = 1,
\end{cases}
$$

then u_f is and only is $\widetilde{F}_m$, i.e.,

$$
\begin{cases}
\mu_{F_i}(\widetilde{F}_m) = 0, & \forall i \neq \pm m \\
\mu_{F_m}(\widetilde{F}_m) = 1,
\end{cases}
$$

The fuzzy representation $\widetilde{F}_m$ of the nonlinear control action u_f is defuzzified using the center of gravity or the fuzzy singleton method. Thus by condition (b), we have

$$
u_f = \frac{\sum\limits_{i=-m}^{m} \mu_{F_i}(s)\widetilde{F}_i}{\sum\limits_{i=-m}^{m} \mu_{F_i}} = \frac{\mu_{F_m}\widetilde{F}_m}{\mu_{F_m}}
$$

$$
= \widetilde{F}_m\Big|_{\mu_{F_m}(\widetilde{F}_m)=1} = -K
$$

Similarly, if $s \leq \xi$, i.e.,

$$
\mu_{A_i}(s) = e^{-\dfrac{(A_i - s)^{\beta}}{\sigma}}
$$

then u_f is calculated base upon rule of inference and defuzzification, i.e.,

$$
u_f = \frac{\sum\limits_{i=-m}^{m} \mu_{F_i}(s)\widetilde{F}_i}{\sum\limits_{i=-m}^{m} \mu_{F_i}} \leq -K
$$

Thus, if we select K as given by (4.38), we have:

$$
|u_f| \geq \eta + |\Delta f|, \quad \forall |s| \geq \xi \tag{13.3}
$$

For the SMC rule, we have:

$$s\dot{s} = s(\Delta f - u_f)$$

$$\leq |s\|\Delta f| - |s|(\eta + |\Delta f|), \quad \forall |s| \geq \xi \tag{13.4}$$

$$= -\eta |s|$$

i.e., the robustness condition (5.4) is satisfied while s moves out of the region $[-\xi, \xi]$, though in the region, smooth control u_f is obtained by fuzzy logic.

Based on the above proof, a further enhancement of the fuzzy SMC control law with a PID equivalent control given by (4.60) in Chapter 4 is adopted. The resultant FSMC control law is given by:

$$u = u_f - K_I \int edt - K_p e - K_D \frac{de}{dt} \tag{13.5}$$

where

$$K_I = \begin{cases} k_{I1}, & \forall s < -\xi \\ k_{I2}, & -\xi \leq s \leq \xi \\ k_{I3}, & \forall s > \xi \end{cases} \tag{13.6a}$$

$$K_P = \begin{cases} k_{P1}, & \forall es < -\xi \\ k_{P2}, & -\xi \leq es \leq \xi \\ k_{P3}, & \forall es > \xi \end{cases} \tag{13.6b}$$

$$K_D = \begin{cases} k_{D1}, & \forall \dot{e}s < -\xi \\ k_{D2}, & -\xi \leq \dot{e}s \leq \xi \\ k_{D3}, & \forall \dot{e}s > \xi \end{cases} \tag{13.6c}$$

and u_f provide the major switching power and is determined by fuzzy means in (13.2). Single dimensional rule-base is used here for u_f, which specifies the relationship between $s(t)$ and fuzzy control actions. Two-dimensional rule-base can also be used with $s(t)$ and $\dot{s}(t)$ as its input variables.

Based on the above derivations, a small difference is identified when comparing the control law of (13.5) to that of a hard-switching scheme with a sliding region given in Chapter 4. The difference is the number of sliding regions used in the nonlinear element in which several of these regions are used in the FSMC system, whilst the sliding region hypothesis uses only three. Therefore, the control law is structurally similar to that of (5.5).

13.2.2 Fuzzy Gain Scheduling for the Switching Element in the SMC System

Based upon the discussions so far, the occurrence of undesirable chattering is usually due to the control actions produced by the nonlinear element. Therefore, it can be induced directly that the switching gains of the nonlinear element play the role of contributing to that chattering. Hence the application of fuzzy logic control in adjusting the values of switching gains would, intuitively, improve the damping ratio of the control system. Ideally, it works in a such way that when $s(t)$ is far away from the sliding hyperplane, the switching gain has a higher value and when $s(t)$ is nearer the sliding hyperplane, the gain is adjusted to a smaller value. The application of fuzzy logic control to adjust the switching gains of the nonlinear element in hard-switching scheme will result to similar controller as discussed in Section 13.2.1. Therefore, the soft-switching control scheme with a sliding region will be used in this fuzzy application. The FSMC system is given by:

$$u = -K_f \tanh(\frac{s}{\xi_1}) - \left(K_I \int edt\right) - \left(K_P e\right) - \left(K_D \frac{de}{dt}\right) \tag{13.7}$$

where K_P, K_I, and K_D are given by (5.6a~c). Here $K_f \tanh(\frac{s}{\xi})$ term represents the major switching action, ξ_1 is a positive constant defining the slope of the *hyperbolic-tangent* function and K_f is the switching gain to be determined by fuzzy means.

In such a fuzzy SMC scheme, the observable sliding surface $s(t)$ forms the input space of the fuzzy implications of the major switching rule. Its switching gain is written in the form of a fuzzy rule, given by:

$$R_j\text{: If } s(t) \text{ is } A_j, \textbf{ Then } \tilde{K}_f \text{ is } B_j, i = -m,..., m, j = 1, ..., r \tag{13.8}$$

where r is the number of rules. With fuzzy implications, K_f is transformed to an adjustable parameter and hence the fuzzy inference mechanism is used as an estimation mechanism for adaptive control.

13.2.3 Fuzzy PD SMC System with Integral Equivalent Control

Fuzzy PD control system has a problem of dealing with steady-state error due to its inherent structure of a conventional PD controller. Hence, this shortcoming has made the controller unfavourable for implementation. Wang and Kwok (1992) had proposed a fuzzy PID controller which uses a fuzzy PD controller in conjunction with a convention integral controller. Such a conventional integral controller only acts upon the accumulation of errors to improve the steady-state response and provides no switching power to it. Major switching power is, thus, concentrated in the fuzzy PD term. To further enhance such a fuzzy SMC system, a switching capability is proposed in this thesis for the conventional I-term.

Based upon the sliding mode design methodology devised from the conventional PID algorithm, a switching integrator is to be used in conjunction with the fuzzy PD controller. Such FSMC control law is described as:

$$u = u_{FPD} - \left(K_I \int edt \right) \qquad (13.9)$$

where

$$K_I = \begin{cases} k_{I1}, & \forall s < -\xi \\ k_{I2}, & -\xi \le s \le \xi \\ k_{I3}, & \forall s > \xi \end{cases} \qquad (13.10)$$

Here $K_I \int edt$ term is used as $\hat{u}_{eq}$ and u_{FPD} is determined by fuzzy means which act upon the error and rate of change of error, namely e and $\dot{e}$. The fuzzy PD control rules are written in the form of:

$$\boldsymbol{R_j}\text{: If } e \text{ is } A_i \text{ and } \dot{e} \text{ is } B_i \text{ Then } \tilde{u}_{FPD} \text{ is } U_i, \ i = -m,..., \ m \ , j = 1, \ ...., \ r \quad (13.11)$$

where m represents the maximum number of linguistic variables, e.g., NB, PB, ZO, etc., and r the number of rules. Hence, a two-dimensional rule-base is used in this case. Such a hybrid system not only provides a fuzzy PID feature but also reduces the computational time (both in design and implementation) and complexities associated with a three-dimensional fuzzy PID controller. Again, like the true fuzzy PID controller, three-terms switching capability can be found in this hybrid system. As we know, the sliding mode switching scheme is a generalized version of a fuzzy switching scheme in terms of the less switching regions.

13.3 Fuzzy SMC System Designs Using a GA

Without a loss of generality, the second-order laboratory-scale nonlinear coupled liquid-level regulation system (SISO) given by Equation (8.9) in Section 8.4.2 is also used in the pilot study of fuzzy sliding mode designs using genetic algorithms. Like the sliding region hypothesis and soft-switching scheme, the objective of all proposed FSMC system designs is to demonstrate the reduction of unwanted chattering. This would, therefore, provide a wider scope for selection of control strategies in switching control system design.

The design step for the FSMC system is, however, a combination of both fuzzy logic and sliding mode system as discussed in Chapters 3 and 4, respectively. In this GA design approach, each design candidate is consisted of two groups of design parameter sets. Mainly, (a) fuzzy logic design parameter sets (the rule-base, membership function designs, scaling factors, etc., (b) SMC design parameter sets (the switching gains, slope of sliding hyperplane, width of sliding region, etc. To reckon with seven fuzzy sets to be used for the fuzzy input decision variable, base-7 coding scheme is advisable instead of base-10. Hence, the computational power is maximized without any unused coding range arising from, for, e.g., a non-direct mapping of base-10 on to seven fuzzy sets. Similar design criteria of low overshoots, fast rise-time and minimum steady-state errors are also adopted here in the FSMC design.

13.3.1 FSMC-I System with a Fuzzy Switching Element

Simulated and real-implemented results of such GA-designed FSMC system of (13.5) are given in Figures 13.3 and 13.6, respectively, whose optimized controller parameters are given below.

$$
\begin{aligned}
h &= 0.122 \\
\xi &= 0.0034
\end{aligned}
\qquad
K_P = \begin{cases}
40.0 & , \quad \forall es < -0.0034 \\
4.7 & , \quad -0.0034 \le es \le 0.0034 \\
28.9675 & , \quad \forall es > 0.0034
\end{cases}
$$

$$
K_I = \begin{cases}
0.136 & , \quad \forall s < -0.0034 \\
0.135 & , \quad -0.034 \le s \le 0.0034 \\
0.125 & , \quad \forall s > 0.0034
\end{cases}
\qquad
K_D = \begin{cases}
0.48 & , \quad \forall \dot{es} < -0.0034 \\
0.44875 & , \quad -0.0034 \le \dot{es} \le 0.0034 \\
0.43 & , \quad \forall \dot{es} > 0.0034
\end{cases}
$$

Rule-base:

$$X_s(t)$$

+B	+M	+S	0	-S	-M	-B
-M	-M	-M	0	+S	+S	+B

Input and output scaling factors:

$$
S_i = \frac{\mathbf{3.0}}{\xi} ; S_o = 0.5
$$

Note: Bold value is manually fixed.

Input membership functions—position parameters:

$$\alpha_{\pm B} = \pm\mathbf{3.0} \; ; \; \alpha_{\pm M} = \pm 1.095 \; ; \; \alpha_{\pm S} = \pm 0.503; \; \alpha_0 = \mathbf{0.0};$$

Note: Bold values are manually fixed.

Output membership functions—position parameters:

$$\alpha_{\pm B} = \pm\mathbf{3.0} \; ; \; \alpha_{\pm M} = \pm 2.314 \; ; \; \alpha_{\pm S} = \pm 2.205; \; \alpha_0 = \mathbf{0.0}$$

Note: Bold values are manually fixed; β_i and b_i are redundant as singleton de-fuzzification is used.

Input membership functions—base-length parameters:

$$b\pm B = 0.97917 ; b\pm M = 0.66667 ; b\pm S = 0.66667 ; b_0 = 0.58333$$

Input membership functions—shape parameters:

$$\beta_{\pm B} = 3.5; \beta_{\pm M} = 1.5; \beta_{\pm S} = 2.0; \beta_0 = 2.0$$

Input membership functions—membership-type:

err: $\pm B \rightarrow N; \pm M \rightarrow N; \pm S \rightarrow L; ZO \rightarrow N$

Note: L represents triangular/trapezoidal-type and N represents bell-type.

As can be observed in Figure 13.3, the unwanted chattering is reduced by using a fuzzy switching element in the control law. However, a larger damping effect is achieved by the FSMC system when comparing Figures 13.3 and 4.17 in terms of the amount of oscillations. A plot of s-trajectory produced by such FSMC system is depicted by Figure 13.4.

In terms of robustness to parameter variations, a 50 per cent increment and 20 per cent decrease of the maximum flow rate, Q_1, is exerted for a duration of 400 seconds interval, similar to that in Chapter 4. Figure 13.5 shows the performances of the FSMC system under such parameter variations. Without any further investigation of the controller performance in simulations, a real implementation of the FSMC system is carried out, whose performance is depicted by Figure 13.6. It can be seen that the good performance of the controller is verified in terms of low overshoots, fast rise-time, minimum steady-state errors and disturbances rejection.

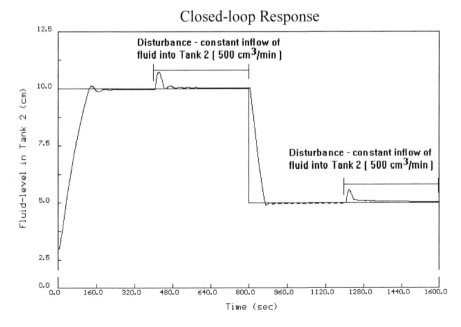

Fig. 13.3 Simulated response of a FSMC-I system designed by GA

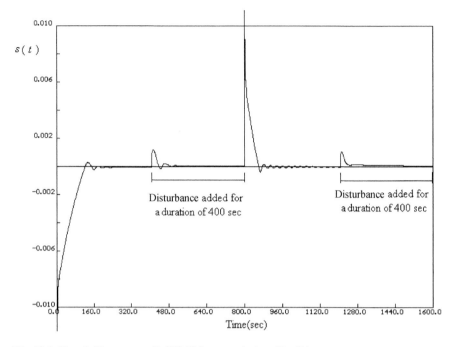

Fig. 13.4 Plot of $s(t)$ response of a FSMC-I system designed by GA

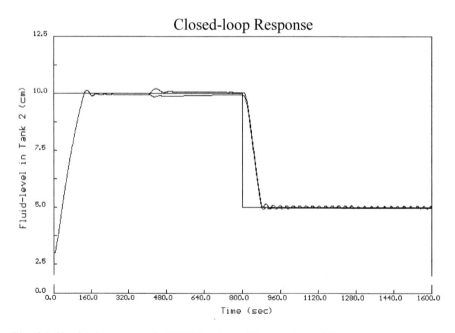

Fig. 13.5 Simulated responses of a FSMC-I system with parameter variations

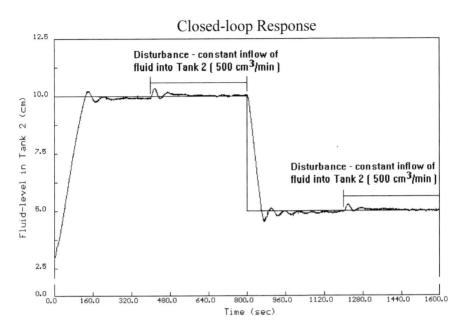

Fig. 13.6 Real-implemented response of a FSMC-I system designed by GA

13.3.2 FSMC-II System with Fuzzy Gain-Scheduling

Simulated and real-implemented results of such GA-designed FSMC system of (5.7) are given in Figures 13.7 to 13.10, respectively, whose optimized controller parameters are given below:

$$h = 0.147$$

$$\xi_1 = 0.0002$$

$$\xi = 0.001$$

$$K_P = \begin{cases} 22.0 & , \quad \forall es < -0.001 \\ 12.493 & , \quad -0.001 \le es \le 0.001 \\ 0.27 & , \quad \forall es > 0.001 \end{cases}$$

$$K_I = \begin{cases} 1.2 & , \quad \forall s < -0.001 \\ 0.13485 & , \quad -0.001 \le s \le 0.001 \\ 0.011 & , \quad \forall s > 0.001 \end{cases} \quad K_D = \begin{cases} 45.0 & , \quad \forall \dot{es} < -0.001 \\ 47.188 & , \quad -0.001 \le \dot{es} \le 0.001 \\ 48.0 & , \quad \forall \dot{es} > 0.001 \end{cases}$$

Rule-base:

$$X_s(t)$$

+B	+M	+S	0	-S	-M	-B
+B	+B	+B	0	+S	+S	+B

Input and output scaling factors:

$$S_i = 372.322; \; S_o = 0.39$$

Input membership functions—position parameters:

$$\alpha_{\pm B} = \pm \mathbf{3.0} \; ; \; \alpha_{\pm M} = \pm 0.961 \; ; \; \alpha_{\pm S} = \pm 0.503; \; \alpha_0 = \mathbf{0.0};$$

Note: Bold values are manually fixed.

Output membership functions—position parameters:

$$\alpha_{\pm B} = \pm \mathbf{3.0} \; ; \; \alpha_{\pm M} = \pm 1.743 \; ; \; \alpha_{\pm S} = \pm 0.548; \; \alpha_0 = \mathbf{0.0};$$

Note: Bold values are manually fixed β_i and b_i are redundant as singleton de-fuzzification is used.

Input membership functions—base-length parameters:

$$b_{\pm B} = \pm 0.41667 \; ; \; b_{\pm M} = \pm 1.08333; \; b_{\pm S} = \pm 1.0; \; b_0 = 0.70833$$

Input membership functions—shape parameters:

$$\beta_{\pm B} = 4.5; \; \beta_{\pm M} = 1.5; \; \beta_{\pm S} = 4.5; \; \beta_0 = 3.0$$

Input membership functions—membership-type:

$$\text{err:} \quad \pm B \rightarrow L; \pm M \rightarrow N; \pm S \rightarrow L; ZO \rightarrow L$$

Note: L represents triangular/trapezoidal-type and N represents bell-type.

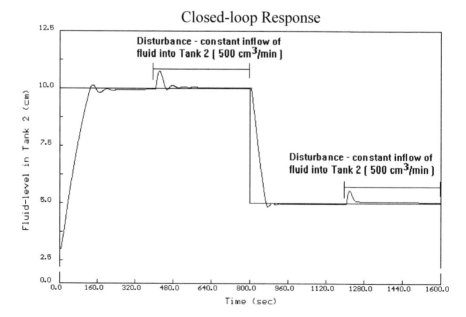

Fig. 13.7 Simulated response of a soft-switching FSMC-II system designed by GA

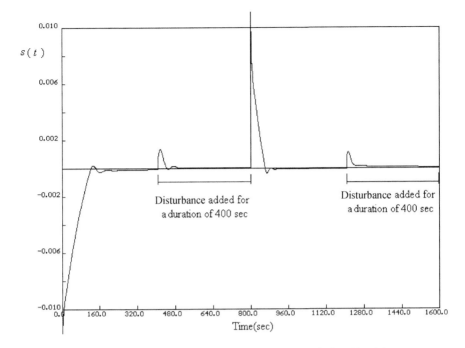

Fig. 13.8 Plot of $s(t)$ response of a soft-switching FSMC-II system designed by GA

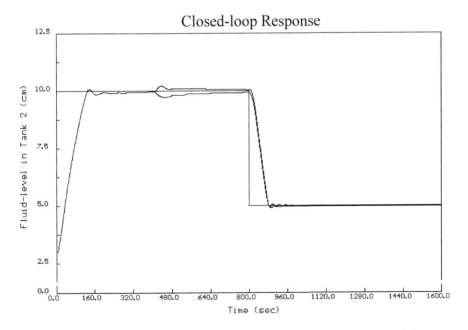

Fig. 13.9 Simulated responses of a soft-switching FSMC-II system with parameter variations

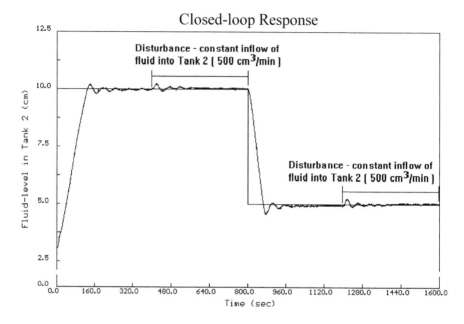

Fig. 13.10 Real-implemented response of a soft-switching FSMC-II system designed by GA

A plot of *s*-trajectory produced by the FSMC system using a fuzzy gain-scheduling is depicted by Figure 13.8. It can be observed in Figures 13.7 and 13.8 that smooth control actions are achieved for reducing the unwanted chattering. However, unlike the soft-switching scheme with a sliding region given by (4.62) in Chapter 4, such an FSMC system has achieved a fast reversal of control actions during the interval of disturbance injections. Such a difference can be seen by comparing Figures 13.8 and 4.22. The superior performance of the FSMC system is due to the use of the fuzzy gain-scheduling. This would contribute to a fast reaction of control actions.

In terms of robustness to parameter variations as mentioned earlier, Figure 13.9 shows the performance of the FSMC system reacting to changes in Q_1. It can be seen that a good performance is maintained in the parameter variation test. In real implementation of the FSMC system, Figure 13.10 shows a similar type of good performance in terms of low overshoots, fast rise-time, minimum steady-state error and disturbance rejections.

13.3.3 Fuzzy PD SMC System with Integral Equivalent Control

Simulated and real-implemented results of such GA-designed FSMC system of (5.9) are given in Figures 13.11 and 13.14, respectively, whose optimized controller parameters are given below:

$$h = 0.112$$

$$\xi = 0.00169$$

$$K_I = \begin{cases} 6.3 & , & \forall s < -0.00169 \\ 0.138 & , & -0.00169 \le s \le 0.00169 \\ 0.12 & , & \forall s > 0.00169 \end{cases}$$

Rule-base:

$$X_{\dot{e}}(t)$$

$X_e(t)$	+B	+M	+S	0	-S	-M	-B
+B	-B	-B	-B	-B	-B	-M	-S
+M	-B	-M	-M	-M	-M	-M	-S
+S	-S	-S	-S	-S	0	+M	+M
0	-B	-M	-S	0	+S	+M	+B
-S	+S	-B	-B	+S	+M	+M	+M
-M	-S	+M	+M	+M	+B	+B	+B
-B	-M	+B	+B	+B	+B	+B	+B

Input and output scaling factors:

$$Ser = 90; \ S\Delta er = 320; \ So = \mathbf{1.0}$$

Note: Bold value is manually fixed.

Input membership functions—position parameters:

err: $\alpha_{\pm B} = \pm\mathbf{3.0}$; $\alpha_{\pm M} = \pm 2.085$; $\alpha_{\pm S} = \pm 1.130$; $\alpha_0 = \mathbf{0.0}$;

Δerr: $\alpha_{\pm B} = \pm\mathbf{3.0}$; $\alpha_{\pm M} = \pm 2.452$; $\alpha_{\pm S} = \pm 1.220$; $\alpha_0 = \mathbf{0.0}$;

Note: Bold value is manually fixed.

Output membership functions—position parameters:

$$\alpha_{\pm B} = \pm\mathbf{3.0} \ ; \ \alpha_{\pm M} = \pm 2.452; \ \alpha_{\pm S} = \pm 1.22; \ \alpha_0 = \mathbf{0.0};$$

Note: Bold value is manually fixed; β_i and b_i are redundant as singleton defuzzification is used.

Input membership functions—base-length parameters:

err: $b_{\pm B} = \pm 1.0$; $b_{\pm M} = \pm 1.59375$; $b_{\pm S} = \pm 1.40625$; $b_0 = 1.28125$

Δerr: $b_{\pm B} = \pm 1.625$; $b_{\pm M} = \pm 0.9375$; $b_{\pm S} = \pm 1.28125$; $b_0 = 0.75$

Input membership functions—shape parameters:

err: $\beta_{\pm B} = 1.5$; $\beta_{\pm M} = 3.5$; $\beta_{\pm S} = 4.5$; $\beta_0 = 4.0$

Δerr: $\beta_{\pm B} = 3.0$; $\beta_{\pm M} = 1.5$; $\beta_{\pm S} = 3.5$; $\beta_0 = 3.0$

Input membership functions—membership-type:

err: $\pm B \rightarrow L$; $\pm M \rightarrow N$; $\pm S \rightarrow N$; $ZO \rightarrow N$

Δerr: $\pm B \rightarrow N$; $\pm M \rightarrow N$; $\pm S \rightarrow N$; $ZO \rightarrow L$

Note: L represents triangular/trapezoidal-type and N represents bell-type.

It can be seen in Figure 13.11 that the common steady-state problem of a fuzzy PD controller is improved by using a sliding mode integrator incorporated into a fuzzy PD control. A zero steady-state response is achieved by such an FSMC system. A plot of s-trajectory produced by the FSMC system is depicted by Figure 13.12. A similar type of trajectory is observed when compared to other SMC or FSMC systems. Figure 13.13 shows the performances of the controller when the parameter Q_1 is subjected to variations as mentioned earlier. Similar to other FSMC systems, a good robustness is achieved in this control scheme. The real implemented results of such FSMC system, as depicted by Figure 13.14, shows a good performance in terms of good transient, steady-state response and disturbance rejections.

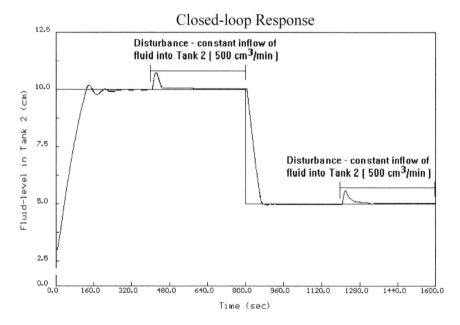

Fig. 13.11 Simulated response of a fuzzy PD SMC system designed by GA

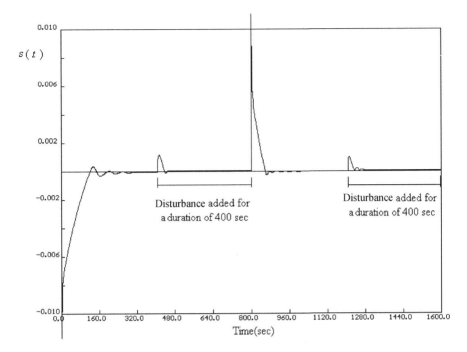

Fig. 13.12 Plot of $s(t)$ response of a fuzzy PD SMC system designed by GA

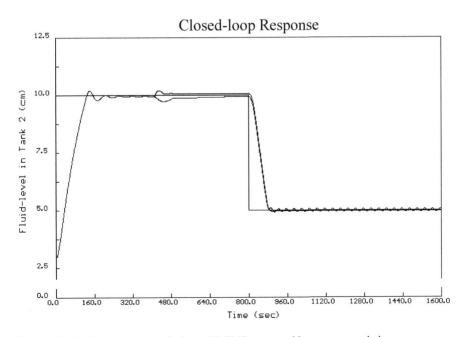

Fig. 13.13 Simulated responses of a fuzzy PD SMC system with parameter variations

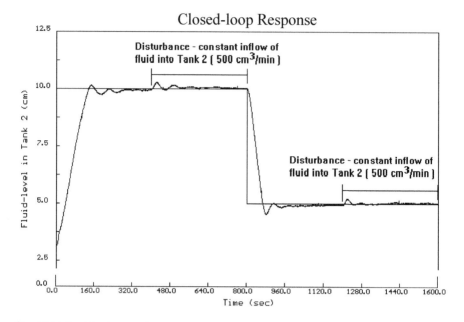

Fig. 13.14 Real-implemented response of a fuzzy PD SMC system designed by GA

13.4 Conclusion

The vast majority of work on the applications of fuzzy logic to sliding-mode control system is concerned with minimizing the undesirable chattering due to the hard-switching feature in the control law. Two main areas where the possible applications of fuzzy logic system are identified are discussed in this chapter. They are (a) the use of fuzzy control scheme to replace the discontinuous/continuous elements in the sliding mode control law and (b) the use of fuzzy control to adjust the switching gains of the discontinuous/continuous elements. Such hybrid systems are termed fuzzy sliding mode control (FSMC) system. Other than these two FSMC systems, the third type is identified and proposed in this thesis. It involves the application of the SMC system to improve the steady-state deficiency of a fuzzy PD controller by providing a switching integral term. Thus the proposed hybrid controller has a full switching capability which is found in the control structure. All proposed FSMC systems have achieved good performances in implementation in terms of good transient and steady-state response, reduced chattering and robustness in rejecting disturbances.

However, all previous work of FSMC system designs are mostly proceeded with a manual design approach. Hence, in the manual design of the fuzzy part of the FSMC system, one would easily encounter similar problems of rules-acquisitions, optimal design of membership functions and their parameters, the choice of scaling factors, availability of expertise, etc., as discussed in Chapter 3. In the sliding mode control part, similar design problems as discussed in Chapter 4 are also encountered—

difficulties in the acquisition of equivalent control, the choice of switching gains and the slope of the sliding hyperplane, deficiency of the gradient-guided optimization techniques, etc. A genetic algorithm-based FSMC design approach is, therefore, proposed here underlying the similar objective of automatic design. In the literature, there is no report of the use of the genetic algorithm in the fuzzy sliding model controller designs. In this genetic algorithm design approach for the FSMC system, the transportation delay, constraints of the actuator's output and a nonlinear model of the plant can be included easily in this design approach.

14

Space Tether for Payload Orbital Transfer

14.1 Introduction

With the original concepts described as a 'space elevator' or 'beanstalk', space tethers were developed to transport payloads vertically without any propellant. A space tether is a long cable ranging from a few hundred meters to many kilometers and uses a series of thin strands of high-strength fiber to couple spacecraft to each other or to other masses. A space tether also provides a mechanical connection that enables the transfer of energy and momentum from one object to the other.

Despite the conception of the space tether being in the 19th century, it has not yet been fully utilized. Space tethers can be used in many applications, including the study of plasma physics and electrical generation in the upper atmosphere, the orbiting or deorbiting of space vehicles and payloads, inter-planetary propulsion, and potentially for specialized missions, such as asteroid rendezvous, as well as the extreme form as the well-publicized space elevator. With the development of space technology, space tethers should be widely used in space exploration [Cartmell and McKenzie (2007), Chen, et al.(2013), Chen, et al. (2014)].

Space tethers have a long history since the original idea was proposed in 1895 and research on space tethers has quickly expanded, especially research on the dynamics and control of space tethers, which are fundamental aspects of this concept. Furthermore, a series of tether missions were conducted for aerospace applications in the last century. The objectives of these missions were mostly for scientific experiments in space tether research [Cartmell and McKenzie (2007), Chen, et al.(2013), Chen, et al. (2014), Misra and Modi (1982), Misra and Modi (1986)a, Eiden and Cartmell (2003), Kumar (2006)].

As shown in Figure 14.1 [Chen (2010a), Logsdon (1997), Curtis (2004), Roy and Clarke (2003)], the geocentric inertial coordinate system is utilized in space tether dynamical modeling, in which the nodes are the points where an orbit crosses on orbital plane and the space tethers are crossing the Earth's equatorial plane. Space tether system dynamics are quite complex because they are governed by a set of ordinary or partial non-linear equations and coupled differential equations, various aspects of which can affect the space tether system be-

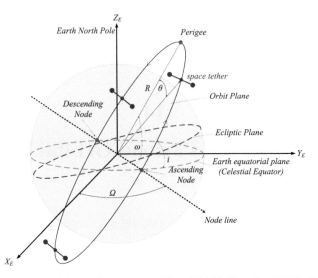

Fig. 14.1 Geocentric inertial coordinate system [Chen (2010a), Logsdon (1997), Curtis (2004), Roy and Clarke (2003)]

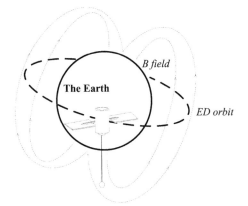

Fig. 14.2 Electrodynamic tether system [Chen (2010a)]

havior and can cause control problems, which could be coupled with other problems. Momentum exchange (MX) tethers and Electrodynamic (ED) tethers are the two principal categories of practical tether systems. There are many types of tether applications, such as Hybrid of Momentum exchange/Electrodynamic Reboost (MXER) tethers, as given in Figure 14.2, and Electrostatic tethers. The research on dynamics and control represents the two fundamentally important aspects of all tether concepts, designs and mission architectures.

Most problems in industrial applications may have several (or possibly conflicting) objectives to be satisfied. It is rare that there is a single point that simultaneously optimizes all the objective functions. Therefore, solutions providing a 'trade-off' need to be found, rather than a single solution when addressing MO optimization problems. To maximize the payload transfer performance (apogee altitude gain and perigee altitude loss) to achieve the most desirable tether motion, while also minimizing tether stress, represents a case of an MO problem with several objectives that need to achieve a compromise. GAs are widely applied for appropriating solutions for MO optimization problems due to their efficiency.

Thus, in this chapter, studies on MO problems in tether payload transfer and tether stress will be conducted using MO GA methods. The four MO GA methods of MOGA, NPGA, NSGA and NSGAII are applied for space tether payload transfer optimization. The task is to search for a set of suitable solutions involving the three conflicting objectives below and achieve acceptable performance in all objective dimensions.

14.2 Motorized Momentum Exchange Tether

The symmetrical motorized momentum exchange tether (MMET) was first proposed by Cartmell in 1998 [Cartmell (1998)]. A more developed MMET model was discussed further [Chen (2010a), Cartmell and Ziegler (2001), Cartmell, et al. (2003), Ziegler and Cartmell (2001), Ziegler (2003)]. This model was motorized by a motor driver and used angular generalized coordinates to represent spin and tilt, together with an angular coordinate for circular orbital motion and a further angular coordinate defining the back-spin of the propulsion motor's stator components.

The basic conceptual schematic of the MMET system is shown in Figure 14.3. The system is composed of the following parts: a pair of propulsion tether subspans (♯A and ♯B), a corresponding pair of outrigger tether subspans (♯C and ♯D), the launcher motor mass within the rotor and that within the stator (♯J and ♯I), the outrigger masses (♯H and ♯G), and the two payload masses (♯E and ♯F), as also shown in Table 14.1.

The MMET system is excited by using a motor and the dynamical model uses angular generalized coordinates to represent spin and tilt, together with the true anomaly for circular orbital motion or the true anomaly and a variable radius coordinate for elliptical orbits. Another angular coordinate defines the backspin of the propulsion motor's stator components. The payload masses are fitted to each end of the tether subspans, and the system orbits a source of gravity in space, in this case, the Earth. The use of a tether generally means that all the constituent parts of the system have the same angular velocity as the overall center of mass (COM). As implied by Figure 14.3, the symmetrical double-ended motorized spinning tether can be applied as an orbital transfer system to exploit MX for propelling and transferring payloads in space. A series of terrestrial-scale model tests of the MMET system were performed on ice by Cartmell and Ziegler in 2001 [Cartmell and Ziegler (2001)] and 2003 [Cartmell, et al. (2003)], as shown in Figure 14.4.

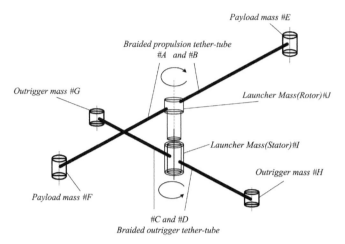

Fig. **14.3** Conceptual schematic of the motorized MX tether [Cartmell, et al. (2003), Ziegler and Cartmell (2001), Chen and Cartmell (2007)MMET]

Table 14.1 The Components of Basic Conceptual Schematic of MMET System

Position	Component
♮ A	Braided propulsion tether subspan
♮ B	Braided propulsion tether subspan
♮ C	Braided outrigger tether subspan
♮ D	Braided outrigger tether subspan
♮ E	Payload mass
♮ F	Payload mass
♮ G	Outrigger mass
♮ H	Outrigger mass
♮ I	Launcher mass (stator)
♮ J	Launcher mass (rotor)

Fig. **14.4** The scale model of the MMET experiment on ice [Cartmell and Ziegler (2001), Cartmell, et al. (2003)]

14.3 Payload Transfer

As shown in Figure 14.3, the symmetrical double-ended motorized spinning tether can be applied as the orbital transfer system to exploit MX to propel and transfer payloads. Tethered payloads, orbiting a source of gravity in space, possess the same orbital angular velocity as the overall COM. As the tethered system's acceleration caused by the motor builds up about the COM, eventually, the tangential velocity of the payloads reaches the required level and the payloads are released on to a desired tangential path.

As the upper payload is released from a spinning tether, it is always aligned along the gravity vector. The upper payload carries more angular velocity than required to stay on that circular orbit; however, because the upper payload does not have sufficient energy to escape the Earth's gravity, it goes into an elliptical orbit with the release point being the perigee of the orbit, as shown in Figure 14.5. In this way, the upper payload can be transferred from low Earth orbit (LEO) to geostationary Earth orbit (GEO).

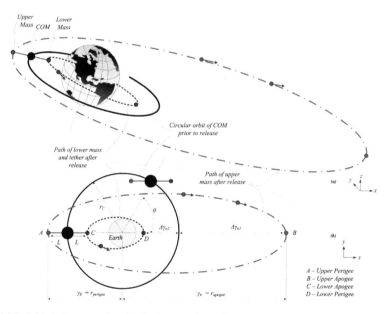

Fig. 14.5 Orbital elements of payload release and transfer

The lower payload remains attached to each remaining free end of the tether. Similarly, the lower payload does not possess sufficient velocity to remain on its circular orbit when it is released, and thus, it also transfers into an elliptical orbit but with the release point being the apogee of the orbit. Half an orbit later, the upper payload reaches its apogee and is hence further from the Earth than it was at the point of release. Upon reaching the perigee of the orbit, the lower payload is closer to the Earth than it was at release. Thus, the upper and lower masses released from a spinning tether are raised and lowered, respectively. When the tether with a payload at each end is undergoing spin-up, libration or rotating about the COM, the resulting stress will be generated in each tether sub-span. The stress of the tether should be kept within the stress limit of the tether.

As shown in Figure 14.5, when a symmetrical MMET is in LEO, the upper payload can be released from a spinning tether, which is always aligned along the local gravity vector. The payload will carry more angular velocity than required to stay on that circular orbit; however, it does not possess sufficient energy to escape the Earth's gravity. The upper payload will transfer into an elliptical orbit, with the release point being the perigee of the orbit. Half an orbit later, the upper payload reaches its apogee and is further from the Earth than it was at the point of release. Similarly, the lower payload does not possess sufficient energy to remain on its circular orbit when it is released, and it transfers into another elliptical orbit, with the release point being the apogee of that orbit. Upon reaching the perigee of its elliptical orbit, the lower payload is closer to the Earth than it was at release. Therefore, the upper and lower payload masses are released from a spinning tether and raised and lowered, respectively.

$$\frac{\Delta r_{\pi 1}}{L} = \frac{\dfrac{(r_c+L)^2 \left[(r_c+L)\dot{\theta}+L\dot{\psi}\right]^2}{2\mu - (r_c+L)\left[(r_c+L)\dot{\theta}+L\dot{\psi}\right]^2} - r_c}{L} \tag{14.1}$$

$$\frac{\Delta r_{\pi 2}}{L} = \frac{\dfrac{(r_c-L)^2 \left[(r_c-L)\dot{\theta}-L\dot{\psi}\right]^2}{2\mu - (r_c-L)\left[(r_c-L)\dot{\theta}-L\dot{\psi}\right]^2} - r_c}{L} \tag{14.2}$$

The radial separation of Δr_π, which is shown in Figure 14.5, gives the distance half an orbit after the tether releases the payload between the payload and the facility's orbit at the time of release. This separation describes how well the tether facility performs at transferring the payload and is defined as the tether's performance index [Ziegler and Cartmell (2001)]. $\Delta r_\pi/L$ is defined as the tether's efficiency index. The upper payload releases at its perigee point. A larger apogee altitude gain of $\Delta r_{\pi 1}/L$ achieves better tether transfer performance, given by Equation (14.1). For the lower payload, released at its apogee point, a smaller perigee altitude loss of $\Delta r_{\pi 2}/L$ also achieves better tether transfer performance, given by Equation (14.2).

Assuming $L \ll r_c$ and $L/r_c \ll 1$ and applying the binomial expansion to Equations (14.1) and (14.2), these equations can be re-written as Equations (14.3) and (14.4). Table 14.2 lists the parameters for MMET payload transfer.

$$\frac{\Delta r_{\pi 1}}{L} = 7 + \frac{30L}{r_c} + \frac{4\dot{\psi}}{\dot{\theta}} + \frac{2Lr_c^2\dot{\psi}\left(18\dot{\theta}+5\dot{\psi}\right)}{\mu} \tag{14.3}$$

$$\frac{\Delta r_{\pi 2}}{L} = -7 + \frac{30L}{r_c} - \frac{4\dot{\psi}}{\dot{\theta}} + \frac{2Lr_c^2\dot{\psi}\left(18\dot{\theta}+5\dot{\psi}\right)}{\mu} \tag{14.4}$$

Table 14.2 Tether Parameters

Symbol	Quantity	Value
A	Tether cross-sectional area	$62.83 \times 10^{-6} m^2$
ρ	Tether density	$970\ kgm^{-3}$
T, U	Kinetic and potential energy	J
M_M, M_P	Motor and payload mass	$5000, 1000\ kg$
Δr_π	Tether performance index	m
$\Delta r_\pi / L$	Efficiency index	
$\frac{\Delta r_{\pi 1}}{L}$	Upper payload efficiency index	
$\frac{\Delta r_{\pi 2}}{L}$	Lower payload efficiency index	
L	Tether length	$50000\ m$
l	Tether length from COM, a point along the tether	m
$\dot{\psi}$	Angular pitch velocity	$0.01\ rad/s$
$\dot{\theta}$	Angular orbital velocity	$\sqrt{\mu/r_c}\ rad/s$
r_c	LEO circular orbit radius of COM at payload release	$6890\ m$
σ_0	Axial stress at end of tether where it connects to payload	GPa
σ	Stress	GPa
σ_{LIM}	Stress limit	$5\ GPa$
μ	Gravitational constant	3.9877848×10^{14} $m^3 s^{-2}$
e	Eccentricity	0

14.4 Tether Strength Criterion

The symmetrical motorized MX tether with a payload at one end will experience centripetal acceleration when rotating about the facility. A resulting tensile force and stress are generated by the rotation. If it is assumed that the gravity gradient effects are negligible, the stress in the tether can be given by Equation (14.5) [Ziegler and Cartmell (2001)], where l is the tether length, $l \in [0, L]$. Equations (14.5) and (14.6) show the symmetrical stress distribution within the symmetrical tether sub-spans, and Figure 14.6 [Chen and Cartmell (2007)MMET] gives the stress distribution within the tether.

$$\sigma = \frac{1}{A} \dot{\psi}^2 l \left(M_P + \frac{1}{2} \rho A l \right) \tag{14.5}$$

$$\sigma_0 = \sigma|_{l=L} = \frac{1}{A} \dot{\psi}^2 L \left(M_P + \frac{1}{2} \rho A L \right) \tag{14.6}$$

As shown in Figure 14.6 and Equations (14.5) and (14.6), the max stress point is where the payloads connect to the tether, and the max stress of the full tether sub-span is given by Equation (14.6). The max tether stress of σ_0 should stay within the safe stress limit of the tether, i.e., $\sigma_0 < \sigma_{LIM}$.

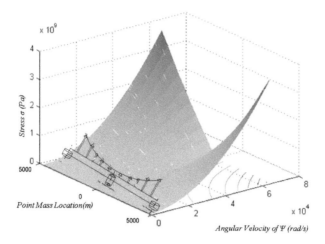

Fig. 14.6 Stress distribution within tether sub-span

14.5 Payload Transfer Objective Definition

The MMET payload transfer and design strength problems are defined as three conflicting objectives that need to be subject to a compromise in the optimization study, and the four MO methods of MOGA, NPGA, NSGA and NSGAII are used as the design optimizers. Then, the practical MO problems of MMET payload transfer can be defined as follows:

- Tether strength

$$Min : y_1 = \sigma_0 \tag{14.7}$$

- Upper payload transfer

$$Max : y_2 = \frac{\Delta r_{\pi 1}}{L} \tag{14.8}$$

- Lower payload transfer

$$Min : y_3 = \frac{\Delta r_{\pi 2}}{L} \tag{14.9}$$

Subject to

$$\sigma_0 \preceq \sigma_{LIM} \tag{14.10}$$

If the GAs are used as the optimizers for practical problems, the practical objectives need to be converted into proper fitness functions for GAs. For MMET payload transfer application, the maximum fitness functions can be defined through Equations (14.11), (14.12) and (14.13), which are transformed from Equations (14.7), (14.8) and (14.9), respectively.

$$f_1(\psi, L, A) = \sigma_{LIM} - \sigma_0 \tag{14.11}$$

$$f_2(\dot{\psi}, L) = \frac{\Delta r_{\pi 1}}{L} \tag{14.12}$$

$$f_3(\dot{\psi}, L) = \frac{1}{\frac{\Delta r_{\pi 2}}{L} + \varepsilon} \tag{14.13}$$

where ε is a small value parameter that helps to avoid $f_3 \longrightarrow \infty$ and $\varepsilon \longrightarrow 0$.

14.6 Simulations

Table 14.3 presents the GA parameter settings for the MMET payload transfer MO optimization simulation. The Simple Genetic Algorithms Laboratory (SGALAB) toolbox and the Smart MATLAB® and MATHEMATICA link laboratory toolbox (SMATLINK) in MATLAB are utilized for the simulation and the MMET payload transfer optimization. More details about the SMATLINK tool can be found in [SMATLINK (2008)].

Table 14.3 GA Parameter Setting

Symbol	Quantity	Value
	Max generations	50
	Crossover probability	0.8
	Mutation probability	0.001
	Population	50
	Selection operator	tournament
	Crossover operator	single point
	Mutation operator	single point
	Encoding method	binary
x_1	$\dot{\psi}$	$[1 \times 10^{-5}, 1]$
x_2	L	$[1 \times 10^3, 2.5 \times 10^5]$
x_3	A	$[5 \times 10^{-6}, 5 \times 10^{-1}]$
y_1	Tether stress objective function	Eq. (7)
y_2	Upper efficiency index objective function	Eq. (8)
y_3	Lower efficiency index objective function	Eq. (9)
f_1	Fitness function 1	Eq. (11)
f_2	Fitness function 2	Eq. (12)
f_3	Fitness function 3	Eq. (13)
ε	A small value parameter that helps to avoid $f_3 \longrightarrow \infty$.	eps:the floating point relative accuracy

Figures 14.7 to 14.10 are the diagrams of the mean value of fitness functions 1, 2 and 3 within the max generation under the MOGA, NPGA, NSGA and NSGAII methods. Performance information is shown in these figures. All the fitness lines achieve steady state at high convergence rates. The number of generations upon convergence for the four MO methods is approximately 10 under the MOGA, four under the NPGA, four under the NSGA and four under the NSGAII. This means

that the MOGA needs more time or generations to reach convergence and that its efficiency is lower than that of the other MO methods while obtaining similar results. The GA optimizer starts from random initial values for the variables ψ, L and A within corresponding specific variable ranges in Table 14.4. This leads to random initial fitness values. Unlike single-objective problems, the MO GAs reach convergence by achieving a compromise for each fitness function, and the fitness lines present fluctuations in Figures 14.7 to 14.10.

Figure 14.7 is the mean value of the fitness diagram for the MOGA, where the convergence point of the number of generations is approximately 10; the fitness lines subsequently achieve steady state. Within 1 to 10 generations, there are some local fluctuations of fitness functions 1, 2 and 3; these local fluctuations are the MOGA's stochastic searching outputs.

Figures 14.8, 14.9 and 14.10 present the mean values of the fitness diagrams under the NPGA, NSGA and NSGAII, and the convergence point of the number of generations is approximately 4. Then, the fitness lines remain in steady state without any local fluctuations. This means that, for the current three objective problems of MMET payload transfer, NPGA, NSGA and NSGAII can converge with a higher performance than can the MOGA.

For the 'trade-off', solutions need to be provided for the three MO MMET payload transfer problems, rather than single-point solutions that can optimize all the objectives simultaneously. Figures 14.11, 14.12 and 14.13 are the Pareto fronts of three MMET payload transfer fitness functions for the MOGA, NPGA, NSGA and NSGAII. The figures present fitness 1 vs. fitness 2, fitness 1 vs. fitness 3, and fitness 2 vs. fitness 3, respectively.

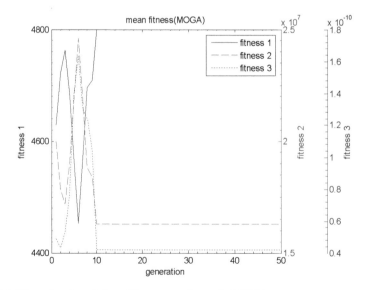

Fig. 14.7 Fitness functions 1, 2, and 3 vs. generation under MOGA

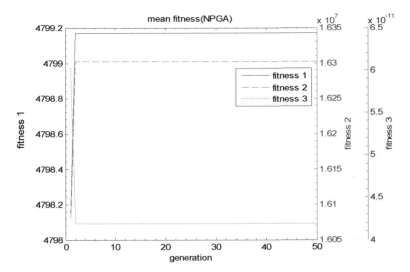

Fig. 14.8 Fitness functions 1, 2, and 3 vs. generation under NPGA

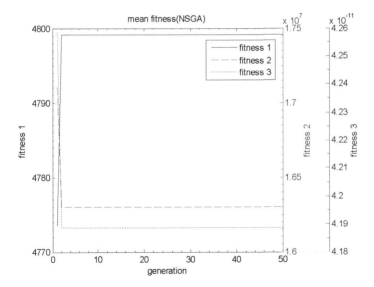

Fig. 14.9 Fitness functions 1, 2, and 3 vs. generation under NSGA

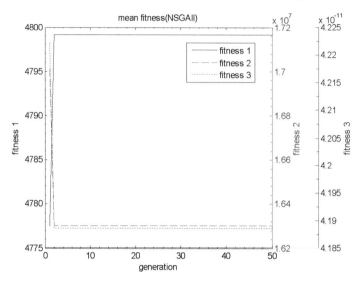

Fig. 14.10 Fitness functions 1, 2, and 3 vs. generation under NSGAII

Table 14.4 Comparison of Simulation Results and Existing Design

Type	f_1	f_2	f_3
MOGA	4.7991×10^3	1.6311×10^7	4.1891×10^{-11}
NPGA	4.7991×10^3	1.6311×10^7	4.1891×10^{-11}
NSGA	4.7999×10^3	1.6311×10^7	4.1891×10^{-11}
NSGAII	4.7999×10^3	1.6311×10^7	4.1891×10^{-11}
Existing Data	4.7982×10^3	1.6302×10^7	4.1887×10^{-11}

Figure 14.11 shows that the four MO methods—MOGA, NPGA, NSGA and NSGAII—have searched all the regions along the 'Fitness 1 ⟶ Max' and 'Fitness 2 ⟶ Max' axes, with the searching policy designed to achieve a 'trade-off' between the maximum of fitness 1 and the maximum of fitness 2. The markers of the MOGA are scattered over a larger area than in other MO methods and will result in the MOGA requiring greater searching time. The three methods can reach similar regions of solutions. A position having a higher scatter data density means that the Pareto-optimal 'trade-off' solutions are near the 'Fitness 1 ⟶ Max' axis.

Figures 14.12 and 14.13 are similar Pareto fronts for the four MO methods. The methods have searched all regions along the 'Fitness 1 ⟶ Max' and 'Fitness 3 ⟶ Max' axes with a searching policy designed to obtain a 'trade-off' between the maximum of fitness 1 and the maximum of fitness 2 and along the 'Fitness 2 ⟶ Max' and 'Fitness 3 ⟶ Max' axes with a searching policy designed to obtain a 'trade-off' between the maximum of fitness 2 and the maximum of fitness 3, respectively. The four methods can reach a similar region of solutions. A position with a higher scatter data density indicates that the Pareto-optimal 'trade-off' solutions are near the 'Fitness 1 ⟶ Max' axis in Figure 14.12. There are three higher density areas in Figure 14.13, and there are more suitable 'trade-offs' that can be obtained between fitness 2 and fitness 3.

The Pareto fronts indicate the corresponding relations for the three fitness functions and their evolution tendencies. The four MO methods can achieve high performance in the evolution process, and a stable dataset can be obtained quickly. The Pareto fronts suggest comparable solutions for MMET payload transfer applications within a safe stress design criterion.

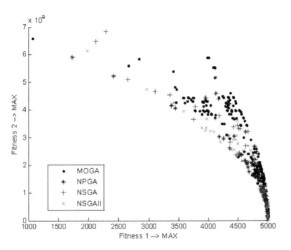

Fig. 14.11 2D Pareto dataset of fitness function 1 vs. fitness function 2

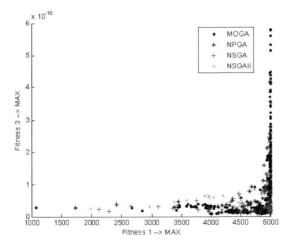

Fig. 14.12 2D Pareto dataset of fitness function 1 vs. fitness function 3

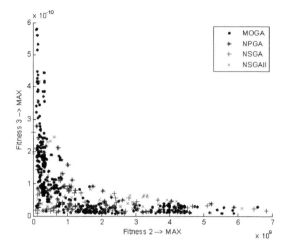

Fig. 14.13 2D Pareto dataset of fitness function 2 vs. fitness function 3

Table 14.4 shows a comparison of the fitness values from the simulation results and existing design. The simulation fitness values are from Table 14.3 and the current existing design fitness values are from Table 14.2. The simulation fitness values of its decision vectors are not dominated by any current existing fitness values of its decision vector, and thus, they can be a set of Pareto-optimal solutions. In addition, there may be special interest in finding or approximating the Pareto-optimal set mainly to gain deeper insight into the problem and knowledge about alternate solutions, respectively. According to practical MMET applications, if there are more specific criteria given, practical results from the Pareto dataset of solutions can be provided.

There are some small differences between the fitness values of the GAs' results and those of existing design in Table 14.4. This helps to validate the current existing design, which performs satisfactory in obtaining a practical solution for optimization. On the other hand, the practical existing solution also helps to prove the practical performance of MO GAs methods, and they can be applied to more complicated MMET payload transfer designs.

14.7 Conclusion and Future Work

MO optimization in MMET payload transfer problems using GAs has been performed. The conflicting problems were to maximize the tether's upper and lower efficiency indexes while minimizing the tether stress or remaining within the tether strength limit. By converting the three practical design criteria into fitness functions for the GAs, four MO methods were applied in the optimization study.

The four MO methods—MOGA, NPGA, NSGA and NSGAII—presented a flexible optimization ability when applied to MMET practical objectives, and the Pareto dataset suggests a series of solutions for MMET designs satisfying strength criteria and payload transfer performance. The simulation example helps to validate the current MMET payload transfer design, and more specific results can be obtained by adding more specific or practical conditions for each MMET application.

Extreme Pareto solutions are found to be physically reasonable and the centers of the Pareto fronts provide a good compromise. The results confirm the feasibility of the MO approaches for MMET tether payload transfer optimization.

Given the advantage of MO GAs, they can be applied as follows:

- Additional MO GA methods will be applied in the current MMET payload transfer study
- Some optimization studies for multiple-orbit MMET payload transfers will be performed
- Intelligent real-time controller design for MMET payload transfer or navigation will be studied
- Dynamical system modeling for MMET will be performed

15

Structural Design for Heat Sinks

15.1 Introduction

Heat sinks have widely been used to enhance heat dissipation from various devices, such as from a hot surface to a cooler ambient environment. With the growing demand for dissipation of high heat fluxes, the use of liquid-cooled heat sinks has increased significantly. In recent years, researchers have worked on the structural design and optimization of the thermal performance of heat sinks for various applications. The structural design and optimization of a heat sink should be such that it can dissipate as much heat as possible, e.g., a heat flux of 750 W/cm^2 [Escher, et al. (2010)] and 270 W/cm^2 [Renfer, et al. (2013)], under given pumping power, volume, or weight limit conditions.

In 1981, Tuckerman and Pease [Tuckerman and Pease (1981)] designed a micro-channel heat sink. In 1992, Knight et al. [Knight, et al. (1992)] presented a dimensionless form of the governing equations of fluid dynamics and heat transfer for both laminar and turbulent flow and used it to determine the geometry of a micro-channel heat sink. In 2000, Perret et al. [Perret, et al. (2000)] presented a cooling device that embeds the heat sink's micro-channels into the silicon wafer. In 2006, Hilbert et al. [Hilbert, et al. (2006)] performed MO design optimization for the blade shape of a heat exchanger, therein considering the coupled solution of the heat transfer processes. Wang et al. [Wang and Wang (2006)] studied a membrane-electrode assembly model of the serpentine cooling channels for polymer electrolyte membrane fuel cells. In 2008 and 2010, Husai et al. [Husain and Kim (2008), Husain and Kim (2010)] performed MO performance optimization of a micro-channel heat sink using surrogate analysis, in which the design variables were the micro-channel width, depth and fin width, and the two objective functions were thermal resistance and pumping power. In 2009, Biswal et al. [Biswal, et al. (2009)] employed an analytic model to optimize a single-phase liquid-cooled micro-channel heat sink. Copiello and Fabbri [Copiello and Fabbri (2009)] investigated the optimization of heat transfer from wavy fins cooled by a laminar flow. In 2012, Turkakar et al. [Turkakar and Okutucu-Ozyurt (2012)] reported their work on the dimensional optimization of silicon micro-channel heat sinks by minimizing the total thermal resistance. Hung et al. [Hung, et al. (2012)] used an optimization procedure comprising a simplified conjugate-gradient method and a three-dimensional fluid flow and heat transfer model to investigate the optimal geometric parameters of a double-layered micro-channel heat sink. In 2014, Xie et al. [Xie, et al. (2014), Xie, et al. (2014)] performed a numerical investigation on microchannel heat sinks to study the laminar fluid flow and thermal performance based on constructal theory.

In designing a heat sink, both thermal resistance and pressure drop should be low. In previous works, the pressure drop was considered as a constraint and parallel heat sinks were modeled. In this research, an MO optimal design of a serpentine channel heat sink has been proposed, therein attempting to reduce the overall thermal resistance and pressure drop. The design includes four design variables—the number of channels, channel width, channel height and inlet velocity. An optimal structural design of a serpentine channel heat sink is presented in this chapter. In the structural modeling of the heat sink, channel width, fin width, channel height and inlet velocity are defined as the design variables, and the 'total thermal resistance' and the 'pressure drop' are the two objectives, subject to constraints of the fixed length and width of the heat sink. In this chapter, an MO artificial swarm fish algorithm with a variable population size using a non-dominated sorting method (MOAFNS), as introduced in Section 7.4.3, is employed to perform the optimization, in which a fast approach of Pareto-optimal solution recommendation using the Pareto risk index is proposed to address the optimal trade-offs between the two conflicting thermal objectives. Then, the optimal solutions are validated by performing related experiments. The Pareto front indicates a trade-off between 'total thermal resistance' and 'pressure drop'. Numerical results and experimental data have produced an agreement and in that the reduction in both the thermal resistance and the pressure drop can be achieved via the determination of the channel configuration and inlet velocity using MOAFNS, resulting in the desired thermal performance of the heat sink.

15.2 Structural Modeling

Compared with parallel channels, serpentine channels exhibit perfect flow distribution uniformity. In a serpentine channel, the flow is interrupted periodically at the bends of the channels, resulting in periodic interruptions at the thermal boundary layers. The impingement, recirculation and flow separation at these sharp bends lead to flow distortion and consequently enhance the heat transfer performance of the serpentine channels.

The structural modeling of a serpentine channel heat sink with one 'inlet' and one 'outlet' is shown in Figure 15.1(a). The model includes n channels with a width W_c and $(n-1)$ fins with a width W_b. Figure 15.1(b) is the 3D model of this heat sink. The heat sink's dimensions are $W \times L \times H$, and the heat sink is subject to a uniform heat flux from the base plate. The insulated 'top plate' is used solely for containing the coolant flow, and the heat is transported away by the coolant.

In this modeling, two factors are considered to represent the performance of the heat sink: (1) the total thermal resistance R_t and (2) the pressure drop of the serpentine channel P_d.

(a) Front View   (b) Bottom View

(c) Isometric View

Fig. 15.1 CAD modeling of a serpentine channel heat sink

15.2.1 Total Thermal Resistance

A serpentine channel heat sink has several sharp bends (Figure 15.2). The total thermal resistance of a serpentine channel heat sink can be written as given in Equation (15.1), which ignores the spreading thermal resistance caused by the temperature difference between the outlet and the inlet and the thermal resistance across the interface between the device and the heat sink.

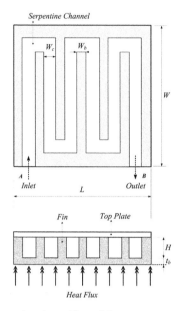

Fig. 15.2 Schematic of a serpentine channel heat sink

$$R_t = \frac{T_{max} - T_{in}}{Q} = R_1 + R_2 + R_3 = \frac{t_b}{k_0 A_0} + \frac{1}{h_c A_s} + \frac{1}{\rho_f W_c H v_a c_p} \tag{15.1}$$

where T_{max} is the maximum temperature of the bottom surface of the heat sink, $^\circ C$; T_{in} is the inlet temperature of the coolant, $^\circ C$; Q is the heat power, W; R_1 is the conductive thermal resistance; R_2 is the total convection thermal resistance; R_3 is the capacitive thermal resistance, which represents the increase in temperature from the inlet to the outlet in bulk fluid; t_b is the thickness of the bottom plate of the heat sink, mm; k_0 is the thermal conductivity of the heat sink, 160 $W/(m{\cdot}K)$; A_0 is the total bottom surface area of the heat sink, mm^2; h_c is the heat transfer coefficient, $W/m^2{\cdot}K$, where $h_c = u_a k_0/D_h$, with u_a as the average Nusselt number; A_s is the surface area available for heat transfer, defined by Equation (15.2); ρ_f is the coolant density, kg/m^3; v_a is the inlet flow velocity, m/s; and c_p is the specific heat of the coolant, 4183 $J/kg{\cdot}K$.

$$A_s = 2n\eta \, aW_c \left(W - 2bW_c - cW_c\right) + nW_c \left(W - 2bW_c\right) \tag{15.2}$$

In Equation (15.2), a is the aspect ratio of the cross-section of the channels, $a = H/W_c$; b is the aspect ratio of the cross-section of the fins, $b = W_b/W_c$; c is the ratio of turn clearance to channel width, $c = l_s/W_c$; l_s is the width of turn clearance, mm; and η is the efficiency of the fins, expressed by Equation (15.3), in which m_p is the fin parameter of approximate equality, defined by Equation (15.4) [Knight, et al. (1992), Li, et al. (2013)].

$$\eta = \frac{\tanh{(m_p H)}}{m_p H} \tag{15.3}$$

$$m_p = \sqrt{\frac{h_c}{k_0 b W_c}} \tag{15.4}$$

There are a few bends in the serpentine channel, which interrupt the hydrodynamic boundary layers. In this context, the average Nusselt number u_a is utilized to describe the thermal profiles considering the bends and the entrance downstream.

$$u_a = \left[\left(2.22 x_1^{-0.33} \right)^3 + u_f \right]^{\frac{1}{3}} \tag{15.5}$$

The average Nusselt number u_a for laminar flow can be calculated using Equation (15.5), in which x_1 is the thermal channel length, $x_1 = x/ReD_h Pr$; Re is the Reynolds number, $Re = v_a D_h / v$; D_h is the hydraulic diameter, $D_h = 2W_c H/(W_c + H)$; v is the coolant viscosity; u_f is the fully developed Nusselt number, $u_f = 8.31G - 0.02$; and G is a geometric factor of the channel defined by Equation (15.6) [Copeland (2000)].

$$G = \frac{\left(\dfrac{W_c}{H} \right)^2 + 1}{\left(\dfrac{W_c}{H} + 1 \right)^2} \tag{15.6}$$

15.2.2 Pressure Drop

As given by Equation (15.7), a pressure drop in the serpentine channel is caused by two components—straight channel friction and bends.

$$P_d = \frac{1}{2} \rho_f v_a^2 \left(4 f_s \frac{L_t}{D_h} + \sum_{i=1}^{n-1} \xi_i \right) \tag{15.7}$$

$$f_s = \frac{\sqrt{\left(3.2 x_2^{-0.57} \right)^2 + (f_l Re)^2}}{Re} \tag{15.8}$$

$$f_l = \frac{4.7 + 19.64G}{Re} \tag{15.9}$$

where f_s is the friction loss coefficient in a straight section defined by Equation (15.8); x_2 is the hydraulic channel length, $x_2 = x/Re_e D_h$; L_t is the total length of the channel; ξ_i is the loss coefficient of bend i, the serpentine channel having a total of $n-1$ U-bends along the fins; and f_l is the friction factor of a fully developed laminar flow given by Equation (15.9).

These bends in series interrupt the hydrodynamic boundary periodically; thus, the effects of laminar flow development and re-development are considered [Copeland (2000), Pharoah (2005), Pharoah (2006), Maharudrayya, et al. (2004)]. The bend loss coefficient ξ_i can be developed as a function of the Reynolds number, the aspect ratio, the curvature, and the width of the fins, as given by Equation (15.10), with $1000 < Re < 2300$, $1 < a < 6$, and $0.25 < b < 1.5$.

$$\xi_i = 8.09 \left(1 - 0.3439a + 0.042a^2\right) \left(1 - 0.3315b + 0.1042b^2\right) \qquad (15.10)$$

15.3 Experimental Setup

Experiments have been performed to compare the experimental results and the simulation results of the heat sink, validate the accuracy of the proposed model, facilitate optimization analysis and provide optimal solutions to improve the heat sink's thermal performance.

A schematic of the heat sink experimental setup is given in Figure 15.3; the upper part of the figure represents the top view of the heat sink, and the lower part represents the side view. In the upper part, A and B are the inlet and outlet of the heat sink, respectively, which are connected by the 'serpentine channel'. As observed in the lower part, the heat sink is placed on the top, with the 'heat sink base' in the middle and the 'heater' with 'insulation'' in the bottom layers. Six platinum thermometer temperature sensors, C_1 to C_6, are sited within the 'contact surface' of the 'heat sink base' and 'plate heater'.

As shown in Figure 15.4, the six temperature sensors are attached to the upper surface of the plate heater to measure the average temperature, which is obtained from the 'temperature data logger' D_1. The 'Z view' of the 'contact surface' given in Figure 15.4 shows the distribution of the six temperature sensors. The periphery of the heater is enveloped with an insulator. In this experiment, a 'plate electrical heater' of the same shape as the 'heat sink' is used as the heat source and is connected to a direct current (DC) power supply with a pre-set power ($Q=15$ W).

The main components of this experiment are shown in Figure 15.5. Two pressure transmitters, PT_1 and PT_2, are connected to inlet A and outlet B, respectively, to measure the pressure drop, which is obtained from the 'pressure transducer', D_3. In addition, two temperature sensors are attached to inlet A and outlet B of the heat sink to measure the flow conditions of T_{in} and T_{out}, which are obtained from the 'temperature data logger', D_2. Table 15.1 gives the experimental data readings in average format, where the temperature data are in ^oC and the pressure data are in Pa.

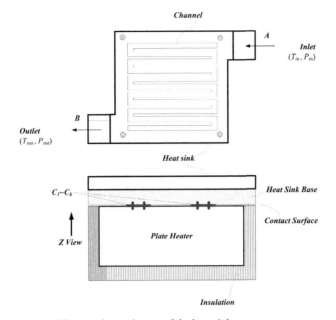

Fig. 15.3 Schematic of the experimental setup of the heat sink

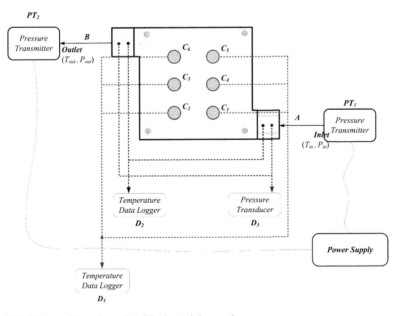

Fig. 15.4 Z view of the schematic of the heat sink experiment

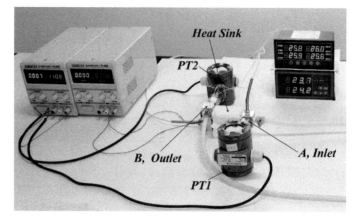

Fig. 15.5 Experimental setup of the heat sink in laboratory

Table 15.1 Experimental Average Data Readings

Design	T_{in}	T_{out}	P_{in}	P_{out}	C_6	C_5	C_4	C_3	C_2	C_1
A	20	21.9	20000	4399.8	21.7	21.8	21.6	20.9	20.4	20.2
B	20	22.1	20000	8972.5	21.9	21.5	21.2	20.8	20.5	20.2
C	20	22.1	20000	14284.9	21.9	22.0	21.9	21.9	21.8	20.2
D	20	22.4	20000	16567.7	22.1	22.2	21.8	21.5	20.9	20.2
E	20	21.8	20000	18219.8	21.4	20.9	20.7	20.6	20.4	20.2

15.4 Optimal Design

A procedure for the optimal design of the serpentine channel heat sink is given in Figure 15.6, which summarizes the 4 steps in the overall process. Step 1 is the experimental data acquisition step, which collects temperature and pressure data. Step 2 is the optimal design of the serpentine channel heat sink, where the two objective fitness functions are defined; then, the MOAFNS method is utilized to perform the MO optimizations and to generate Pareto-optimal solutions. Step 3 conducts PRI assessment on the Pareto-optimal solutions. Step 4 presents the results and is employed for advising engineers and data-enabled applications.

15.5 Fitness Functions

In this context, the 'total thermal resistance' and 'pressure drop', which significantly affect the heat transfer performance and fluid flow characteristics of the sink, are considered as the objectives of structural design for the serpentine channel heat sink. Therefore, based on the FPR approach, this multi-objective optimization has two fitness functions, as given by Equations (15.11) and (15.12), which are determined

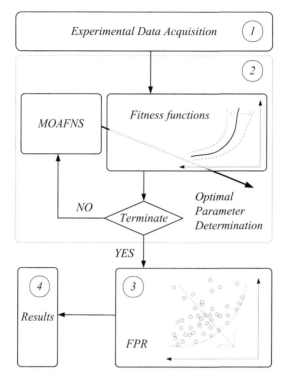

Fig. 15.6 Optimal design procedure

by the number of channels n, the width of the channel W_c, the height of the channel H, and the inlet flow velocity v_a, as stated in Section 15.2.

$$F_1 = minimize\left\{mAP\left(R_t\left(n,W_c,H,v_a\right)\right)\right\} \qquad (15.11)$$

$$F_2 = minimize\left\{mAP\left(P_d\left(n,W_c,H,v_a\right)\right)\right\} \qquad (15.12)$$

The goal of this study is to find the optimal geometric parameters that simultaneously minimize both the thermal resistance R_t and the pressure drop P_d in a serpentine channel heat sink, whose length L, width W and thickness of the bottom plate t_b are constant.

Since the width of the heat sink is constant, the sum of the widths of the channels and fins is also a constant, which provides the constraint function given by Equation (15.13). In addition, the coolant flow should be in the laminar regime $Re < 2300$.

$$nW_c + (n+1)W_b = W \qquad (15.13)$$

15.6 Empirical Results

15.6.1 Simulations

The minimization of the fitness functions results in a minimized thermal resistance R_t and pressure loss P_d. The fitness functions involve the following four design parameters—the number of channels n, the width of the channel W_c, the height of the channel H and the velocity at the inlet v_a. The optimizations of the MO structural design of a serpentine channel heat sink are performed using *SwarmFish* [Swarmfish (2011)] and *SECFLAB* [SECFLAB (2012)], which are toolboxes for MATLAB®. The initial parameters of the heat sink optimization, based on previous research and engineering applications, are given in Table 15.2. Here, the max generation of 50 is the termination condition of each round of testing; the total test number is 100; the visual factor is 2.5; the crowd factor is 0.618; the population is 50, in which the non-replaceable population P_N and replaceable population P_R are 40 and 10, respectively; the attempt number is 5; and the ranges of n, W_c, H and v_a are [4,20], [1,4]$\times 10^{-3}$, [2,5]$\times 10^{-3}$ and [0,2], respectively. The remaining structural constants are also listed in Table 15.2.

Table 15.2 Parameters of the Heat Sink Optimization Using *SwarmFish*

Max generation	50
Test number	100
Visual factor	2.5
Crowd factor	0.618
Population	50
Non-replaceable population	40
Replaceable population	10
Attempt number	5
Number of channels	[4,20]
Width of channel (m)	[1,4]$\times 10^{-3}$
Height of channel (m)	[2,5]$\times 10^{-3}$
Velocity at inlet $(m \cdot s^{-1})$	[0,2]
Length of heat sink (m)	32×10^{-3}
Width of heat sink (m)	32×10^{-3}
Thickness of base plate (m)	1×10^{-3}
Ambient temperature (K)	293.15
Thermal conductivity of water $(W \cdot m^{-1} \cdot K^{-1})$	0.595
Thermal conductivity of heat sink $(W \cdot m^{-1} \cdot K^{-1})$	160
Density of water $(kg \cdot m^{-3})$	1000
Water kinematic viscosity $(m^2 \cdot s^{-1})$	1×10^{-6}
Water Prandtl coefficient	7
Water heat capacity $(J \cdot kg^{-1} \cdot K^{-1})$	4.183×10^3

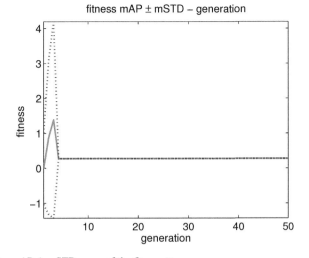

Fig. 15.7 The mAP ± mSTD curve of the fitness F_1

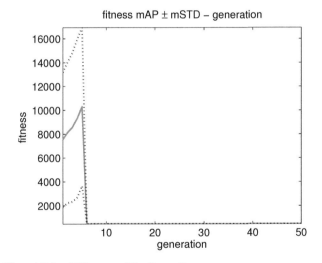

Fig. 15.8 The mAP ± mSTD curve of the fitness F_2

Figures 15.7 to 15.10 illustrate the fitness mAP and mSTD curves of two fitness functions, which are defined in Section 10.5. As shown in the figures, the fitness curves decrease very quickly after generation 1 and reach a plateau (before generation 10), and then, they remain steady from generation 10 to 50. The mAP and mSTD curves move closer and converge, indicating the high efficiency and accuracy of this thermal modeling and the MO optimization.

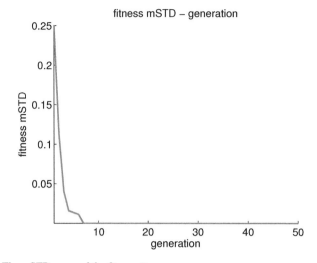

Fig. 15.9 The mSTD curve of the fitness F_1

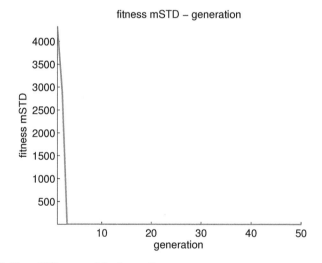

Fig. 15.10 The mSTD curve of the fitness F_2

Figures 15.7 and 15.8 present the mAP $\pm$ mSTD diagrams for F_1 and F_2 over the full simulation generations and indicate that a better solution of F_1 can be optimized without worsening a solution of F_2, which is not dominated by any other solution in the search space. Figures 15.9 and 15.10 show the mSTD curves of F_1 and F_2, which decrease quickly at approximately generations 8 and 4 and continue until the end of the simulation.

15.6.2 Verification

The Pareto-optimal solutions are shown in Figure 15.11, which clearly reveals that these two objectives are in conflict. Any changes in design parameters that decrease the thermal resistance lead to an increase in the pressure drop and vice versa. Designers can select any optimal solution on the Pareto-optimal curve in accordance with the available pressure drop to drive the coolant. The five designs, A, B, C, D, and E, are listed from left to right in Figure 15.12 and are constructed to validate the optimization using the experimental setup of the heat sink introduced in Section 15.3. The validation results are given in Table 15.3.

As shown in Table 15.3, the recommended solutions are listed by the solution number in the row 'β_2', with the values of β_2 ascending from left to right (RANK = 1 is the most recommended). It is reasonable to consider the following four design parameters for a heat sink—the number of channels n, the width of the channel W_c, the height of the channel H and the velocity at the inlet v_a. The errors (*error$_1$* and *error$_2$*, %) between the analytical modeling and the experimental results are relatively small for the two fitness functions—*error$_1$* for F_1 and *error$_2$* for F_2. It can be observed that the *error$_2$* of design D obtains maximum values among all the errors because b of design D is 0.13, which is beyond the range of the bend loss coefficient correlation obtained in Section 15.2.2, and this deviation causes the maximum error in the pressure drop measurement.

Table 15.3 also indicates that the channel number n, the inlet velocity v_a and the width of the channel W_c more strongly affect the design objectives than does the channel height H. When the values of v_a and n increase, F_1 decreases, and F_2 increases; in contrast, when W_c increases, F_1 increases, and F_2 decreases. The height of the channel H has the smallest effect on the design objectives because the optimal value of H remains constant at 5.

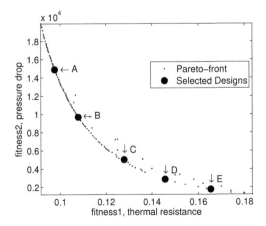

Fig. 15.11 Pareto-front and optimal solutions

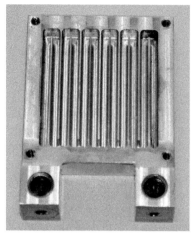

(a) Design A

(b) Design B

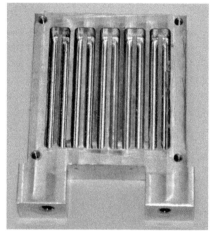

(c) Design C

(d) Design D

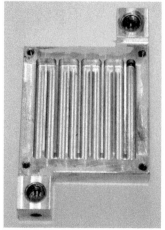

(e) Design E

Fig. 15.12 Five proposed designs of A to E (from left to right)

Table 15.3 Optimal and Experimental Results of the Heat Sink Design

Design	A	B	C	D	E
Rank	1	2	3	4	5
β_2	2.3242	3.4352	3.6322	4.0234	4.2427
n	12	11	10	10	9
W_c	2.0	2.2	2.6	2.8	2.8
H	5	5	5	5	5
v_a	0.7858	0.6742	0.4967	0.3885	0.3022
Simulation Results					
F_1	0.0976	0.1079	0.1279	0.1457	0.1654
F_2	14894.5	9678.6	5026.4	2869.5	1780.2
Experimental Results					
F_1	0.1065	0.1132	0.1216	0.1376	0.1755
F_2	15600.2	11027.5	5715.1	3432.3	2024.1
$error_1(\%)$	8.3568	4.6820	5.1809	5.8866	5.7550
$error_2(\%)$	4.5237	12.2321	12.0505	16.3972	12.0466

15.7 Conclusions and Future Work

In this study, the thermal resistance R_t and pressure drop P_d of a serpentine channel heat sink were obtained using a serpentine channel structural model. A fitness function was defined with four variables, namely, the number of channels n, the width of the channel W_c, the height of the channel H and the inlet flow velocity v_a. Then, the model was optimized using the newly developed MO artificial swarm fish algorithm with the assistance of the PRI β_2 and the trend indices mAP and mSTD.

Based on the above results, the following conclusions can be drawn:

- Simulations of the serpentine channel heat sink model considering the thermal resistance R_t and the pressure drop P_d obtain a good agreement with the experimental validations of the five selected cases
- The FPR was introduced to provide a ranking list of Pareto-optimal solutions to offer a feasible method for facilitating decision making. The evolutionary trends were indicated by the mAP $\pm$ mSTD indices with variable uncertainty tolerances.

16

Battery Capacity Prediction

16.1 Introduction

Estimating the remaining useful lifetime (RUL) of a lithium-ion battery is one of the most important requirements for mechatronics systems, such as portable devices, satellites, deep-space probes, robotic systems, and hybrid vehicles, in which prognostics and health management (PHM) technologies are used to assess the reliability and performance of a mechatronics product and, under its actual life cycle conditions, to determine the advent of failure and mitigate system risk.

Substantial amounts of research on estimating the remaining useful life of batteries has been performed in recent years ([Ramadass, et al. (2003), Kozlowski, et al. (2001), Kozlowski (2003), Goebel, et al. (2008), Saha, et al. (2009), Burgess (2009), He, et al. (2011), Ecker, et al. (2012)] for examples). Ramadass et al. [Ramadass, et al. (2003)] developed a capacity fade prediction model of a semi-empirical approach for Li-ion cells, in which a diffusion coefficient of the lithium electrode was taken as the accounting parameter for the rate capacity losses during cycling. Kozlowski et al. [Kozlowski, et al. (2001), Kozlowski (2003)] discussed data fusion prognostic approaches using feature vectors, auto regressive moving average (ARMA), neural networks and fuzzy logic for the assessment of the state of charge (SOC), state of health (SOH) and state of life (SOL) of batteries. Goebel and Saha [Goebel, et al. (2008), Saha, et al. (2009)] presented a comparative study for the estimation of the RUL of batteries, in which the autoregressive integrated moving average (ARIMA), extended Kalman filtering (EKF), relevance vector machine (RVM) and particle filter (PF) were used. Burgess [Burgess (2009)] proposed a method to assess the RUL of valve-regulated lead-acid batteries using capacity measurements and Kalman filtering. He et al. [He, et al. (2011)] proposed an empirical model using Dempster-Shafer theory (DST) and Bayesian Monte Carlo (BMC) for state of health (SOH) and RUL estimations with available battery data. Ecker et al. [Ecker, et al. (2012)] proposed a parametrized semi-empirical aging model for a multi-variable analysis of accelerated lifetime experiments of a hybrid electric vehicle under typical operating conditions.

In this paper, the AFSAVP is employed as the optimization tool for fast quantitative analysis of a battery residual capacity estimation using an ABF-based fitness function. The adaptive bathtub-shaped function (ABF)'s highly flexible ability for parametric representation and the newly developed highly efficient AFSA method with a variable population size (AFSAVP), which was introduced in Section 7.4, are utilized.

This paper is organized as follows. Section 16.1 introduces the background of the quantitative approach for battery RUL and the artificial fish swarm algorithm method. Section 16.2 introduces the definition of the adaptive bathtub-shaped function. Section 16.3 discusses the modeling of battery capacity prediction. Section 16.4 defines the fitness function of battery capacity prediction using an index of the mean average precision of the coefficient of determination. Section 16.5 discusses the parameter determination for the battery capacity prediction using experimental data. In Section 16.6, the feasibility of the AFSAVP-driven battery capacity behavior approach using the adaptive bathtub-shaped function is discussed.

16.2 Adaptive Bathtub-Shaped Functions

The ABF was first proposed by Chen [Chen, et al. (2011)] to provide a parametric representation of failure behaviors for universal applications. The ABF has been employed for a complementary metal-oxide semiconductor (CMOS) life data prediction study [Chen, et al. (2012)]. In that case, a CMOS device acceleration test was devised and the quantitative representation of the failure behaviors was well demonstrated, indicating a flexible parametric performance of the ABF function with the given CMOS experimental data.

As demonstrated in Figure 16.1, the ABF $\lambda(x)$ is defined by Equation (16.1), where α is the support factor, β is the boundary factor, γ is the internal scale factor, ζ is the shape factor, η is the external scale factor, and δ is the shift factor.

$$\lambda(x) = \begin{cases} \delta + \eta \left(1 - e^{-\dfrac{(|x-\alpha|-\beta)^{\zeta}}{\gamma}} \right) & \text{if } |x-\alpha| \geq \beta, x \geq 0 \\ \\ \delta & \text{if } |x-\alpha| < \beta, x \geq 0 \end{cases} \quad (16.1)$$

In the case of $\delta = 0$ and $\eta = 1$, Equation (16.1) can be re-written as Equation (16.2), which is taken as the basic formula of the fitness function definition in Section 16.4.

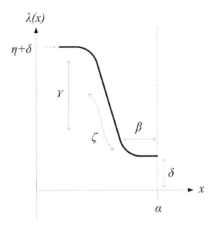

Fig. 16.1 The adaptive bathtub-shaped function [Chen, et al. (2011)]

$$
\lambda(x) = \begin{cases} 1 - e^{-\dfrac{(|x-\alpha|-\beta)^{\zeta}}{\gamma}} & \text{if } |x-\alpha| \geq \beta, x \geq 0 \\[2ex] 0 & \text{if } |x-\alpha| < \beta, x \geq 0 \end{cases} \tag{16.2}
$$

16.3 Battery Capacity Prediction

Table 16.1 summarizes the battery test conditions. The test batteries have a graphite anode and a lithium cobalt oxide cathode, verified by electron dispersive spectroscopy, with a rated capacity of 0.9 *Ah*. The tests were conducted using multiple charge-discharge cycles with the Arbin BT2000 battery test equipment at room temperature [He, et al. (2011)]. The discharge current was 0.45 *A*, the batteries' charging and discharging cut-off voltage was 2.5 *V*, guided by the manufacturer's specifications, and the capacity of the tested batteries was assessed by the Coulomb counting method with full charge-discharge cycles.

Figure 16.2 gives the battery capacity experimental setup, which consists of a set of Lithium-ion battery cases, chargers, loads, a data acquisition (DAQ) system, switches, a computer for control and analysis and various other modules such as a set of sensors for voltage, current and temperature. As shown in Figure 16.3, the battery capacity raw data (observed data) are given in four groups—A3, A5, A8 and A12.

Inspired by the shapes and trends of the raw data curves, the ABF is employed to build a parametric model for the experimental data based on prediction analysis.

Let us define a vector X_j with values $\{x_1, x_2, ..., x_i, ..., x_n\}$ as the observed battery capacity data and a vector $\hat{X}_j$ with the values $\{\hat{x}_1, \hat{x}_2, ..., \hat{x}_i, ...\hat{x}_n\}$ as the estimated battery capacity data, where $i = 1, 2, ..., n$, $j = 1, 2, ..., N$, n is the length of the vector

Table 16.1 Summary of Battery Test Conditions

No.	Test Equipment	Temperature	Cycles	Discharge Current
A3	Arbin BT2000	20–25°C	300	0.45A
A5	Arbin BT2000	20–25°C	300	0.45A
A8	Arbin BT2000	20–25°C	300	0.45A
A12	Arbin BT2000	20–25°C	300	0.45A

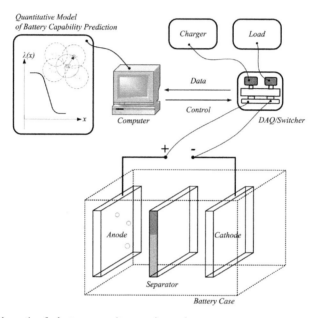

Fig. 16.2 Schematic of a battery capacity experimental setup

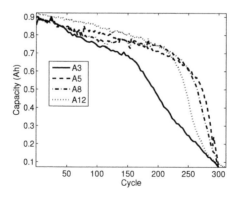

Fig. 16.3 Battery raw data: A3, A5, A8 and A12

X_j, and the population of the vectors X_j is N. The flow chart of the battery capacity prediction analysis is given in Figure 16.4, in which there are five steps, as described below.

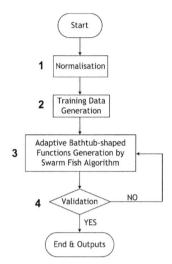

Fig. 16.4 Battery capacity prediction flow chart

Step 1: Normalization: Because the four groups of data are not the same length, all the raw data need to be normalized by scale and shift operations before the next step. The data shift and scale operations are stated in Equations (16.3) and (16.4), respectively [Chen (2010a)]. The charge-discharge cycles (X-axis, unit: cycle) and the battery capacity (Y-axis, unit: Ah) are subject to normalization.

As given by Equation (16.3), the scale operation first calculates the scale factor according to the input range $[x_{min}, x_{max}]$ of the raw data x_i, and then, all the input data are scaled to the range of $[c_l, c_u]$. Specifically, the variables are mapped from the practical value range $[x_{min}, x_{max}]$ to the normalized value range $[c_l, c_u]$, which is $[0,1]$ in this context.

$$x_{i_c} = (c_u - c_l) \times \frac{x_i - x_{min}}{x_{max} - x_{min}} \tag{16.3}$$

In the shift operation, the scaled data x_{i_c} are shifted to the unit range of $[c_l, c_u]$, as given by Equation (16.4).

$$x_{i_h} = c_l + x_{i_c} \tag{16.4}$$

Step 2: Training Data Generation: There are four groups a_k for training data generation. Each of the groups is calculated from the mean values of three out of the four raw datasets—A3, A5, A8 and A12, as stated in Equation (16.5). Table 16.2 lists the raw data grouping for each training data generation. The validation data are b_k.

$$T_k = \frac{a_k}{3} \tag{16.5}$$

Table 16.2 Data Grouping for Training and Validation, Capacity (Ah)

j	Training Samples a_j	Validation Samples b_j
1	A3+A5+A8	A12
2	A3+A5+A12	A8
3	A3+A8+A12	A5
4	A5+A8+A12	A3

Step 3: ABF Generation and Optimization: As defined by Equation (16.1), the ABF $\lambda(x)$ was employed to approximate the experimental data using the predicted data $\hat{x}_i$, in which four parameters (α, β, ζ and γ) of the ABF need to be determined by the AFSAVP.

Step 4: Validation and Error Analysis: The error analysis of the estimated battery capacity data $\hat{x}_i$ and the experimental data x_i validate the optimization process.

16.4 Fitness Function

In this section, an index of the mean average precision (mAP) of the coefficient of determination (R^2) is defined as the fitness to evaluate the optimization process.

As given by Equation (16.6), R^2 is a quantity utilized to measure the proportion of the total fluctuation in the response trend prediction analysis, where a larger R^2 indicates a closer similarity between the predicted data $\hat{x}_i$ and the experimental data x_i [Montgomery and Runger (2003), Gujarati (2004)]. $\bar{x}_i$ is the mean value of the observed data x_i.

$$R_j^2(x_i, \hat{x}_i) = \left(1 - \frac{\sum\limits_{i=1}^{n}(x_i - \hat{x}_i)^2}{\sum\limits_{i=1}^{n}(x_i - \bar{x}_i)^2}\right)_{|x_j} \tag{16.6}$$

The mAP is the mean of the average precision scores for each vector X_j, as expressed by Equation (16.7), in which N is the population of the dataset and $\text{Avg}(*)$ is the average value of each data sequence.

$$\text{mAP}(X_j) = \frac{\sum\limits_{j=1}^{N} \text{Avg}(X_j)}{N} \tag{16.7}$$

The fitness function F of the battery capacity prediction is given by Equation (17.14).

$$F = \text{mAP}\left(R_j^2(x_i, \hat{x}_i)\right) \tag{16.8}$$

16.5 Simulation Results and Discussion

The simulation results for the AFSAVP-driven hybrid modeling of the battery capacity behaviors are performed by *SwarmFish*, which is a toolbox for MATLAB® developed by Chen [Swarmfish (2011)]. According to previous research and engineering applications, the parameters for the battery capacity prediction are initialized as shown in Table 16.3, in which a max generation of 100 is the termination condition of each round of testing; the total test number is 100; the visual factor is 2.5; the crowd factor is 0.618; the population is 30, in which the non-replaceable population P_N and the replaceable population P_R are 20 and 10, respectively; the attempt number is 5 [Chen, et al. (2012), Chen, et al. (2011)]; and the ranges of α, β, γ and ζ are [-10,10], [-10,10], [300,500] and [1,6], respectively.

As can be observed in Figure 16.5, the fitness mAP curves of the four tests are represented by four types of lines, where the fitness curves increase very quickly from generation 1 to 20 to reach a plateau (at approximately generation 20), and then, from generation 20 to 100, the curves remain steady over the remaining generations, and the fitness mAP approaches convergence.

The 4×2 cross-validation (CV) is designed to validate the prediction method. Here, the experimental data, A3, A5, A8 and A12, are split into two parts—one for training (training samples) and the other for validation (validation samples), as listed in Table 16.2.

Table 16.4 gives the 4×2 CV comparison of the four tests with the four groups of datasets for battery capacity prediction in the [0,0.9] normalized cycle range. The mean and standard deviation (MEAN±STD) of α, β, ζ and γ of the four tests are listed and indicate the optimized locations of the parameter settings within their initial ranges; *error* is the absolute error of the normalized battery capacity between the predicted data and the experimental data for validation. The *error* index indicates the acceptable errors of the battery capacity prediction using ABF over the optimization process, in which the *error* values of the training tests are slightly larger than the values of the validation samples. The optimization processes of Test 1 to Test 4 are presented in Figures 16.6 to 16.9.

Table 16.3 Parameters of the SwarmFish Optimization

Max generations	100
Test number	100
Experiment number	100
Visual factor	2.5
Crowd factor	0.618
Population	30
Attempt number	5
α	[-10,10]
β	[-10,10]
γ	[300,500]
ζ	[1,6]

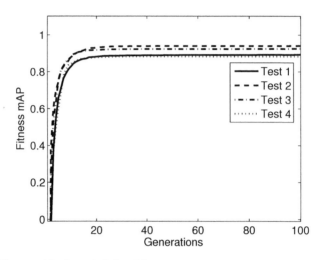

Fig. 16.5 Fitness mAP of tests 1, 2, 3 and 4

As can be observed in Table 16.4, the different raw data sizes could cause different prediction errors. The means of the validation errors are 8.51 per cent > 8.04 per cent > 7.42 per cent > 5.45 per cent, which indicates that closer datasets can result in lower battery capacity predictions. In addition, as shown in Figure 16.3, a larger data size provides better battery capacity prediction results. Specifically, if the given raw data scale is closer to the real battery life-time scale, the prediction result is better. In contrast, a small data scale and data size would result in poor predictions.

Figures 16.6–16.9 are the plots for Test 1 to Test 4. There are two subfigures, (a) and (b), in Figures 16.6–16.9. The calculation curves in subfigure (a) of Figures 16.6–16.9 demonstrate the approximation curves of differences and similarities in Test 1 to Test 4 over the prediction process driven by the AFSAVP, which can be employed as the indicators in battery capacity prediction.

Figures 16.6(a)–16.9(a) compare the results with two curves—the validation data (b_j, dash-dot line) and the calculated battery prediction data (solid line). In addition, the two dashed curves are the ± 5 per cent error upper and lower boundaries of the calculation curve, and there are three vertical dotted lines located at normalized cycle = 0.8, 0.85 and 0.9 as reference lines on the X-axis. Without losing significance, the normalized cycle range [0, 0.9] for the analysis is chopped from the full range of [0,1] to avoid the numerical singularity at normalized cycle = 1 and the errors mainly caused by the singularity in the range of [0.9, 1]. In addition, there are three horizontal dotted reference lines at 5 per cent, 10 per cent and 15 per cent on the Y-axis.

Subfigure (b) compares the absolute errors of the validation. As shown in Figures 16.6 and 16.9, subfigure (b) and Table 16.4, the mean errors (MEAN) are less than 10 per cent, which indicates that this battery capacity prediction method provides

Table 16.4 The 4×2 Cross-validation of Battery Capacity Prediction (MEAN±STD), (Ah)

NO.	Test 1		Test 2		Test 3		Test 4	
	Training	Validation	Training	Validation	Training	Validation	Training	Validation
Datasets	(A3,A5,A8)	A12	(A3,A5,A12)	A8	(A3,A8,A12)	A5	(A5,A8,A12)	A3
α	0.97±0.32		0.62±0.91		0.72±0.37		0.58±0.04	
β	0.39±0.11		0.47±0.82		0.58±0.33		0.33±0.93	
γ	712.72±47.23		515.43±25.22		467.81±84.52		500.75±64.09	
ζ	4.38±0.53		4.62±0.38		4.55±0.35		4.68±0.37	
$e_{A|max}$	2.8571%	5.9428%	3.7143%	3.4286%	8.5714%	1.1429%	4.0973%	15.1429%
RMS	0.0937	0.1219	0.0621	0.1486	0.0724	0.1452	0.0836	0.1757

acceptable accuracy for practical cases. The results also indicate that the proposed method can handle raw data of different sizes, providing robustness and applicability to field data acquisition systems.

To summarize, the results indicate that the developed ABF-based prediction method can be employed to estimate the lithium-ion battery capacity and as a quantitative indicator for assisting users in their real-time applications.

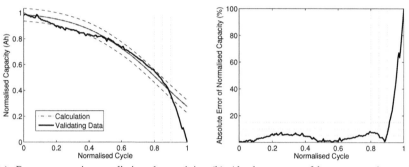

(a) Battery capacity prediction by training (A3,A5,A8) and validating (A12)

(b) Absolute errors of battery capacity prediction

Fig. 16.6 Data comparison of test 1

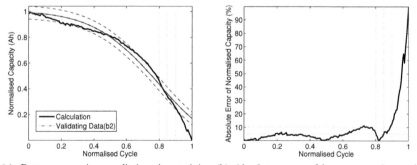

(a) Battery capacity prediction by training (A3,A5,A12) and validating (A8)

(b) Absolute errors of battery capacity prediction

Fig. 16.7 Data comparison of test 2

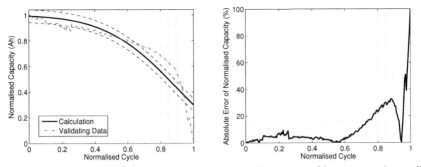

(a) Battery capacity prediction by training (A3,A8,A12) and validating (A5)

(b) Absolute errors of battery capacity prediction

Fig. 16.8 Data comparison of test 3

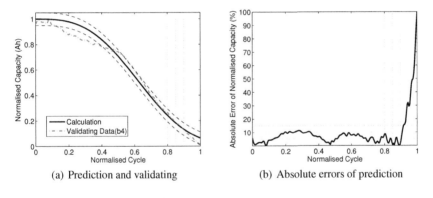

(a) Prediction and validating

(b) Absolute errors of prediction

Fig. 16.9 Data comparison of test 4

16.6 Conclusion and Future Work

In this chapter, the AFSAVP has been employed to quantitatively predict battery RUL based on four groups of historical test data. Designed in the form of an ABF, the index of mAP has been introduced as the fitness function of the R^2 of the observed battery data and estimated battery data for the optimization process. The 4×2 CV method has been deployed to validate the developed model. Under this approach for battery capacity prediction, the four tests have shown agreement with the experimental results.

Quantitative analysis is utilized to aid users in power source planning in fully utilizing available statistical battery data. The ABF formed a prognostic model that can capture the dynamic behaviors of battery capacity. The contributions of this paper include consideration of the dynamic behaviors of historical battery capacity and the development of both a quantitative prediction analysis and a clear indicator of the normalized battery capacity of battery RUL for decision making.

Aiming at on-board battery RUL prediction, this value-added approach has potential applications in protection from battery failure in robotic systems (e.g., exoskeletons, robotic space tethers, humanoid robots and industrial robotics), mobile devices and other systems with electrochemical power sources.

17
Parameter Determination for Fuel Cells

17.1 Introduction

Due to the increasing need for an efficient and clean energy supply, great importance has been placed on the advancement of and fundamental research on polymer electrolyte fuel cell (PEFC) technology. Among the components of PEFCs, the cathode electrode plays a vital function in its operation, in which an oxygen reduction reaction (ORR) occurs and generates heat. Platinum (Pt) loading, ionic conductivity, and the reaction's exchange current density are among the factors that may affect performance. Separate research has been conducted to develop models and approaches that are essential to battery performance and optimization.

Springer et al. [Springer, et al. (1991)] presented an isothermal, one-dimensional, steady-state model for a PEFC with a 117 Nafion membrane, in which the water diffusion coefficients, electro-osmotic drag coefficients, water sorption isotherms, and membrane conductivities were employed. Bernardi and Verbrugge [Bernardi and Verbrugge(1992)] developed a mathematical model of a solid-polymer-electrolyte fuel cell, which was utilized to investigate factors affecting the fuel cell's performance and elucidate the mechanism of species transport in a complex network of gas, liquid and solid phases. Amphlett et al. [Amphlett, et al. (1995)] reported a parametric model to predict the performance of a solid PEFC considering mass transport properties. Bevers et al. [Bevers, et al. (1997)] presented a one-dimensional dynamic model of a gas diffusion electrode with various effects brought about by changes in parameters. Kulikovsky [Kulikovsky, et al. (1999)] developed a two-dimensional model of the cathode compartment of a PEFC with gas channels. Rowe and Li [Rowe and Li(2001)] proposed a one-dimensional non-isothermal model of a PEFC after investigating the effects of various design and operating conditions on the cell performance. Baschuk and Li [Baschuk and Li (2000)] formulated a mathematical model for the performance and operation of a single PEFC. Song et al. [Song, et al. (2001)] utilized the AC impedance method to optimize the thickness and composition of the supporting PEFC layer. Ramadass et al. [Ramadass, et al. (2003)] developed a semi-empirical approach for capacity fade prediction for Li-ion cells, which considers active material and rate capability losses. Wang et al. [Wang, et al. (2003)] studied the effects of different operating parameters on the performance of a PEFC through an experiment with pure hydrogen on the anode side and air on the cathode side. Yerramalla et al. [Yerramalla, et al. (2003)] developed a mathematical model for investigating the dynamic performance of a PEFC with a number of single cells combined into a fuel cell stack. Song et al. [Song, et al. (2004)] investigated one- and two-parameter numerical optimization analyses of PEFC cathode catalyst layers, which consider Nafion content, Pt loading, catalyst layer thickness and porosity. Grujicic and Chittajallu [Grujicic (2004)] developed a model for determining air-inlet pressure, cathode thickness and length, and the width of the shoulders in the inter-digitized air distributor. Weber and Newman [Weber and Newman (2004)] reviewed models of PEFCs, general modeling methodologies and some related summaries. Pathapati et al. [Pathapati, et al. (2005)] reported a mathematical model for simulating the transient phenomena in a PEFC, which can predict the transient responses of cell voltage, temperature, hydrogen/oxygen out-flow rates and cathode and anode channel temperatures/pressures under sudden changes in load current. Wang and Feng [Wang and Feng (2008), Wang and Feng (2009)] presented a one-dimensional study on electrochemical phenomena within the electrode cathode. Wang et al. [Wang, et al. (2011)] published a review on recent PEFC technical progress and applications, the role of fundamental research in fuel-cell technology and major challenges to fuel-cell commercialization. Chen et al. [Chen, et al. (2013)] proposed a quantitative approach for the prediction of RBL using an adaptive, bathtub-shaped function. Considering thermoelectric and thermoeconomic objectives, Sayyaadi and Esmaeilzadeh [Sayyaadi and Esmaeilzadeh (2013)] studied a methodology for optimal PEFC control, in which the net power density and energetic efficiency are maximized. Pathak and Basu [Pathak and Basu (2013)] discussed a mathematical model for the anode and the cathode with an anion exchange membrane in an attempt to predict the performance of the fuel cell considering reaction kinetics and ohmic resistance effects. Noorkami et al. [Noorkami, et al. (2014)] investigated the temperature uncertainty as a key parameter in determining the performance and durability of a PEFC. Molaeimanesh and Akbari [Molaeimanesh and Akbari (2014)] proposed a three-dimensional lattice Boltzmann model of a PEFC cathode electrode, in which electrochemical reactions occur on the catalyst layer. This model is able to simulate single- and multi-species reactive flow in a heterogeneous, anisotropic gas diffusion layer.

Despite a large number of previous studies, the available literature on the analytical modeling of cathode electrodes fails to address two concerns—first, such work does not capture the coupling effects on PEFC performance due to the interactions among design variables; second, few effective methods have been developed that allow for quantitative analysis, model verification and parameter optimization. To fill this void, this paper proposes a swarm bat algorithm with variable population (BAVP) for constructing and optimizing the quantitative cathode electrode model, which will be embedded into the CIAD [Chen, et al. (2014)] framework. Because this new CIAD framework provides an expanded capability to accommodate a variety of CI algorithms, it presents three advantages: (1) mobilized computational resources, (2) the utilization of multiple CI algorithms, and (3) reduced computational costs. This has been demonstrated by some of our previous work in various areas, including applied energy [Chen, et al. (2013)], new drug development for public health care [Xu, et al. (2012), Liu et al. (2012)], economics and finance [Chen and Zhang (2013)], sustainable development [Chen, et al. (2013), Chen, et al. (2012), Chen and Song (2012)], aerospace engineering [Chen and Cartmell (2007)], automotive engineering [Chen, et al. (2012)], public security [Chen, et al. (2011)] and engineering modeling and design [Chen, et al. (2012)].

Inspired by the echolocation behavior of bats and first proposed by Yang [Yang (2010)] in 2010, the swarm bat algorithm (BA) allocates computational resources by adjusting its population and accelerating the calculation speed. By using echolocation, schooling bats can quickly respond to changes in the direction and speed of their neighbors during activities such as detecting prey, avoiding obstacles, and locating roosting crevices in dark surroundings. Useful behavioral information is passed among bats and guides them to move from one configuration to another as one unit. By borrowing this intelligence of social behavior, the BAVP is made parallel, independent of initial values, and able to achieve a global optimum.

Following the introduction section, the remainder of the chapter is organized as follow: Section 17.2 gives the analytical modeling of the cathode electrode; Section 17.3 defines the fitness function for optimizing the analytical model using the model proposed in Section 17.2; Section 17.4 provides the empirical results and further verifies the optimal design and Section 17.5 concludes the chapter.

17.2 Analytical Modeling

A schematic diagram of a dual-layer configuration of a cathode electrode is shown in Figure 17.1, where five specific areas are numbered and explained below. The left side of the electrode attaches to the polymer electrode membrane (PEM), whereas the right side connects to the diffusion media [Wang and Feng (2008), Wang and Feng (2009)].

① includes the assumption that the oxygen concentrations, temperatures, electronic phase potentials and equilibrium potential are the same between the two layers and uniform within each layer. The electrodes are thin layers ($\leq 10\ \mu m$) coated on the PEM surface containing a catalyst (typically, $\mathbf{Pt}$) or carbon ($\mathbf{C}$), an ionomer electrolyte and void space. Generally, there are three phases in the electrode: (i) the void space for gaseous reactant transport, (ii) ionomer content for proton transfer and (iii) carbon support for electronic current conduction. In addition to the electrochemical catalyst, which is essential for all functions, Equation (17.1) is given as follows:

$$O_2 + 4e^- + 4H^+ \to 2H_2O + Heat \tag{17.1}$$

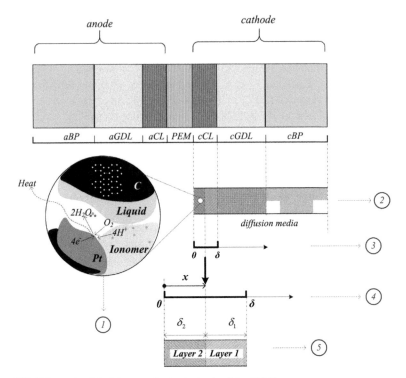

Fig. 17.1 Schematic of dual-layer cathode electrode of the PEFC

② The two sub-layers are denoted as 'Layer 1' and 'Layer 2', where five parameters are considered in this model—the ionic conductivity σ_m, the catalyst specific area a, the exchange current density i, the ionic resistance R_δ, the current density I_δ, the thickness of sub-layer δ, and the interface location of the two sub-layers l, etc.

The factors of the ionic conductivity σ_m of 'Layer 1' and 'Layer 2' are denoted as σ_1 and σ_2, which are determined by the electrolyte water content λ, the ionomer tortuosity τ_m, the Nafion content ε_m and temperature T, as given by Equation (17.2). The ratio of the ionic conductivity factors of 'Layer 1' and 'Layer 2' is given by Equation (17.3).

$$\sigma_m = \varepsilon_m^{\tau_m}(0.5139\lambda - 0.326)\exp\left[1268\left(\frac{1}{303} - \frac{1}{T}\right)\right] \qquad (17.2)$$

$$r_\sigma = \frac{\sigma_1}{\sigma_2} \qquad (17.3)$$

The catalyst specific area a describes the active catalyst surface area per unit volume. The exchange current density i depends on factors like temperature and the catalyst electrochemical characteristics. As given by Equation (17.4), the factor of the catalyst specific area and the exchange current density multiplication ai is determined by factors such as the structural feature of the electrode, including the reaction interface roughness and the mean radius of the catalyst particles, as well as the most important factor for catalyst cost reduction, Pt loading. The ai ratio of 'Layer 1' and 'Layer 2' is given by Equation (17.5).

$$ai = ai_0 (1 - s) \exp \left[-\frac{E_a}{R_g} \left(\frac{1}{T} - \frac{1}{353.15} \right) \right] \qquad (17.4)$$

$$r_{ai} = \frac{ai_1}{ai_2} \qquad (17.5)$$

③ A lumped variable ΔU is defined by Equation (17.6), in which $R_\delta = \delta/\sigma_m$ is the overall ionic resistance across the cathode electrode and $I_\delta = -j_\delta \delta$ is the current density based on the transfer current density j_δ at the interface between the electrode and cathode.

$$\Delta U = R_\delta I_\delta \qquad (17.6)$$

④ The relative location of the interface location of the two sub-layers is defined by Equation (17.7), in which δ is the total thickness of the dual-layer electrodes.

$$l = \frac{x}{\delta} \qquad (17.7)$$

⑤ The thickness ratio of the two sub-layers r_δ is defined by Equation (17.8):

$$r_\delta = \frac{\delta_1}{\delta_2} \qquad (17.8)$$

Considering the cathode electrode in one dimension (the x direction), the two indices (η_1 and η_2) of the over-potential difference of 'Layer 1' and 'Layer 2' are given by Equations (17.9) and (17.10), respectively, where Π, Ψ and Ω are defined by Equations (17.11) to (17.13) [Wang and Feng (2009)].

$$\eta_1 (\Delta U, l) = \frac{R_g T}{\alpha_c F} \ln [\Pi (\Delta U, l) + 1] \qquad (17.9)$$

$$\eta_2 (\Delta U, r_\sigma, r_{ai}, r_\delta, l) = \frac{R_g T}{\alpha_c F} \ln [\Psi (\Delta U, r_\sigma, r_{ai}, r_\delta) (\Omega (\Delta U, r_\sigma, r_{ai}, r_\delta, l) + 1)] \qquad (17.10)$$

$$\Pi (\Delta U, l) = \tan^2 \left(\pm \left(\Delta U \frac{\alpha_c F}{2 R_g T} \right)^{\frac{1}{2}} \cdot (1 - l) \right) \qquad (17.11)$$

$$\Psi (\Delta U, r_\sigma, r_{ai}, r_\delta) = \Pi \left(\Delta U, \frac{1}{1 + r_\delta} \right) (1 - r_\sigma r_{ai}) + 1 \qquad (17.12)$$

$$\Omega\left(\Delta U, r_\sigma, r_{ai}, r_\delta, l\right) =$$
$$\tan^2\left(\frac{\sqrt{\Delta U \dfrac{\alpha_c F}{2R_g T} \cdot \dfrac{r_\sigma}{r_{ai}} \cdot \Psi\left(\Delta U, r_\sigma, r_{ai}, r_\delta\right) \cdot \left(l - \dfrac{1}{1+r_\delta}\right)}}{-\tan^{-1}\sqrt{\dfrac{\Pi\left(\Delta U, \dfrac{1}{1+r_\delta}\right)}{\Psi\left(\Delta U, r_\sigma, r_{ai}, r_\delta\right)} r_\sigma r_{ai}}}\right) \tag{17.13}$$

17.3 Fitness Function

To determine the optimal parameters for the over-potential difference η, this section employs two trend indices, *mmAP* and *mmVAR*, for evolutionary optimization, as given in Section 10.5.2.

The fitness function F of the optimal over-potential difference function is given by Equation (17.14). As the fitness function is defined as the *mmAP* values of the over-potential difference function η, the goal of this model is to determine the optimal combination of five parameters, ΔU, r_σ, r_{ai}, r_δ and l, that simultaneously maximizes the objective of η.

$$F\left(\Delta U, r_\sigma, r_{ai}, r_\delta, l\right) = maximize\left\{\text{mmAP}\left(\eta\left(\Delta U, r_\sigma, r_{ai}, r_\delta, l\right)\right)\right\} \tag{17.14}$$

The over-potential difference function η is given by Equation (17.15), in which ΔU is the lumped variable, given by Equation (17.6); r_σ is the ratio of the ionic conductivity of the two sub-layers, given by Equation (17.3); r_{ai} is the *ai* ratio, given by Equation (17.5); l is the location factor, given by Equation (17.7); and r_δ is the ratio of the thicknesses, defined by Equation (17.8).

$$\eta\left(\Delta U, r_\sigma, r_{ai}, r_\delta, l\right) = \eta_1 \cdot \left(\frac{1}{1+r_\delta} \le l \le 1\right) + \eta_2 \cdot \left(0 \le l \le \frac{1}{1+r_\delta}\right) \tag{17.15}$$

17.4 Empirical Results and Discussion

Maximizing the fitness function F yields the maximum *mmAP* of η, which is obtained by the specially designed toolboxes *SwarmBat* [SwarmBat] and *SECFLAB* [SECFLAB (2012)]. The computer used for the simulations has a 2.1 GHz Intel dual-core processor, Windows XP Professional v5.01 Build 2600 service pack 3, 2.0 GB of 800 MHz dual-channel DDR2 SDRAM, and MATLAB® R2008a.

The initial parameters are listed in Table 17.1, in which the max-generation number is 100, which serves as the termination condition in each test. The test number is also 100. The frequency range is set to [20000,500000] Hz; the reduction factor α is 0.9; the population is 50, in which the non-replaceable population P_N and replaceable population P_R are 40 and 10, respectively; the random step is 0.01; and the ranges of $\Delta U, r_\sigma, r_{ai}, r_\delta$, and l are [0,10], [0,10], [0,10], [0,2] and [0,3], respectively.

Table 17.2 gives the optimal combinations (MEAN±VAR) of $\Delta U, r_\sigma, r_{ai}, r_\delta$ and l and indicates that the over-potential is both non-uniform within the cathode and at particularly high values of the lumped parameter ΔU and sensitive to the spatial variation l.

Table 17.1 Parameters of the Swarmbat Optimization

Max generations		100
Test number		100
Frequency range		[20000,500000]Hz
Reduction factor	α	0.9
Population		50
Random step		0.01
Lumped variable	ΔU	[0,0.1]
Ratio of ionic conductivity	r_σ	[0,2]
Ratio of ai	r_{ai}	[0,2]
Ratio of thickness	r_δ	[0,4]
Location factor	l	[0,1]

Table 17.2 Optimal Results of 'MEAN±VAR'

Parameters	Results
ΔU	0.07754 ± 0.01034
r_σ	1.03191 ± 0.12681
r_{ai}	0.85241 ± 0.13736
r_δ	2.05281 ± 0.16181
l	0.40709 ± 0.15818

Figure 17.2 shows the *mmAP* curves, with the upper and lower *mmVAR* boundaries given, in which the fitness increases very quickly. In addition, the fitness reaches a plateau from generations 1 to 60. It is noted that all lines converge at generation 100.

Figure 17.3 depicts the fitness *mmVAR* over the entire simulation. The curves decline quickly within the first 60 generations and finally reach 0 in generation 100. These two figures indicate that the proposed optimization algorithm is efficient and accurate.

As also listed in Table 17.3, to demonstrate the coupled effects of the five variables on η, Figures 17.4 to 17.10 provide seven '3D' sub-figures for evaluation. Figures 17.4 and 17.5 show that r_σ and r_{ai} have similar positive effects on η; when these values increase, η also increases. Figure 17.6 shows that η is sensitive to $r_\delta \in$

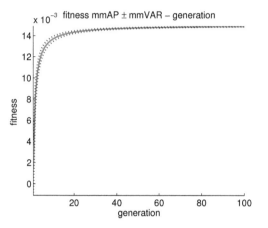

Fig. 17.2 Fitness curves of mmAP $\pm$ mmVAR over the simulation

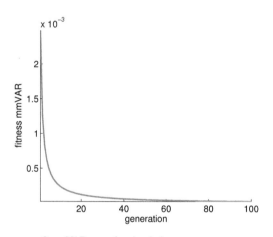

Fig. 17.3 Fitness curves of mmVAR over the simulation

Table 17.3 Coupled Effects of Five Variables on η in 3D Figures

Figure No.	$X - Y - Z$ axis
Figure 17.4	$\Delta U - r_\sigma - \eta$
Figure 17.5	$\Delta U - r_{ai} - \eta$
Figure 17.6	$\Delta U - r_\delta - \eta$
Figure 17.7	$\Delta U - l - \eta$
Figure 17.8	$r_\delta - l - \eta$
Figure 17.9	$r_\sigma - l - \eta$
Figure 17.10	$r_{ai} - l - \eta$

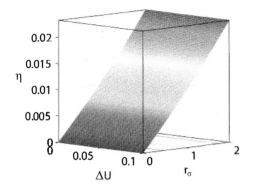

Fig. 17.4 Effects of variables on η: η vs. ΔU, r_σ

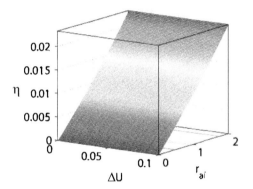

Fig. 17.5 Effects of variables on η: η vs. ΔU, r_{ai}

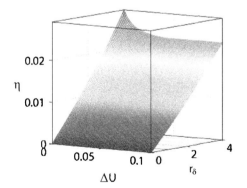

Fig. 17.6 Effects of variables on η: η vs. ΔU, r_δ

[3,4] and $\Delta U < 0.05$. Figure 17.7 indicates that η increases faster with $l > 0.5$ and $\Delta U < 0.05$, and an improved η is obtained with larger l and smaller ΔU.

Figure 17.8 shows that η is sensitive to a smaller r_δ, and in Figure 17.9, there is a plateau for $r_\sigma < 1$. Figure 17.10 shows that there are a few η peaks when both r_{ai} and l are found in their lower range, which implies that η exhibits instability with small values of r_{ai} and l.

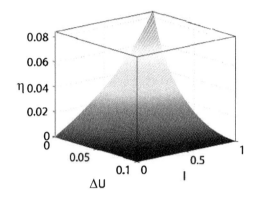

Fig. 17.7 Effects of variables on η: η vs. $\Delta U, l$

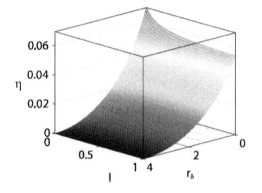

Fig. 17.8 Effects of variables on η: η vs. r_δ, l

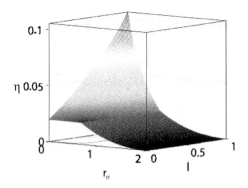

Fig. 17.9 Effects of variables on η: η vs. r_σ, l

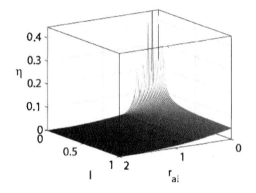

Fig. 17.10 Effects of variables on η: η vs. r_{ai},l

17.5 Conclusion and Future Work

In this study, an analytical model incorporating five parameters is proposed to explore the transport and electrochemical phenomena in dual-layered cathode electrodes of polymer electrolyte fuel cells. These parameters include the lumped variable ΔU, the ratio of ionic conductivity of two sub-layers r_σ, the ai ratio of the two sub-layers r_{ai}, the ratio of thickness r_δ and the relative location factor l. Moreover, a theoretical study on the spatial distribution of reaction rates across the electrode is presented.

The proposed model is utilized to define a design objective, which determines the optimal combinations of the five parameters to maximize the over-potential difference η. Based on trend indices mmAP and mmVAR, a fitness function was constructed with the five variables as discussed above, which are optimized by the swarm bat algorithm with a variable population.

The numerical solutions obtained by this study have been applied to optimize the electrode performance through a set of optimal dual-layer configurations; the research findings can be summarized with the following three points:

- A dual-layered cathode electrode modeling for optimal parameter determination that provides a strong argument for implementing the solutions to explore the effects of each layer's properties on their performance is proposed
- Based on the developed dual-layered cathode electrode model, a swarm bat algorithm with variable population is developed. The algorithm directly affects the optimal parameter determination due to its high efficiency and accuracy
- The two trend indices of *mmAP* and *mmVAR* have been utilized to smooth out short-term fluctuations and highlight longer term trends until the maximum generation fitness point is reached. This facilitates the measurement of the computational performance of the BAVP or deployment of other algorithms

Notations

ΔU	Lumped variable
r_σ	Ratio of ionic conductivity of two sub-layers
r_{ai}	*ai* Ratio of two sub-layers
r_δ	Ratio of thickness of two sub-layers
l	Relative location factor
η	Over-potential difference
σ_m	Ionic conductivity
R_δ	Ionic resistance
I_δ	Current density
δ	Thickness of sub-layer
λ	Electrolyte water content
τ_m	Ionomer tortuosity
F	Fitness function
ε_m	Nafion content
T	Temperature
ai	Catalyst specific area and the exchange current density multiplication
PEFC	Polymer electrolyte fuel cell
ORR	Oxygen reduction reaction

18

CIAD Towards the Invention of a Microwave-Ignition Engine

18.1 Introduction

This chapter illustrates how CIAD may be used to search for a new invention. Similar to the principles of computational intelligence algorithms, microwave ignition was also first proposed in the 1950s [Linder (1952)]. It was to replace the spark ignition for a petrol or gasoline internal combustion engine (ICE) with volumetric ignition. An electromagnetic field, instead of an electric spark, is generated by microwave resonance, instead of a high voltage between electrodes of the sparkplug. Academic research into the potential invention of such an engine has then studied both engine and microwave aspects. Recently, University of Glasgow [Li, et al. (2013)] have developed a homogeneous charge microwave ignition (HCMI) system. The homogeneous charge air-fuel mixture can burn more thoroughly and faster with improved thermal efficiency and reduced emissions.

The success of HCMI primarily relies on the adequacy of the resonant power and the resonant frequency of the electromagnetic (EM) wave emitted into the engine cylinder. Theoretical analysis and simulations have shown that 100 W input power is enough to generate an EM field of which 82 per cent volume is above the required ignition field strength [Schöning (2014)]. An extra complication is the unmatched impedance and hence, unwanted microwave reflection, which reduces the EM field intensity and could also harm the microwave source. In [Chen (2008)], emitter parameters are isolated and optimized in simulation for impedance matching. On the other hand, a change in the emitter, as well as the piston position, air-fuel-ratio (AFR), cylinder diameter, etc., also changes the resonant frequency [Schöning (2014), Sun (2010)]. Hence the emitter design in HCMI system should consider both the resonant frequency and impedance matching.

There are usually multiple parameters that define an HCMI emitter. Adjusting the parameters by a human engineer based on a usual trial-and-error approach would cost an impractical amount of time and scope. Using an automated exhaustive search with a computer-based simulation module will save physical prototyping tests, but requires an exponential time and hence is intractable in reality. However, evolutionary computation can help find globally near-optimal parameters and their structure in a nondeterministic polynomial time. Such heuristic virtual prototyping allows the exploration of HCMI with a shortened design cycle.

There exist various heuristic methods that have been widely studied for various optimization problems. The optimization problem arising from the design of an HCMI system is significantly challenging and different from many other applications. The evaluation of optimal parameters depends on the resonant mode and frequency, leading to constraints coupled with arguments. Further, the sensitivity of resonant frequency implies that the search resolution on the frequency should be small, which poses a significant challenge to a generic GA.

In this chapter, the problem of HCMI design with cylinder and emitter models are first described in Section 2. In Section 3, heuristic search methods adaptable to the presented problem are discussed first, including the Nelder-Mead (NM) simplex method and the conventional GA, and then a 'predefined genetic algorithm' (PGA) is developed. Two case studies are presented in Section 4, with the three methods applied and compared in Section 0. Conclusions are drawn in Section 5.

18.2 HCMI Design Evaluation and Virtual Prototyping Through Simulation

18.2.1 Models of the Emitter and Cylinder

The geometry model in Figure 18.1 is a basic model of cylinder with a single antenna in the center of the cylinder. This model consists of four parts: the cylinder head, the cylinder body, the emitter and a coaxial transmission line. The only changeable variable of the emitter is the length of the emitter, which is denoted by a_1. There are four different materials used for this emitter design. The material inside the cylinder head and the cylinder body represents the homogeneous air-fuel. The inner material of the emitter, in this case part 3, is copper with a radius of 2.25 mm. The bound on the outside is steel with a fixed width of 2 mm for this model. A dielectric material fills the gap between the steel and copper, according to a standard coaxial cable.

The antenna is extended by additional antenna designs in Figure 18.2 with two more variables, emitter width as a_2 emitter height as a_3. The additional antenna is made of steel and connected to the outer shield of the emitter. The emitter width is limited, due to the dimensions of the cylinder and the emitter radius; in this case, to a maximum of 7.35 mm. The emitter height starts at the top of the plug and can be in the range of 0.5 mm to 15 mm (the height of the cylinder head).

In order to obtain best propagation performance inside the combustion chamber, search methods are applied for emitter design. Ideally, the EM field inside the cylinder needs to reach a maximum electrical field. The simulation is built with COMSOL software and EM field intensity is calculated through the FEM solver.

The major challenge in the design process is the search resolution in the simulations. In the past, trial and error method was used to find optimal parameters—try one

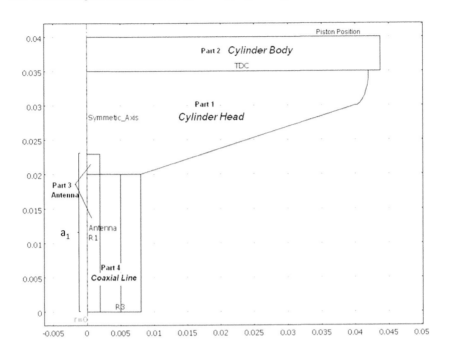

Fig. 18.1 Default geometry model of the emitter and the cylinder

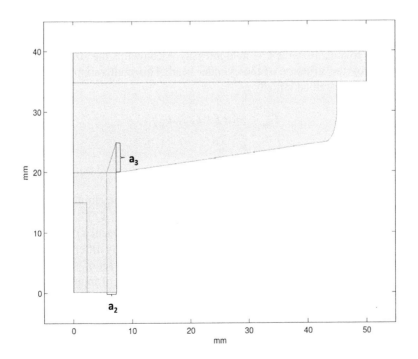

Fig. 18.2 Extended model of an emitter

set of parameters and run the simulation to get resonant frequency and maximum EM intensity, then try another set of parameters to search for improvement, and so on. In this chapter, an auto search based on optimization algorithm in MATLAB® is implemented. Values of parameters are the output of MATLAB while input of simulation models is COMSOL. And maximum EM field intensity is feedback to optimization algorithm from simulation in COMSOL to get the next set of parameters. This optimization algorithm needs to be able to communicate with the simulation model to exchange data and perform a detailed search.

At each search iteration, once the parameters are determined, the frequency search is adopted in COMSOL to search for resonant frequency with the emitter parameters from optimization algorithm. In frequency search, different frequency microwave would generate EM field with different intensity. By searching, the maximum EM field intensity, the corresponding frequency is the resonant frequency. For this case, FEM simulation for EM field would be implemented multiple times for one set of parameters to determine the maximum EM filed intensity and resonant frequency. For optimal parameters design, the FEM simulation could be the major cost of optimizer overhead.

18.2.2 Coupled Constraint Optimization Problem

The HCMI and its emitter design is a multivariable optimization problem. Their performance evaluation is mainly the EM field intensity. For optimal emitter parameters, the EM field intensity would need to be the maximum. Meanwhile, for each fixed set of parameters, resonant state of electromagnetic waves is reached if and only if the input frequency is a resonant frequency. The objective of this evaluation is to maximize the EM field intensity. Meanwhile, the size of the emitter is restricted by the structure of the cylinder head. This multivariable optimization problem can be described initially as:

$$
\begin{aligned}
E_{max} &= max\ E\left(a_1, a_2, a_3, f\right) \\
s.t.\quad & g_i\left(a_i\right) \geq 0, \qquad i = 1, 2, 3 \\
& g_4\left(f\right) \geq 0
\end{aligned}
\tag{18.1}
$$

where E is the intensity of the EM field, a_i, $i = 1,2,3$, are the emitter parameters, f is the frequency of the input microwave, and $g_i\left(a_i\right) \geq 0$ and $g_4\left(f\right) \geq 0$ are boundary conditions, depending on the size of the cylinder and the frequency range.

For different sets of a_i, $i = 1,2,3$, the resonant frequency is different. In order to evaluate emitter parameters, the resonant frequency for each set of parameters must be determined first. For each set of a_i, $i = 1,2,3$, frequency of input microwave is resonant frequency if and only if the intensity of EM field is at the maximum. Thus f in optimization problem (18.1) is the implicit function of a_i, $i = 1,2,3$. Problem (1) is hence expressed as a 'coupled constraint optimization problem':

$$E_{max} = max \; E \; (f \,|\, a_1, a_2, a_3)$$

$$s. \; t. \; g_4(f) \geq 0 \tag{18.2}$$

$$E_{max} = max \; E \; (a_1, a_2, a_3)$$

$$s. \; t. \quad g_i(a_i) \geq 0, \qquad i = 1, 2, 3 \tag{18.3}$$

To solve problem (3), problem (2) needs to be solved first. The EM field intensity, E_{max}, is calculated using a finite element method (FEM). Hence the computational cost depends on searching times and solving the process of FEM. For a cylinder with a pre-determined shape, the solving time of FEM for different emitter parameters varies little. In (2), the search range of frequency is given as $g_4(f) \geq 0$.

In the application to an HCMI system, it is desirable to have as few function evaluations as possible due to the high computing cost in evaluating the EM virtual prototyping. If the search range of frequency is too small, the search result might reach a local extreme. If the search range of frequency is too broad, the computational cost would be high. The common search methods are determined with search range, which becomes a dilemma for this specific application. It is necessary to improve the current search method for tractable performance. To proceed; however, models of emitter and cylinder are introduced first.

18.3 Heuristic Methods and Improved GA Search

18.3.1 Existing Heuristic Methods Tested

The optimization problem described in Section 0 can be divided into two sub-optimization problems: the search for resonant frequencies with fixed emitter parameters and the search for optimal emitter parameters. Both of the sub-problems have the same evaluation of intensity of the EM field. Hence, the solution to such an optimization problem is:

Step 1: Generate emitter parameters as input variables by the search algorithm

Step 2: Evaluate HCMI field strengths on the FEM models

Step 3: Go back to Step 1 if the search is not finished

Step 4: Output the resonant frequency, emitter parameters and field intensity of the final set of candidates.

Figure 18.3 gives a diagram of a general solution to this optimization problem. The frequency is considered as input variables. There are various heuristic search methods that have been widely employed in various fields targeting similar optimization problems, whether they are deterministic or nondeterministic.

A typical deterministic example is the NM simplex method, which is a posteriori algorithm first proposed by Nelder and Mead in 1965 [Nelder and Mead (1965)].

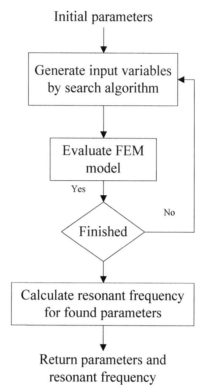

Initial parameters

Fig. 18.3 Optimal heuristic search

This method makes it possible to search and calibrate several design parameters at once. It finds a locally optimal solution to a problem with N variables if the objective function varies smoothly [Haupt and Haupt (2004)]. Though the NM search method is a local search method, it only requires a few function evaluations per iteration, in comparison to an evolutionary algorithm (EA). The NM search requires a rough knowledge of the solution range along multi-optimization. When resonant frequency is involved in optimizing the HCMI system, the performance of the NM method is inadequate.

An EA, on the other hand, is a non-deterministic search algorithm, with its idea originating from the 1950s. The EA includes three operators—selection, crossover and mutation. The EA begins with the creation of a random population of a defined number of individuals. The search performance also depends on how initial parameters are assigned, as they can affect the convergence towards global extrema.

18.3.2 Improved GA Search

The definition of initial parameters includes initial values and the search range of the parameters. If the search range is too small, the results could be local extreme; not a global one. However, if the search range is too wide, the search would cost a lot of time. In the application to HCMI, the time of solving the FEM relates to the time of search iterations. Thus, a suitable definition of initial parameters is important.

To address this issue, a 'predefined genetic algorithm' is developed to narrow down the search range for each search of the emitter parameters. The PGA will generate an initial population, as well as calculate the frequency range before starting the search for emitter parameters. To pre-search the frequency range, the algorithm will select characteristic values out of the initial population and locate the resonance frequency for these values. Additional to the characteristic values, the algorithm will also select a defined number of random parameters out of the initial population and evaluate their corresponding resonant frequency. The frequency range will be calculated from the found minimum and maximum and expanded by a defined boundary. Therefore, the defined frequency range will be appropriate to the minimum required range for the given search parameters. Once the search range of frequency is determined, the GA is applied to search for optimal emitter parameters. Detailed in Figure 18.4, a full implementation procedure of PGA is presented as follows.

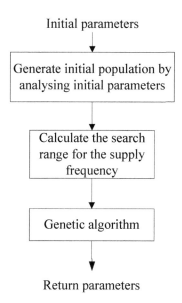

Fig. 18.4 Flow chart of the PGA

A. **Initialization:** Generate an initial population $\{A_0^i, i = 1,2, \cdots ,N\}$, which are drawn from the given input parameters. N is the population size. $A_0^i = \{a_j, j = 1,2,3\}$, a_j are the emitter parameters and $g_j(a_j) \geq 0$, A_0^i is regarded as an individual.

B. **Calculate the frequency range:** Obtain the frequency range $[f_{min}, f_{max}]$.

   1) Select characteristic values λ out of $\{A_0^i, i = 1,2, \cdots ,N\}$ and locate the resonance frequency f_{r_1} for λ;

2) Select a defined number of random parameters out of $\{A_0^i, i = 1, 2, \cdots, N\}$ and evaluate their corresponding resonance frequency f_{r2};

3) Calculate the frequency range: $f_{min} = \min(f_{r1}, f_{r2})$, $f_{max} = \max(f_{r1}, f_{r2})$, and $g(f) \geq 0$.

C. **Execute the GA**: Obtain optimal emitter parameters $\{â_j, j = 1, 2, 3\}$.

1) Evaluate the fitness of each individual with regard to the given objective function. The given objective function is the evaluation of EM field intensity (18.2), which is calculated by using a FEM;

2) Select the individual with higher fitness as the parent individuals, the roulette wheel method is chosen;

3) Perform the crossover and mutation operation to produce the offspring individuals;

4) Repeat the whole process until the termination condition is reached.

D. **Return parameters**.

Compared with the NM and the generic GA, the impact of the initial parameters of PGA can be reduced by decoupling the frequency range. Further, the PGA improves the search efficiency and optimality of the GA.

18.4 Case Studies

The NM, the generic GA and the PGA methods are all applied to solve this optimization problem with three case studies. For both NM and GA, it is easy to be trapped into local extrema and fail. However, in PGA, the initial range of frequency has been predefined, which helps not only shorten the search time but also make sure it reaches global extrema.

Case 1: Coupled Resonant Frequencies and Emitter Lengths

For deterministic optimization as the NM algorithm, the initial value of parameters is given as a starting step for the search. The starting conditions for the NM search are given in Table 18.1. This optimization algorithm is available from the MATLAB Optimization Toolbox. The optimization settings and stopping criteria for the algorithm used were set as in Table 18.2.

For a non-deterministic search, like the GA or PGA, initial search ranges are given as initial conditions.

Table 18.3 gives the initial ranges of input variables for the GA and Table 18.5 for the PGA. It is not required to define a starting value for the supply frequency here because the minimum necessary frequency range can be search for in a PGA. The default settings of the NM and GA are adopted from the MATLAB Optimization Toolbox. For the GA, the optimization setting and stopping criteria for the used algorithm were set as per Table 18.4. For the PGA, the optimization settings and stopping criteria for the NM part of the combined search algorithm were set the same as in Table 18.2. The optimization setting and stopping criteria for the GA part of the combined search algorithm were set the same as in Table 18.4.

Table 18.1 Starting Conditions of NM Searches

Search	Frequency (GHz)	Emitter Length (mm)
1st	2.5	2
2nd	2.59	2
3rd	2.5	10
4th	2.59	10

Table 18.2 Algorithm Setting in MATLAB Optimization Toolbox for NM Search

Display	Final
MaxFunEvals	500
MaxIter	500
TolFun	1E-4
TolX	1E-4

Table 18.3 Starting Conditions of GA Searches

Search	Frequency (GHz)	Emitter Length (mm)
1st	2.55–2.65	0–10
2nd	2.5–2.7	0–20

Table 18.4 Algorithm Setting in MATLAB Optimization Toolbox for GA Search

Display	Final
CrossoverFcn	crossoverscattered
CrossoverFraction	0.8
MutationFraction	0.2
EliteCount	2
Generations	40
PopulationSize	101
SelectionFcn	selectionstochunif
StallGenLimit	20
TimeLimit	Inf
TolFun	1E-6

Table 18.5 Starting Condition of PGA Searches

Search	Frequency (GHz)	Emitter Length (mm)
1st	Not required	0–10
2nd	Not required	0–20

Case 2: Coupled Frequencies, Emitter Lengths, Emitter Heights and Emitter Widths

Two variables are added to the extended emitter models, i.e., the emitter heights and emitter widths. The default values of the height and width of the simplest emitter are set at zero.

Similar to Case 1, the starting conditions of NM searches are given in Table 18.6. The optimization algorithm is available from the MATLAB optimization toolbox.

Table 18.7 gives the initial range of the input variables for the GA and Table 18.8 gives the initial range of the input variable for the PGA. It is not required to define a starting value for the supply frequency here because the minimum necessary frequency range will be calculated during the search process.

The default settings of the NM and the GA are adopted from the MATLAB optimization toolbox with the same settings as in Case 1.

Table 18.6 Starting Conditions of NM Searches

Search	Frequency (GHz)	Emitter Length (mm)	Emitter Height (mm)	Emitter Width (mm)
1st	2.55	5	5	2
2nd	2.59	10	7	5
3rd	2.55	5	5	2
4th	2.59	10	7	5

Table 18.7 Starting Conditions of GA Searches

Search	Frequency (GHz)	Emitter Length (mm)	Emitter Height (mm)	Emitter Width (mm)
1st	2.55–2.6	0–5	0–6	0–5
2nd	2.55–2.6	0–10	0–6	0–5

Table 18.8 Starting Conditions of PGA Searches

Search	Frequency (GHz)	Emitter Length (mm)	Emitter Height (mm)	Emitter Width (mm)
1st	Not required	0–10	0–6	0–5
2nd	Not required	0–20	0–6	0–5

18.5 Virtual Prototyping Results and Comparison

18.5.1 Virtual Prototyping for Case 1 with a Default Emitter

Search Results of the NM Method

As the NM search algorithm is a deterministic heuristic method, it is unnecessary to run multiple simulations to achieve a reliable performance of the convergence speed. Figure 18.5 shows the search trace of the maximum EM field intensity with the search of the first set of initial conditions. It takes about 100 times of searches to reach the maximum point. Table 18.9 lists the gathered search results for different initial parameters. These results show that the initial values are highly relevant for the search to attain acceptable results. Furthermore, this confirms that the NM algorithm can only reach a local extreme value. In the results, resonant frequency from four searches remains at approximately 2.596 GHz, while the emitter length varies for every single search.

Search Results of the Generic GA

Figure 18.5 shows a typical result of the search trace of the maximum EM field intensity. It takes about 15 generations for the GA to reach an acceptable result, with

Table 18.9 Search Results of NM Searches

Search	Frequency (GHz)	Emitter Length (mm)	Maximum EM Field Intensity (V/m)
1st	2.596058	2.407	1.335×10^8
2nd	2.596060	4.422	0.757×10^8
3rd	2.596058	6.477	0.339×10^8
4th	2.596058	6.503	0.335×10^8

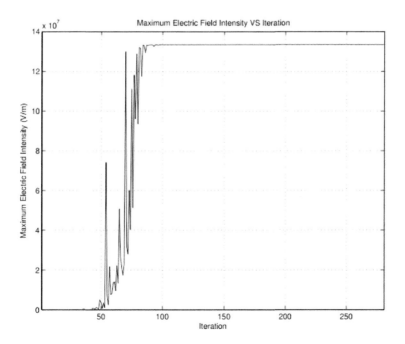

Fig. 18.5 NM search trace for the maximum EM field intensity in Case 1

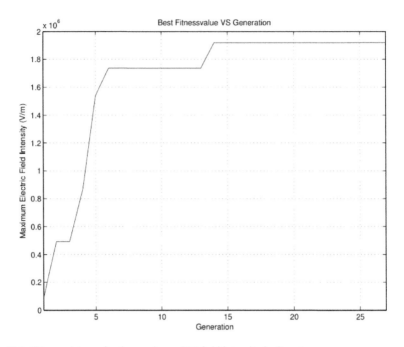

Fig. 18.6 GA search trace for the maximum EM field intensity in Case 1

Table 18.10 Search Results of GA Searches

Search	Frequency (GHz)	Emitter Length (mm)	Maximum EM Field Intensity (V/m)
1st	2.596120	9.881	0.882×10^6
2nd	2.595678	13.446	1.920×10^6

the second search of the initial condition, which has the maximum EM intensity in GA search for Case 1. Table 18.10 gives the results of the different searches. The resulting frequency of both optimization searches is located around 2.596 GHz. However, the emitter lengths vary between the searches, which influence the EM field distribution inside the cavity. Compared with the results of the NM method, the EM field intensity is much smaller, which could imply that the nondeterministic GA is an inferior heuristic search method for such an application.

Search Results of the PGA

It also takes about 15 generations for the PGA to reach an acceptable result, with the second search of the initial condition, as a typical result shown in Figure 18.7. Compared with the results of the GA, the convergence speed was similar. However, in Table 18.11, the results of the different searches show a consistency in PGA

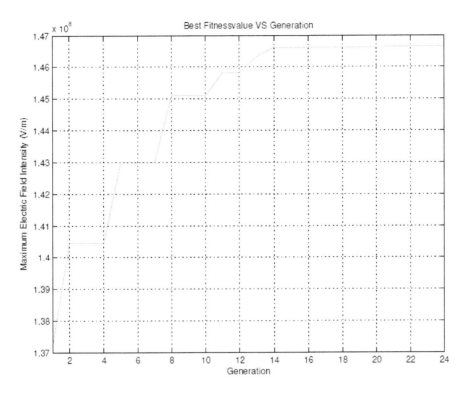

Fig. 18.7 PGA search trace for the maximum EM field intensity in Case 1

Table 18.11 Search Results of PGA Searches

Search	Frequency (GHz)	Emitter Length (mm)	Maximum EM Field Intensity (V/m)
1st	2.596059	1.154	1.466×10^8
2nd	2.596059	1.401	1.473×10^8

search. The resulting frequency of the optimal searches is found at 2.596059 GHz. The best emitter length is found to be 1.154 mm in first and 1.401 mm in second searches, with the maximum EM field intensity being approximately 1.47×10^8 V/m. Compared with the results of PGA with NM and GA, it is the highest of all.

18.5.2 Virtual Prototyping for Case 2 with an Extended Emitter

Search Results of the NM Simplex

Figure 18.8 shows the search trace of the maximum EM field intensity with the first set of initial condition. It takes about 420 times of searches to reach the maximum

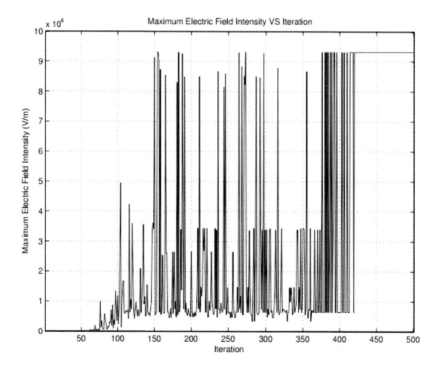

Fig. 18.8 NM search trace for the maximum EM field intensity in Case 2

Table 18.12 Search Results of NM Searches

Search	Frequency (GHz)	Emitter Length (mm)	Emitter Height (mm)	Emitter Width (mm)	Maximum EM Field Intensity (V/m)
1st	2.537480	5.32	4.77	1.98	9.318×10^6
2nd	2.462754	10.68	7.03	5.00	1.067×10^6
3rd	2.535684	5.25	4.84	2.03	3.413×10^6
4th	2.532301	18.11	8.99	-1.07	1.459×10^6

point. As shown in Figure 18.8, the EM field intensity varies much during the search. That's because the EM field intensity is quite sensitive to frequency. In Case 2, the number of input variables increases to 3. It would take more time to complete the search than in Case 1. Table 18.12 lists the gathered search results for the different initial parameters. These results are all local extremes. In the results, both the frequency and emitter parameters are divergent.

Search Results of the GA

The GA search is a non-deterministic optimization method and therefore the results will be inconsistent between different optimization searches. Figure 18.9 shows that

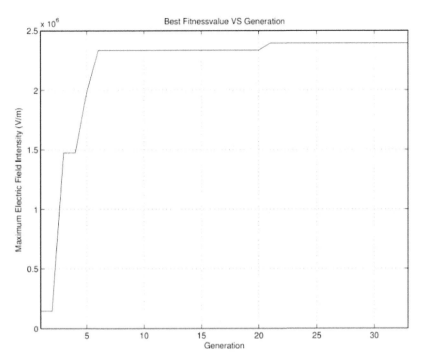

Fig. 18.9 GA search trace for the maximum EM field intensity in Case 2

Table 18.13 Search Results of GA Searches

Search	Frequency (GHz)	Emitter Length (mm)	Emitter Height (mm)	Emitter Width (mm)	Maximum EM Field Intensity (V/m)
1st	2.594545	4.85	0.64	0.53	7.294×10^5
2nd	2.583839	9.78	1.96	1.63	2.394×10^6

it takes about 40 generations for the GA to reach an acceptable result, with the 2nd search of the initial condition due to the higher complexity of Case 2.

Table 18.13 gives the typical results of the different searches. Both resonant frequency and maximum EM field intensity are divergent, which proves that the results are not a global extremum. Furthermore, the EM field intensity is much smaller here than using the NM search.

Search Results of the PGA

It takes about 15 generations for the PGA to reach an acceptable result with the first search of the initial condition in Figure 18.10. In Table 18.14, the results of the different searches show a consistency, unlike with the GA. The resulting frequency and the emitter length are the same as in Case 1. The emitter height is 0.0 mm,

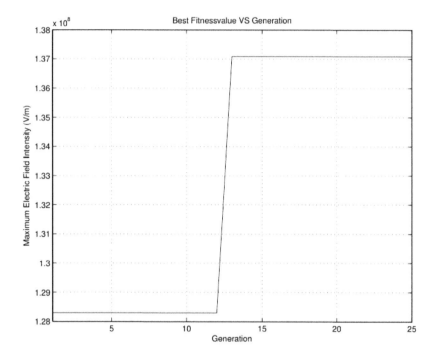

Fig. 18.10 PGA search trace for the maximum EM field intensity in Case 2

Table 18.14 Search Results of PGA Searches

Search	Frequency (GHz)	Emitter Length (mm)	Emitter Height (mm)	Emitter Width (mm)	Maximum EM Field Intensity (V/m)
1st	2.596060	1.154	0.0	1.42	1.371×10^8
2nd	2.596059	1.401	0.0	3.33	1.274×10^8

while the emitter width is 1.42 mm and 3.33 mm in two searches, also delivering the highest filed strength as shown in the table.

The PGA heuristic search combines a deterministic and a non-deterministic methods, hence the results being inconsistent with multiple searches. The PGA exhibits a significant better performance than the GA. The found maximum EM field intensity is nearly identical to that of the default emitter model in Case 1.

18.6 Conclusion

This chapter has developed an improved GA heuristic method for a challenging real-world application—the potential invention of an HCMI engine. During the optimization for this coupled constraint problem, the NM simplex search method and the conventional GA failed due to the high influence of the incident EM frequency. It has been found that improved GA, i.e., the PGA, offers a higher convergent speed and reaches global extrema in various tests. Furthermore, the selection of initial values has little impact on the final results of the PGA. When the complexity of the problem increases with the number of input variables, the PGA also offers a consistent performance but despite that, the NM and the GA yield divergent results.

19

Control for Semi-Active Vehicle Suspension System

19.1 Introduction

Ride comfort, which is concerned with the comfort of passengers during operating conditions, is one of the most important characteristics of a vehicle suspension system and is mainly affected by various sources of vibration of the vehicle body. Usually, a major source of vibrations is road surface irregularities, which range from standard test roads to the random variations in the road surface elevation profile through the vehicle suspension system. An example of a vehicle suspension is shown in Figure 19.1 [Chen (2004)], which includes tyres, springs, dampers and some other accessory components.

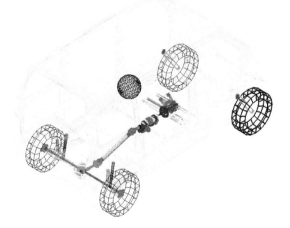

Fig. 19.1 An example of a vehicle suspension system [Chen (2004)]

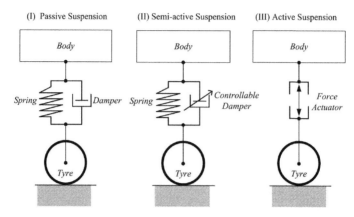

Fig. 19.2 Passive, semi-active and active suspension systems

As shown in Figure 19.2, the main components of a vehicle suspension system include the vehicle body, springs, dampers and tyres. Here, (I) is a passive suspension system, in which the spring stiffness and damping coefficient values are fixed and cannot be adjusted during the suspension system's working process; (II) is a semi-active suspension system, which can only change the damping coefficient and does not invoke any energy inputs into the vehicle suspension system; and finally, (III) is an active suspension system, which uses an actuator to exert an independent force on the suspension system to improve the ride characteristics and the need for extra energy inputs.

Active suspension systems have been investigated since the 1930s; however, due to the bottleneck of complexity and high hardware costs, it has been difficult to achieve wide practical usage, and it is only available on sports vehicles, military vehicles and premium luxury vehicles [Yi, et al. (1993)]. Active suspension is designed to use independent actuators to improve the suspension system's ride comfort performance. By reducing vibration transmission and ensuring proper tyre contact, active and semi-active suspension systems are designed and developed to achieve better ride comfort performance than passive suspension systems. The semi-active (SA) suspension system was introduced in the early 1970s [Karnopp, et al. (1974)], and it has been considered a good alternative to active and passive suspension systems. The concept behind SA suspension is to replace active force actuators with continually adjustable elements, which can vary or shift the rate of the energy dissipation in response to an instantaneous condition of motion. An SA suspension system can only change the damping coefficient of the shock absorber; it will not add additional energy to the suspension system. An SA suspension system is also less expensive and energy consumptive than an active suspension system in operation [Jalili (2002)]. In recent years, research on SA suspension systems has continued to advance with respect to their capabilities, therein narrowing the gap between SA and active suspension systems. An SA suspension system can achieve the majority of the performance characteristics of an active suspension system, thus providing a wide class of practical applications. Magnetorheological/electrorheological (MR/ER) [Stanway (1996), Spencer, et al. (1997), Caracoglia and Jones (2007), Zhou (2008)] dampers are the most widely studied and tested components of an SA suspension system. MR/ER fluids are materials that respond to an applied magnetic/electrical field with a change in rheological behavior. A dynamical model for an SA suspension system will be discussed as a control plant, in which the damping coefficient is adjustable.

The variable structure control (VSC) with sliding mode was introduced in the early 1950s by Emelyanov and was published in the 1960s [Emelyanov (1967)]; further work has been performed by several researchers [Itkis (1976), Utkin (1978), Slotine and Li (1991), Hung, et al. (1993)]. An SA suspension system is one of the widely used VSC systems because the structural parameters continue to vary during a motion process. Sliding mode control (SMC) has been recognized as a robust and efficient control method for complex, high-order, nonlinear dynamical systems and also has been applied to MR/ER damper control for SA suspension systems. The major advantage of sliding mode control is the low sensitivity to the changes in a system's parameters under various uncertainty conditions and such control can decouple system motion into independent partial components of lower dimension, thus reducing the complexity of the system control and feedback design. The major drawback of traditional SMC is chattering, which is the high-frequency oscillations of the system outputs irritated by the discontinuous control switchings across sliding surfaces.

For addressing the chattering phenomenon, one of the best-known methods is fuzzy logic theory. Fuzzy logic theory was first proposed by Zadeh in 1965 [Zadeh (1965a)] and is based on the concept of fuzzy sets. Fuzzy logic control (FLC) has been used in a wide variety of applications in engineering, such as in aircraft/spacecraft, automated highway systems, autonomous vehicles, washing machines, process control, robotics control, decision-support systems and portfolio selection. Practically speaking, it is not always possible to obtain a precise mathematical model for nonlinear, complex or ill-defined system. FLC is a practical alternative for a variety of challenging control applications because it can provide a convenient method for constructing nonlinear controllers via the use of heuristic information (or knowledge). The heuristic information may originate from an operator that acts as a 'human-in-the-loop' controller and from whom experimental data are obtained. In recent years, substantial research has been generated in the area of fuzzy sliding mode control (FSMC) [Ishigame, et al. (1991), Ishigame, et al. (1993), O'Dell (1997)], which concerns the chattering phenomenon of traditional SMC design. The involvement of FLC in the design of an FSMC-based controller can be harnessed to help reduce the chattering problem. The smooth control feature of fuzzy logic can help in overcoming the disadvantages of chattering, which is why it can be useful to combine the FLC method with the SMC method and thus to create the FSMC method. The involvement of FLC in the design of the FSMC-based controller can be harnessed to help avoid the chattering problem. A FSMC with skyhook surface scheme will be discussed, based on which an improved control method supervised using a polynomial function will be proposed.

The skyhook control strategy was introduced in 1974 by Karnopp et al. [Karnopp, et al. (1974)]. It has been applied to reduce vertical vibration of SA suspension systems because of its mathematical simplicity. The basic idea is to link the vehicle body spring mass to the stationary sky by a controllable 'skyhook' damper, which can reduce the vertical vibrations caused by road disturbances of all types. In practice, no such stationary sky can be found as a mathematical assumption or in an engineering product, but this concept can be borrowed to design the sliding surface of an SMC with the assistance of FLC. This is one of the motivations for the new control method that we will discuss in the sections below. The controller's parameter selection and optimization are another problem that we need to solve considering the objectives of passenger ride comfort, suspension deformation and tyre loads. The optimizer that we will use in this context is a GA.

In this chapter, the micro-GA method will be utilized as the optimizer for the parameters of the newly proposed polynomial function supervising FSMC, which will then be applied to ride comfort control for an SA suspension system.

19.2 Two-Degree-of-Freedom Semi-Active Suspension System

The role of the vehicle suspension system is to support and isolate the vehicle body and payload from road disturbances and maintain the traction force between the tyres and the road surface. An SA suspension system can provide both reliability and versatility, such as in terms of a passenger's ride comfort, with decreased power demands. To achieve a basic understanding of the passenger's response to the vehicle's vibrational behavior, as given in Figure 19.3, a two-degree-of-freedom (2-DOF) vehicle ride model that focuses on the passenger's ride comfort performance is applied to an SA suspension system. Here, m_1 and m_2 are the unsprung mass and the sprung mass, respectively; k_1 is the tyre stiffness coefficient; k_2 and c_2 are the suspension stiffness coefficient and the suspension damping coefficient, respectively; c_e is the SA suspension damping coefficient, which can generate the SA damping force f_d through the MR/ER absorber, as given by Equation (19.2); z_1, z_2 and q are the displacements of the unsprung mass, the sprung mass and the road disturbance, respectively; v_0 is the vehicle speed, which is one of the input parameters for the road disturbance q; and g is the acceleration of gravity.

$$\begin{cases} m_1\ddot{z}_1 + k_2(z_2 - z_1) + (c_2 + c_e)(\dot{z}_2 - \dot{z}_1) - k_1(z_1 - q) + m_1g = 0 \\ m_2\ddot{z}_2 - k_2(z_2 - z_1) - (c_2 + c_e)(\dot{z}_2 - \dot{z}_1) + m_2g = 0 \end{cases} \tag{19.1}$$

$$f_d = c_e(\dot{z}_2 - \dot{z}_1) \tag{19.2}$$

Using Newton's second law, the 2-DOF SA suspension model can be given by Equation (19.1), where f_d is the damping force, as stated by Equation (19.2).

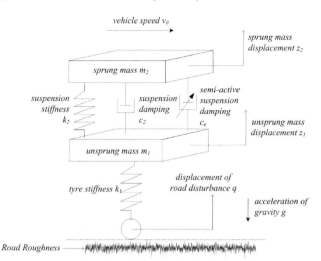

Fig. 19.3 2-DOF SA suspension system

$$\begin{cases} \dot{X} = AX + BQ + EU \\ Y = CX + DQ + FU \end{cases} \tag{19.3}$$

$$X = \begin{Bmatrix} x_1 \\ x_2 \\ x_3 \\ x_4 \end{Bmatrix} = \begin{Bmatrix} z_1 - q \\ z_2 - z_1 \\ \dot{z}_1 \\ \dot{z}_2 \end{Bmatrix} \tag{19.4}$$

$$Y = \begin{Bmatrix} y_1 \\ y_2 \\ y_3 \end{Bmatrix} = \begin{Bmatrix} \ddot{z}_2 \\ z_1 - q \\ z_2 - z_1 \end{Bmatrix} \tag{19.5}$$

$$U = \left\{ f_d \right\} \tag{19.6}$$

$$f_{tyre} = k_1 \left(z_1 - q \right) \tag{19.7}$$

$$Q = \begin{Bmatrix} \dot{q} \\ g \\ 0 \end{Bmatrix} \tag{19.8}$$

$$A = \begin{bmatrix} 0 & 0 & 1 & 0 \\ 0 & 0 & -1 & 1 \\ \dfrac{k_1}{m_1} & -\dfrac{k_2}{m_1} & \dfrac{c_2}{m_1} & -\dfrac{c_2}{m_1} \\ 0 & \dfrac{k_2}{m_2} & -\dfrac{c_2}{m_2} & \dfrac{c_2}{m_2} \end{bmatrix} \tag{19.9}$$

$$B = \begin{bmatrix} -1 & 0 & 0 \\ 0 & 0 & 0 \\ 0 & -1 & -\dfrac{1}{m_1} \\ 0 & -1 & \dfrac{1}{m_2} \end{bmatrix} \tag{19.10}$$

$$C = \begin{bmatrix} 0 & \dfrac{k_2}{m_2} & -\dfrac{c_0}{m_2} & \dfrac{c_0}{m_2} \\ 1 & 0 & 0 & 0 \\ 0 & 1 & 0 & 0 \end{bmatrix} \tag{19.11}$$

$$D = \begin{bmatrix} 0 & -1 & \dfrac{1}{m_2} \\ 0 & 0 & 0 \\ 0 & 0 & 0 \end{bmatrix} \tag{19.12}$$

$$E = \left\{ \begin{array}{c} 0 \\ 0 \\ -\dfrac{1}{m_1} \\ \dfrac{1}{m_2} \end{array} \right\} \tag{19.13}$$

$$F = \left\{ \begin{array}{c} \dfrac{1}{m_2} \\ 0 \\ 0 \end{array} \right\} \tag{19.14}$$

To observe the status of the 2-DOF SA suspension system, Newton's second law, as given by Equation (19.1), can be re-written as state-space equations in Equation (19.3). The state-space analysis concerns three types of variables (input variables, output variables and state variables) [Slotine and Li (1991), Ogata (1996)], as shown by Equations (19.4), (19.5) and (19.6) [Chen (2009)] in the form of vectors (state vectors). Here, X is the state vector for the 2-DOF SA suspension system, which includes the tyre deformation ($x_1 = z_1 - q$), the suspension deformation ($x_2 = z_2 - z_1$), the unsprung mass velocity ($x_3 = \dot{z}_1$) and the sprung mass velocity ($x_4 = \dot{z}_2$), as given by Equation (19.4); Y is the output vector with three state variables for the 2-DOF SA suspension system, which includes the vehicle body acceleration ($y_1 = \ddot{z}_2$), the tyre deformation ($y_2 = z_1 - q$), the suspension deformation ($y_3 = z_2 - z_1$), as given by Equation (19.5) and U is the input vector (control force vector) in Equation (19.6). According to the tyre deformation (y_2), the tyre load can be written as in Equation (19.7); Q is the external road disturbance vector in Equation (19.8), which contains two external disturbance signals of the road velocity profile and the acceleration of gravity; and A, B, C, D, E, and F are the coefficient matrices in Equations (19.9) to (19.14).

In this context, it is convenient to select measurable quantities as state variables using a block diagram because the full-state feedback control law requires the feed-

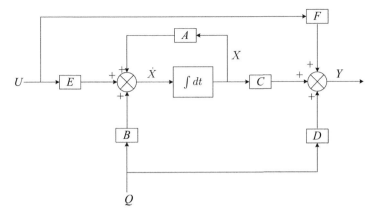

Fig. 19.4 Block diagram for the 2-DOF SA suspension system

back of all selected state variables with suitable weightings. The block diagram for the state-space equations is given in Figure 19.4, which represents the 2-DOF SA suspension system for further controller design.

19.3 Sliding Mode Control with Skyhook Surface Scheme

As shown in Figure 19.5, the skyhook control method is known as one of the most effective methods in terms of the simplicity of the control algorithm. The skyhook control can reduce the resonant peak of the sprung mass quite significantly and thus achieves a good ride quality by adjusting the skyhook damping coefficient when the vehicle body velocity and other conditions are changing.

By applying this concept to reduce the sliding chattering phenomenon, a soft switching control law is introduced for the major sliding surface switching activity in Equation (19.19), as shown in Figure 19.6. This is used to reduce chattering and achieve good switch quality for a sliding mode control with skyhook surface scheme (SkyhookSMC).

As shown in Figure 19.7, when designing a SkyhookSMC, the objective is to consider the 2-DOF suspension system as the control plant, which is defined by the state-space equations, as stated in Equation (19.3). s is the sliding surface of the hyperplane, which is given by Equation (19.15), where λ is a positive constant that defines the slope of the sliding surface. Because the 2-DOF SA suspension system is a second-order system, it can be given $n = 2$, in which s defines the position and velocity errors, and Equation (19.15) can be re-written as Equation (19.16).

$$s(e,t) = \left(\frac{d}{dt} + \lambda\right)^{n-1} e \qquad (19.15)$$

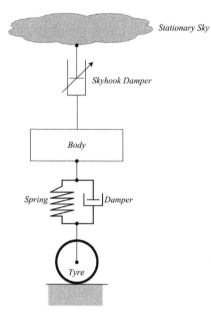

Fig. 19.5 Ideal skyhook damper definition, adopted from [Karnopp, et al. (1974)]

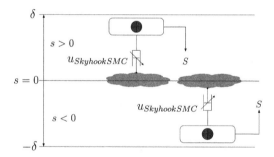

Fig. 19.6 Sliding mode surface with skyhook scheme [Chen (2009), Chen and Cartmell (2009), Chen and Cartmell (2009), Chen and Cartmell (2010), Chen (2010b)]

$$s = \dot{e} + \lambda e \tag{19.16}$$

$$V(s) = \frac{1}{2}s^2 \tag{19.17}$$

$$\dot{V}(s) = s\dot{s} \tag{19.18}$$

According to Equation (19.16), the second-order tracking problem of the 2-DOF SA suspension system is now being represented by a first-order stabilization problem, in which the scalar s is set to zero using a governing condition

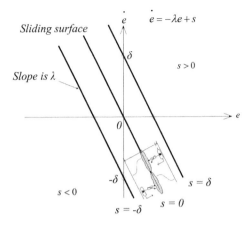

Fig. 19.7 Sliding surface generation with skyhook scheme [Slotine and Li (1991), Chen and Cartmell (2009), Chen and Cartmell (2009), Chen and Cartmell (2010), Chen (2010b), Slotine (1982)]

[Slotine and Li (1991)]. Obtained by using the Lyapunov stability theorem, the Sky-hookSMC is designed such that the origin is a globally asymptotically stable equilibrium point for the control system. The Lyapunov candidate function V and its time derivative function $\dot{V}$ are given by Equations (19.17) and (19.18), respectively. The energy-like Lyapunov candidate function V is positively definite. If the derivative function $\dot{V}$ satisfies the negative-definite condition, the plant is in a stable status; if the derivative function $\dot{V}$ is positive definite, the plant is unstable and the controller thus needs be activated, as defined by Equation (19.19).

$$u_{SkyhookSMC} = \begin{cases} -c_0 \tanh\left(\dfrac{s}{\delta}\right) & s\dot{s} > 0 \\ 0 & s\dot{s} \leq 0 \end{cases} \qquad (19.19)$$

The smooth control time function generated by the SkyhookSMC is expressed by Equation (19.19), where c_0 is an assumed positive damping ratio for the switching control law. The SkyhookSMC law needs to be chosen in such a way that the existence and reach of the sliding mode are both guaranteed. It is noted that δ is an assumed positive constant, which defines the thickness of the sliding mode boundary layer [Slotine (1982)].

19.4 Fuzzy Logic Control

Generally, in the FLC design methodology, the human operator needs to write down a set of rules on how to control the process. This is called the 'rule base'. Then, a fuzzy controller can emulate the decision-making process of the human, using the rule base, in which the heuristic information (knowledge) may come from a control engineer who has performed extensive mathematical modeling, analysis and development of control algorithms for a particular process. Again, such expertise is loaded into the fuzzy controller to automate the reasoning processes and actions of the expert. Regardless of where the heuristic control knowledge comes from, fuzzy control provides a user-friendly formalism for representing and implementing ideas that can help to achieve high-performance control.

As shown in Figure 19.8, the fuzzy controller has four main components—The 'rule base' (a set of 'IF-THEN' rules) contains a fuzzy logic quantification of the expert's linguistic description of how to achieve good control. The 'inference mechanism' emulates the expert's decision making in interpreting and applying knowledge about how efficiently to control the plant. A set of 'IF-THEN' rules are loaded into the rule base, and an inference strategy is chosen. Then, the system is ready to be tested, and closed-loop specifications are needed. The 'fuzzification' interface converts 'crisp' inputs into 'fuzzy' information that the inference mechanism can interpret and compare to the rules in the rule base. The 'defuzzification' interface converts the conclusions by the inference mechanism into the FLC crisp (actual) outputs as the control inputs for the plant.

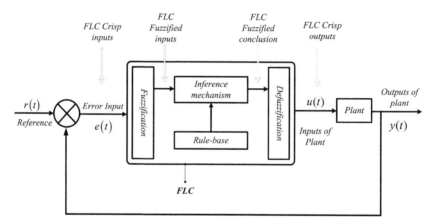

Fig. 19.8 Fuzzy logic controller architecture [Chen (2010b)]

Briefly, fuzzy control systems can be designed in the following steps: $\langle 1 \rangle$ choosing the fuzzy controller inputs and outputs, $\langle 2 \rangle$ choosing the preprocessing that is needed for the controller inputs and possibly the postprocessing that is needed for the outputs, and $\langle 3 \rangle$ designing each of the four components of the fuzzy controller: fuzzification, the inference mechanism, the rule base and defuzzification, as shown in Figure 19.8.

Fuzzification is the process of decomposing the system inputs into fuzzy sets. Specifically, this process maps variables from the crisp space to the fuzzy space. The process of fuzzification allows the system inputs and outputs to be expressed in linguistic terms so that the rules can be applied in a simple manner to express a complex system. In the FLC for the 2-DOF SA suspension system, the velocity and acceleration of the vehicle body are selected as the crisp error (e) and the crisp change-in-error (ec) feedback signals for the 2-DOF SA suspension system control. There are 7 linguistic terms in the fuzzy sets for two inputs of the fuzzified error (E) and the fuzzified change-in-error (EC) as well as one output of the fuzzified force (U): $\langle$ NL, NM, NS, ZE, PS, PM, PL $\rangle$, as stated in Table 19.1. In addition, their linguistic values are also included for further numerical simulations with proper ranges of [-5,5] and [-2,2]. Defuzzification is the opposite process of fuzzification and maps variables from the fuzzy space to the crisp space.

Table 19.1 Fuzzy Linguistic Values

Fuzzy Linguistic Value	Description	E	EC	U
NL	Negative Large	-5	-5	-2
NM	Negative Middle	-4	-4	-1.5
NS	Negative Small	-3	-3	-1
ZE	Zero	0	0	0
PS	Positive Small	3	3	1
PM	Positive Middle	4	4	1.5
PL	Positive Large	5	5	2

A membership function (MF) is a concept that defines how each point in the input space is mapped to a membership value between 0 and 1. The MF for the 2-DOF SA suspension system is a triangular-shaped membership function. The inputs of E and EC are interpreted from this fuzzy set and the degree of membership is interpreted. The structure of the FLC for the 2-DOF SA suspension system is a standard 2-in-1-out FLC, and the 'If-Then' rule base is then applied to describe the experts' knowledge. The FLC rule base is characterized by a set of linguistic description rules based on conceptual expertise that arises from typical human situational experience. In particular, the 2-in-1-out FLC rule base for the ride comfort of the 2-DOF SA suspension system is given in Table 19.2 [Chen (2009)]. Given two inputs and seven linguistic values for each input, there are at most $7^2 = 49$ possible rules, as given by the following list:

$\langle 1 \rangle$ IF **E** = *NL*, AND **EC** = *NL*, THEN **U** = *PL*;
$\langle 2 \rangle$ IF **E** = *NL*, AND **EC** = *NM*, THEN **U** = *PL*;
$\langle 3 \rangle$ IF **E** = *NL*, AND **EC** = *NS*, THEN **U** = *PM*;

$\vdots$

$\langle 49 \rangle$ IF **E** = *PL*, AND **EC** = *PL*, THEN **U** = *NL*;

Table 19.2 defines the relationship between 2 inputs of the fuzzified error (E) and the fuzzified change-in-error (EC) with 1 output of the fuzzified control force (U), which originated from the previous experience gained for the SA damping force control during body acceleration changes for ride comfort. Briefly, the main linguistic control rules are the following: $\langle 1 \rangle$ when the body acceleration and velocity

Table 19.2 The 2-in-1-out FLC Rule Table for the 2-DOF SA Suspension System [Chen (2009)]

U		EC						
		NL	NM	NS	ZE	PS	PM	PL
	NL	PL	PL	PM	PS	PS	PS	ZE
	NM	PL	PM	PS	PS	PS	ZE	NS
E	NS	PM	PS	ZE	ZE	ZE	NS	NM
	ZE	PM	PS	ZE	ZE	ZE	NS	NM
	PS	PM	PS	ZE	ZE	ZE	NS	NM
	PM	PS	ZE	ZE	ZE	NS	NM	NL
	PL	ZE	NS	NS	NS	NM	NL	NL

increase, the SA damping force decreases, and $\langle 2 \rangle$ when the body acceleration and velocity decrease, the SA damping force increases.

Fuzzy inference is the process of formulating the mapping from a given input to an output using fuzzy logic. The mapping then provides a basis from which decisions can be made or patterns discerned. The process of fuzzy inference involves all the pieces that are described in the previous sections—Membership Functions, Logical Operations, and 'IF-THEN' Rules. Mamdani's fuzzy inference method [Mamdani (1975), Mamdani (1976), Mamdani (1977)], which was proposed in 1975 by Mamdani, as an attempt to control a steam engine and boiler combination by synthesizing a set of linguistic control rules obtained from experienced human operators, is the most commonly utilized fuzzy methodology and was among the first control systems built using fuzzy set theory. Mamdani's effort was based on Zadeh's research [Zadeh (1973a)] on fuzzy algorithms for complex systems and decision processes in 1973. The Fuzzy Inference System (FIS) of Mamdani-type inference for the 2-in-1-out FLC is shown in Figure 19.9.

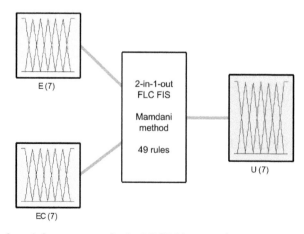

Fig. 19.9 The fuzzy inference system for the 2-DOF SA suspension system

19.5 Fuzzy Sliding Mode Control with Switching Factor α-FαSMC

An FSMC with switching factor α (FαSMC) [Chen and Cartmell (2009), Chen and Cartmell (2009)] was introduced to combine FLC with the SkyhookSMC to address the chattering phenomenon and has been applied to reduce the 2-DOF SA suspension system ride comfort control with proper parameter selection. A flow diagram for the FαSMC, applying the SkyhookSMC approach, is given in Figure 19.10. The control effects of the FLC and the SkyhookSMC are combined through Equation (19.20). In Equation (19.20), α is a switching factor that balances the weight of the FLC to that of the SkyhookSMC. Clearly, α = 0 represents SkyhookSMC, and α = 1 represents FLC, α ∈ [0,1]; the SkyhookSMC was discussed in Section 19.3, and FLC was discussed in Section 19.4.

$$u_{F\alpha SMC} = \alpha u_{FLC} + (1 - \alpha) u_{SkyhookSMC} \qquad (19.20)$$

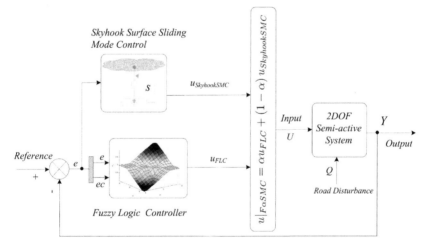

Fig. 19.10 FαSMC control flow diagram

19.6 Polynomial Function Supervising FαSMC—An Improvement

To make the necessary improvement to the FαSMC method, a hybrid real-time polynomial function [Chen, et al. (2004)] supervising FαSMC with skyhook surface (PSFαSMC) is proposed and will be applied to ride comfort control for the 2-DOF SA vehicle suspension system. The basic idea of PSFαSMC is to generate a series of supervising functions (SF) in polynomial function (PF) form that are optimized and produced by a micro-GA in an offline step [Chen, et al. (2004)] through a training process for the PSFαSMC parameter selection. The SFs are used to generate parameters for FαSMC in an online real-time control step.

Briefly, there are two steps in PSFαSMC controller design, the offline step and the online step, where $\langle 1 \rangle$ the offline step takes the micro-GA as the optimizer to generate PFs for each parameter in FαSMC, including K_e, K_{ec}, K_u, α, c_0, δ, and λ. In the micro-GA optimization process, each loop will take more time than practical real-time control allows and that is why the PFs can be taken as the practical real-time control parameter generators for the online step. $\langle 2 \rangle$ The online step generates proper parameters using the PFs, which come from the offline step. In the online step, the polynomial functions are the real-time parameter generators, which supervise the FαSMC controller by adjusting its parameters.

The parameter selection for FαSMC needs substantial manual testing and is time consuming. To reduce the working time of parameter selection for the hybrid control method PSFαSMC, the micro-GA is to be applied as the optimizer to generate proper results for parameter selection, which has been widely applied to industrial applications [Coello and Pulido (2001), Lo and Khan(2004), Tam, et al. (2006), Davidyuk, et al. (2007), Andr, et al. (2009)].

19.6.1 *Multi-objective Micro-GA for the Offline Step*

The population size in the GA usually ranges from tens to hundreds and sometimes thousands. Such a large number of individuals can lead to formidable calculation times. It is important to design highly efficient GAs for MO optimization problems. One of the popular methods in this direction is the micro-GA, with a very small internal population (3→6 individuals) [Coello and Pulido (2001)]. As shown in Figure 19.11, generally, there are two loops in the MO micro-GA (MOμGA) process: an internal cycle and an external cycle. In addition, there are two groups of population memories—the internal population memory, which is used as the source of diversity of the micro-GA internal loop and the external population memory, which is used to archive individuals of the Pareto-optimal set. In the internal loop, the internal population memory is divided into two parts—a replaceable part and a non-replaceable part. The percentages corresponding to each part can be adjusted by the user, as discussed in Section 2.4.

For the small internal population, mutation is an optional operator for a micro-GA. In practice, there are three parts to micro-GA optimization: $\langle 1 \rangle$ fitness function definition, $\langle 2 \rangle$ encoding and decoding definition, and $\langle 3 \rangle$ genetic operator definition, including selection, crossover and mutation. Once the three parts have been well defined, the micro-GA can then create a population of solutions and apply genetic operators, such as mutation and crossover, to evolve the solutions to achieve better results.

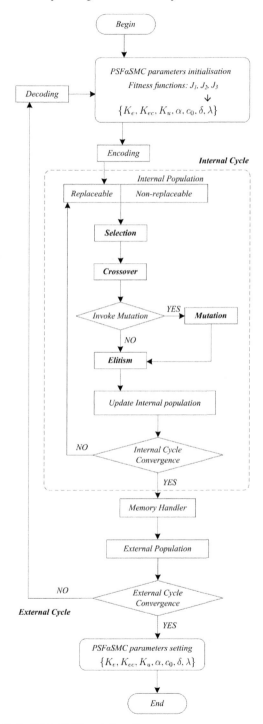

Fig. 19.11 Micro-GA for PSFαSMC work flow diagram

As defined by Equation (19.5), the output matrix Y contains three state variables for the 2-DOF SA suspension system, which are related to the ride comfort performance—vehicle body acceleration (y_1), tyre deformation (y_2) and suspension deformation (y_3). $H(Y)$ is the error state function, as defined by Equation (19.21), which can generate the error state variables (e_1, e_2 and e_3) for three output state variables (y_1, y_2 and y_3). $y_1|_{ref}$, $y_2|_{ref}$ and $y_3|_{ref}$ are the reference state variables for the PSFαSMC.

$$H(Y) == \left\{ \begin{array}{c} e_1 \\ e_2 \\ e_3 \end{array} \right\} = \left\{ \begin{array}{c} y_1 - y_1|_{ref} \\ y_2 - y_2|_{ref} \\ y_3 - y_3|_{ref} \end{array} \right\} = \left\{ \begin{array}{c} \ddot{z}_2 - y_1|_{ref} \\ z_1 - q - y_2|_{ref} \\ z_2 - y_3|_{ref} \end{array} \right\} \tag{19.21}$$

As shown in Figure 19.12, in the PSFαSMC offline step, the micro-GA optimizes and generates PFs for each parameter (K_e, K_{ec}, K_u, α, c_0, δ, and λ), and there are three fitness values for three objectives as J_i, as stated in Equation (19.22). $F(*)$ is the fitness function and $RMS[*]$ is the function for the root mean square values.

$$J_i = F(e_i) = MIN\{RMS[ITAE(e_i)]\}, i = 1, 2, 3 \tag{19.22}$$

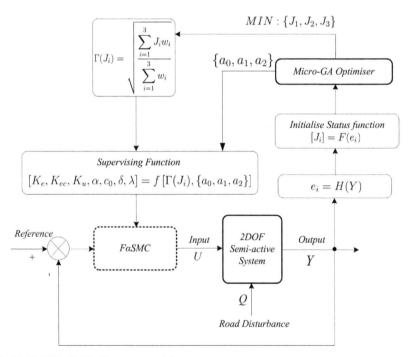

Fig. 19.12 PSFαSMC offline step - training

According to International Standard Organization (ISO) 2631 [International Organization for Standardization (1997)], ride comfort is specified in terms of *RMS* acceleration over a frequency range. The fitness functions (J_i) for the 2-DOF SA suspension system are functions of the error state variables (e_i), which are the values of the root mean square ($RMS[*]$) of the integral over time multiplied by the absolute error ($ITAE$) performance indexes, that is, J_1, J_2 and J_3 are the *RMS* of the *ITAE* indexes for the following error state variables—body acceleration (e_1), suspension deformation (e_2) and tyre loads (e_3), respectively. The *ITAE* of the error state variables (e_i) is expressed by Equation (19.23).

$$ITAE\,(e_i) = \int_0^\infty t\,|e_i(t)|\,dt \qquad (19.23)$$

As stated by Equation (19.22), the micro-GA's optimizing criterion is to minimize the fitness functions (J_i) and to generate a set of solutions to achieve improved supervising function parameters (K_e, K_{ec}, K_u, α, c_0, δ, and λ) via the interval arguments a_1, a_2, a_3 and Γ, which will be discussed in Section 19.6.2. Then, the micro-GA can provide the optimality of a set of solutions for the MO applications of ride comfort control in the online step and engineers can attempt each of the solutions or select a solution based on proper policy, e.g., the outer range, the inner range or the average of the Pareto set, as an engineering solution. In the optimization process using the micro-GA, binary encoding/decoding, roulette-wheel selection and single-point crossover are applied in the micro-GA evolutional process. In the micro-GA optimization process, the initial conditions need to be given for the seven design variables, as shown in Table 24.2.

19.6.2 Offline Step

As shown in Figure 19.12, the offline step is a training process that optimizes and generates a series of PFs for further use in the online step, and the micro-GA is the optimizer, as discussed in Section 2.4.

$$f\,[\Gamma(J_i),\{a_0,a_1,...,a_N\}] = a_N\Gamma(J_i)^N + a_{N-1}\Gamma(J_i)^{N-1} + ... + a_2\Gamma(J_i)^2 + a_1\Gamma(J_i) + a_0 \qquad (19.24)$$

As stated in Equation (19.24), $f\,[\Gamma(J_i),\{a_0,a_1,...,a_N\}]$ is the supervising function in polynomial form, which is fitted by the least square principle based on output data from the micro-GA optimization block, where N is a positive integer and a_0, a_1, ..., a_N are constant coefficients.

$$\Gamma(J_i) = \sqrt{\frac{\sum\limits_{i=1}^{3} J_i w_i}{\sum\limits_{i=1}^{3} w_i}} \tag{19.25}$$

$\Gamma(J_i)$ is a component of the polynomial supervising function $f[\Gamma(J_i),\{a_0,a_1,...,a_N\}]$, as defined by Equation (19.25), which is an index weighted by the optimized 2-DOF SA suspension ride comfort indexes J_i.

$$f[\Gamma(J_i),\{a_0,a_1,a_2\}] = a_2 J_i^2 + a_1 J_i + a_0 \tag{19.26}$$

Basically, a smooth supervising function curve is required in the SA suspension system ride comfort control, and $N = 2$ is chosen as the highest degree of the supervising PFs. The PF with $N = 2$ can provide acceptably accurate outputs with relatively low central processing unit (CPU) time consumptions, that is, $N = 2$ PF provides a relatively high accuracy *vs.* time-consumption ratio in engineering applications [Chen, et al. (2004)]. Equation (19.26) is for generating the parameter K_e and similar processes will be applied to other parameters. There are seven supervising functions for the seven PSFαSMC parameters (K_e, K_{ec}, K_u, α, c_0, δ, and λ), as given by Equation (19.27). As shown in Figure 19.12, the supervising functions will be applied to the FαSMC block.

$$\begin{cases} K_e &= f[\Gamma(J_i),\{a_0,a_1,a_2\}]|_{K_e} \\ K_{ec} &= f[\Gamma(J_i),\{a_0,a_1,a_2\}]|_{K_{ec}} \\ K_u &= f[\Gamma(J_i),\{a_0,a_1,a_2\}]|_{K_u} \\ \alpha &= f[\Gamma(J_i),\{a_0,a_1,a_2\}]|_{\alpha} \\ c_0 &= f[\Gamma(J_i),\{a_0,a_1,a_2\}]|_{c_0} \\ \delta &= f[\Gamma(J_i),\{a_0,a_1,a_2\}]|_{\delta} \\ \lambda &= f[\Gamma(J_i),\{a_0,a_1,a_2\}]|_{\lambda} \end{cases} \tag{19.27}$$

19.6.3 Online Step

As shown in Figure 19.13, the online step applies the PFs to supervise the FαSMC for the 2-DOF SA suspension system. There are seven design variables (K_e, K_{ec}, K_u, c_0, δ, λ, and α) in PSFαSMC in the online step, which are given by Equation (19.27). As stated in Equation (19.27), the supervising functions are functions of $\Gamma(J_i)$, a_0, a_1, and a_2, which can be used to generate the parameters for the FαSMC during real-time simulation. In Figure 19.13, the PFs are the optimized data source for the FαSMC parameters (K_e, K_{ec}, K_u, α, c_0, δ, and λ), which have been produced in the offline step and include the following: Ke are the FLC scaling gains for e; Kec are the FLC scaling gains for ec; Ku are the FLC scaling gains for u; c_0 is the SkyhookSMC damping coefficient; δ is the thickness of the sliding mode boundary layer; λ is the slope of the sliding surface; and α is the switching factor of FαSMC in PSFαSMC.

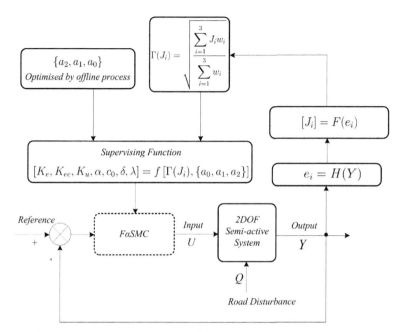

Fig. 19.13 PSFαSMC online step—control for SA suspension system

19.7 Road Surface Profile—Modeling of the Source of Uncertainty

To simulate road excitations for the vehicle suspension system, a road profile is defined as the cross-sectional shape of a road surface under the given conditions, which can be expressed by statistical procedures [Sayers (1996), Wei, et al. (2005)]. There are a few types of excitations for the road surface profile and include sine waves, step functions and triangular waves, which can provide a basis for comparative studies under some simple road surfaces. However, these excitations can hardly serve as a valid and general basis for a practical road roughness analysis of ride behavior. As shown in Figure 19.13, it is more realistic to describe a road surface profile as a random function or data sequence—the road roughness. Various methods, such as the International Roughness Index values and the Fourier transform-based sequence, described in ISO 8608:1995 'Mechanical Vibration-Road Surface Profiles-Reporting of Measured Data' [International Organization for Standardization (1995)], have been developed; however, both methods only give an average condition for a relatively long section of the pavement.

A widely used statistical method of generating road excitations is to describe the road roughness using the power spectral density (PSD). When the road surface profile is regarded as a random function, it can be characterized by a PSD function [Wong (2001)]. To classify the roughness (irregularities) of road surfaces, the ISO has proposed a road roughness classification, roughness-A (very good) to roughness-H (very poor), based on the PSD, in which the relationships between the PSD function $S_g(\Omega)$ and the spatial frequency Ω for different classes of road roughness can be approximated by two straight lines with different slopes on a $log - log$ scale. This can be expressed as Equation (19.28) [International Organization for Standardization (1982)], and the values of N_1 and N_2 are 2.0 and 1.5, respectively. In this case, to generate the road profile of a random base excitation for the 2-DOF SA suspension simulation, a spectrum of a geometrical road profile with road class 'roughness-C' is considered, and Ω_0 is the reference spatial frequency. The vehicle is traveling at a constant speed v_0, and the historical road irregularity is given by the PSD method [Elbeheiry and Karnopp (1996), Ramji, et al. (2004), Liu, et al. (2008)].

$$
\begin{cases}
S_g(\Omega) = S_g(\Omega_0) \left(\dfrac{\Omega}{\Omega_0} \right)^{-N_1}, \Omega \leq \Omega_0 = \dfrac{1}{2\pi} cycles/m \\[4mm]
S_g(\Omega) = S_g(\Omega_0) \left(\dfrac{\Omega}{\Omega_0} \right)^{-N_2}, \Omega > \Omega_0 = \dfrac{1}{2\pi} cycles/m
\end{cases}
\tag{19.28}
$$

19.8 Uncertainty Studies

As shown in Figure 19.14 [Rocquigny, et al. (2008)], an uncertainty analysis with a feedback loop is proposed corresponding to the PSFαSMC online step control process, as discussed in Section 19.5. The goal of the 2-DOF SA suspension system modeling is primarily to demonstrate compliance with uncertainty criteria embodied by the target values in the guidelines (Goal-U, A, S and C), as given in Table 19.3 [Rocquigny, et al. (2008)]. According to Figure 19.14 and Table 19.3, a metrological chain is compared to investigate which model best complies with the proper criterion guidelines (Goal-A, S and C). Through the establishment of industrial emission practices, it also appears that understanding the importance of the various sources of uncertainty will become more essential in improving the metrological options in the long term (Goal-U).

Primarily, two uncertainty variables of interest are considered, as listed in Figure 19.14—2-DOF SA suspension vehicle body acceleration Y and the sliding surface s. Based on the International Standard in Metrological Uncertainty (GUM) [International Organization for Standardization (2008)], the target values are specified by the uncertainty decision criteria—a quantity of interest representing the relative uncertainty is compared to a maximal percentage of relative uncertainty.

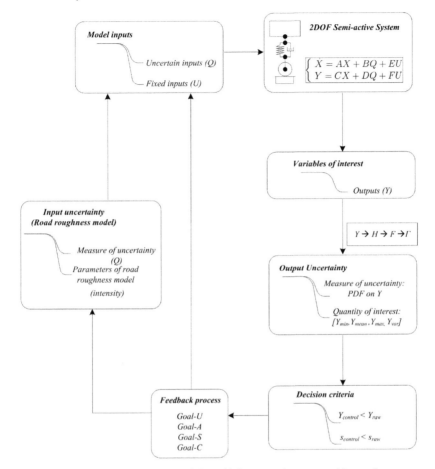

Fig. 19.14 Uncertainty analysis framework for vehicle suspension system ride comfort

Table 19.3 The Goals of Quantitative Uncertainty Assessment [Rocquigny, et al. (2008)]

Type	Goal
Goal-U (Understand)	To understand the importance of uncertainty and to establish measurements and models.
Goal-A (Accredit)	To give credit to a method of measurement and to reach an acceptable quality level for its use. This may involve calibrating sensors and estimating the parameters of the model inputs.
Goal-S (Select)	To compare relative performance and optimize the choice of objective policy, operation or design of the system.
Goal-C (Comply)	To demonstrate compliance of the system with an explicit criterion.

19.9 Simulations

All the results for the ride comfort are obtained using the parameters for the 2-DOF SA vehicle suspension system and PSFαSMC in Table 19.4. The numerical results are obtained using a specially devised simulation toolkit for the Micro-GA for *MATLAB*®, known henceforth as *SGALAB* [SGALAB (2009)]. Unless stated otherwise, all the results are generated using the parameters of the GAs as listed in Table 19.5, in which binary encoding/decoding, tournament selection, single-point crossover and mutation are utilized by the Micro-GA evolutionary process, and a set of PF coefficients a_i for the Micro-GA are given in Table 19.6.

Table 19.4 2-DOF SA Vehicle Suspension System Parameters

m_1	Unsprung mass	$36\,kg$
m_2	Sprung mass	$240\,kg$
c_2	Suspension damping coefficient	$1400\,Ns/m$
k_1	Tyre stiffness coefficient	$160000\,N/m$
k_2	Suspension stiffness coefficient	$16000\,N/m$
g	Gravity acceleration	$9.81\,m/s^2$
Ω_0	Reference spatial frequency	$0.1\,m^{-1}$
$S_g(\Omega_0)$	Degree of roughness	$128 \times 10^{-6}\,m^2/cycles/m$
v_0	Vehicle speed	$72\,km/h$
w_1	Body acceleration weight factor	0.9
w_2	Suspension deformation weight factor	0.05
w_3	Tyre load weight factor	0.05

Table 19.5 Micro-GA Parameters

External cycle	100
Internal cycle	4
External population	50
Internal population	6
Replaceable population	2
Crossover probability	0.9
a_0 Initial range	$[-100,100]$
a_1 Initial range	$[-100,100]$
a_2 Initial range	$[-100,100]$

Table 19.6 A Set of PF Coefficients for a_i by MOμGA

$\{a_2,a_1,a_0\}\|_{K_e}$	a_i Coefficients of K_e	$\{-3.3,2.11,0.31\}$
$\{a_2,a_1,a_0\}\|_{K_{ec}}$	a_i Coefficients of K_{ec}	$\{0.08,-0.19,-10.12\}$
$\{a_2,a_1,a_0\}\|_{K_u}$	a_i Coefficients of K_u	$\{5.34,0.61,15.42\}$
$\{a_2,a_1,a_0\}\|_\alpha$	a_i Coefficients of α	$\{-0.09,-0.22,0.92\}$
$\{a_2,a_1,a_0\}\|_{c_0}$	a_i Coefficients of c_0	$\{0.04,1.56,4999.04\}$
$\{a_2,a_1,a_0\}\|_\delta$	a_i Coefficients of δ	$\{4.34,1.86,25.15\}$
$\{a_2,a_1,a_0\}\|_\lambda$	a_i Coefficients of λ	$\{0.46,0.26,9.64\}$

As discussed in Section 19.2, there are three performance indexes for the vehicle suspension system—body acceleration y_1, tyre deformation y_2 and suspension deformation y_3. In this context, the results for the three indices are applied to evaluate the performance in terms of ride comfort of the 2-DOF SA vehicle suspension system.

The PSFαSMC parameters require a judicious choice as follows: $\langle 1 \rangle$ The FLC scaling gains of Ke and Kec are used for fuzzification of e and ec, where Ku is the defuzzification gain factor. $\langle 2 \rangle$ The SkyhookSMC damping coefficient c_0, as stated in Equation (19.19), is required to expand the normalized controller output force into a practical range. The thickness of the sliding mode boundary layer is given by δ and the slope of the sliding surface is given by λ. Data for both δ and λ come from the design step in the Micro-GA in the offline step. $\langle 3 \rangle$ In the PSFαSMC, α is required to balance the control weight between the FLC and SkyhookSMC. It is easy to switch the controller between the SkyhookSMC and FLC with a proper value of α. $\langle 4 \rangle$ The coefficients $\{a_2, a_1, a_0\}$ for the polynomial supervising functions, which are fitted by the least mean squares (LMS) algorithm based on data from the offline step in the Micro-GA optimization, are given in Table 19.6.

The PFs for K_e, K_{ec}, K_u, α, δ, λ and c_0 are shown in Figures 19.15 to 19.21 with a set of coefficients $\{a_2, a_1, a_0\}$, which are listed in Table 19.6. The coefficients $\{a_2, a_1, a_0\}$ for the PFs are optimized by the Micro-GA in the offline step and then, they are applied to the supervising functions for ride comfort control in the online step. As shown in Figures 19.15 to 19.21, the coefficients $\{a_2, a_1, a_0\}$ can directly affect the shapes of the PFs.

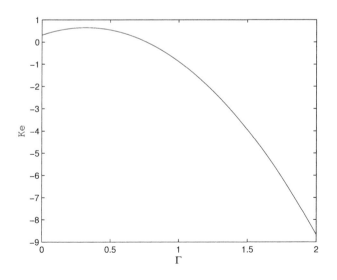

Fig. 19.15 Polynomial supervising functions of PSFαSMC parameters: K_e

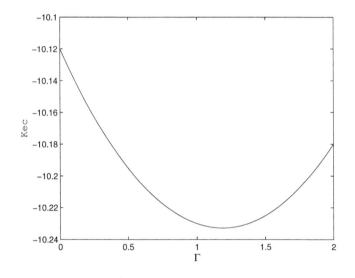

Fig. 19.16 Polynomial supervising functions of PSFαSMC parameters: K_{ec}

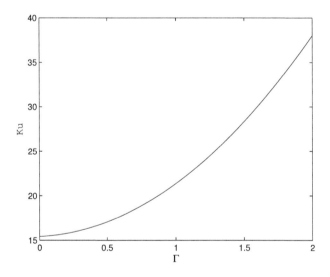

Fig. 19.17 Polynomial supervising functions of PSFαSMC parameters: K_u

With the initial conditions for the Micro-GA listed in Table 19.6, the evolutionary process for each fitness function is listed in Figures 19.22 to 19.25, in which Figures 19.22, 19.23 and 19.24 are the Pareto dataset for J_1 vs. J_2, J_1 vs. J_3 and J_2 vs. J_3. Figure 19.25 is a 3-D surface for the relationships among J_1, J_2 and J_3, in which a position with greater scattered data density represents the Pareto optimal of the

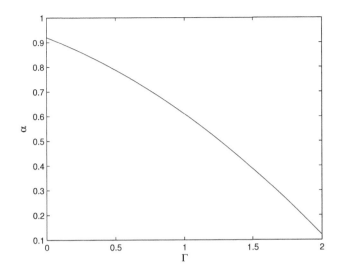

Fig. 19.18 Polynomial supervising functions of PSFαSMC parameters: α

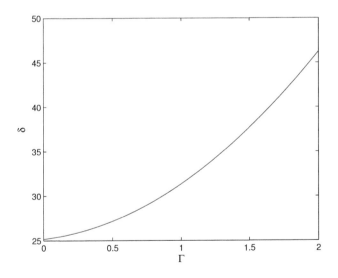

Fig. 19.19 Polynomial supervising functions of PSFαSMC parameters: δ

'trade-off' solutions. A set of PF coefficients a_i for one of the selected Pareto fronts is given in Table 19.6.

Figure 19.26 gives the suspension vertical behavior of the body accelerations. PSFαSMC presents a better control effect than does FLC and SkyhookSMC on the vehicle body acceleration, and both the PSFαSMC and SkyhookSMC methods can

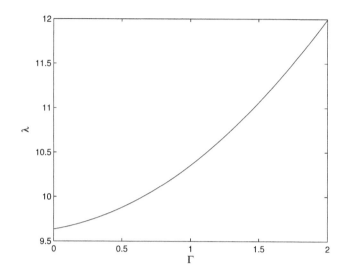

Fig. 19.20 Polynomial supervising functions of PSFαSMC parameters: λ

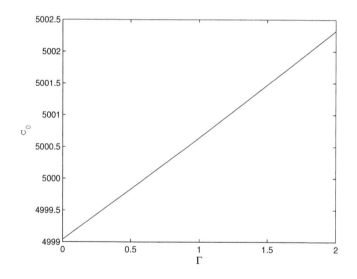

Fig. 19.21 Polynomial supervising functions of PSFαSMC parameters: c_0

provide better ride comfort control effects than can the FLC method on the 2-DOF
SA suspension system.

Figure 19.27 shows the tyre load response that the PSFαSMC and SkyhookSMC
control methods produce under similar tyre load levels, which is smaller (better)
than the FLC control method for the 2-DOF SA suspension system. As shown in

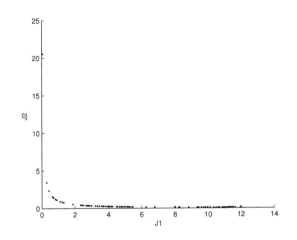

Fig. 19.22 J_1 vs. J_2

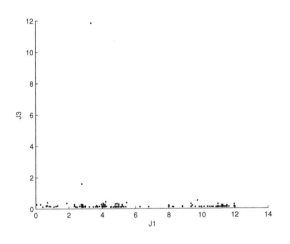

Fig. 19.23 J_1 vs. J_3

Figure 19.27, the tyre load responses start from $0\,N$ @ 0 seconds and increase to approximately $4000\,N$ @ approximately 0.5 seconds and approximately $3000\,N$ @ 1.0 second. The 0 to 1.0 second duration is the transient phase with gravity's effect on the initially loose (uncompressed) SA suspension; then, starting at approximately 1 second, the tyre load responses step into a steady phase until the simulation is terminated. The transient phase (0 to approximately 1.0 second) and the steady phase (1.0 second to the end of the simulation) demonstrate the tyre load responses' dynamical behavior under the initial gravity impact and the road roughness, respectively, which validate the control effects for the SA suspension system with discrete and continuous disturbances.

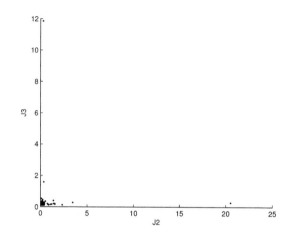

Fig. 19.24 J_2 vs. J_3

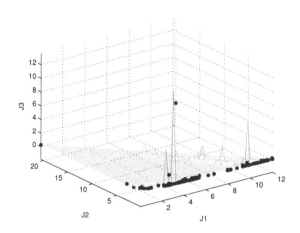

Fig. 19.25 J_1, J_2 and J_3

Figure 19.28 shows the relative displacement (suspension deformation) between the vehicle sprung mass and the vehicle unsprung mass. Compared with passive suspension deformation, the PSFαSMC, SkyhookSMC and FLC control methods can reduce the 2-DOF SA suspension deformation, and the PSFαSMC and SkyhookSMC control methods present similar suspension deformation levels, all of which achieve suspension deformations that are smaller than those of the FLC and passive suspension system. Specifically, PSFαSMC can provide a better ride comfort performance for the 2-DOF SA suspension system.

Figure 19.29 is the body acceleration in the frequency domain, which shows that the control methods under PSFαSMC, SkyhookSMC and FLC can reduce the

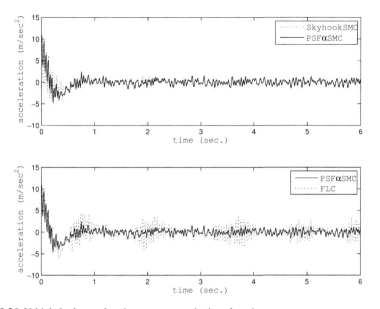

Fig. 19.26 Vehicle body acceleration y_1 response in time domain

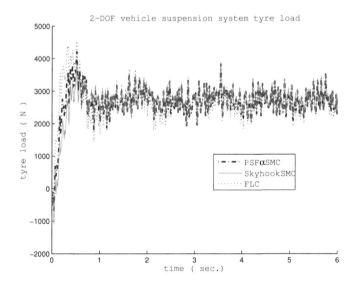

Fig. 19.27 Tyre load response y_2 in time domain

amplitudes at two of the key resonance points (10^0 Hz and 10^1 Hz). The figure also shows that the PSFαSMC can produce better control effects on the 2-DOF SA suspension system ride comfort than can the FLC and SkyhookSMC control

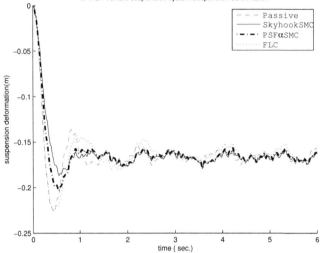

Fig. 19.28 Suspension deformation y_3 response in time domain

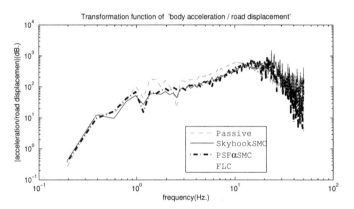

Fig. 19.29 Vehicle body acceleration response in frequency domain

methods for the 2-DOF SA suspension system. In addition, in the higher frequency range (> 10 Hz), PSFαSMC achieves a better performance than the other controllers to some extent.

The phase plot (body velocity vs. body acceleration) is shown in Figure 19.30 as the limit cycles, which represent the improved ride comfort performance of the 2-DOF SA suspension body vertical vibration with controllers. The curves, which corroborated the 2-DOF SA suspension system's interpretations of steady state, started from the initial value point of $(0, g)$ and gathered at the stable area near $(0,0)$ in the

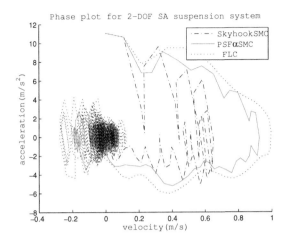

Fig. 19.30 Vehicle body response phase plot

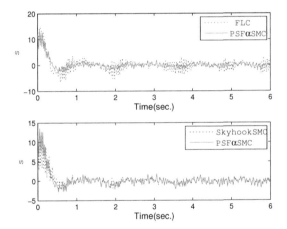

Fig. 19.31 Sliding surface switching plot

clockwise direction. The PSFαSMC reaches the steady-state area faster than does FLC and in a smoother manner than SkyhookSMC.

Figure 19.31 shows that all the 2-DOF SA suspension system's sliding surfaces are switching and going around $s = 0$, and the PSFαSMC exhibits a smaller and smoother switching behavior than does FLC or SkyhookSMC for ride comfort control.

The characteristics of the vehicle suspension system's uncertainty measures are given in Tables 19.7, 19.8 and 19.9, in which the measures of the maximum (*Max*), minimum (*Min*), average (*Mean*, *Median* and *Mode*), dispersion (*STD*,

Table 19.7 Parameters for Uncertainty Analysis - PSFαSMC (α=0.5, FαSMC)

Statistics	y_1	y_2	y_3	s
Max	11.06	4210.26	0.00	13.58
Min	-4.64	-671.54	-0.20	-2.79
Mean	0.0093	2669.26	-0.16	0.24
Median	0.0074	2699.63	-0.16	-0.061
Mode	-4.64	-671.54	-0.20	-2.79
STD	1.31	531.63	0.026	2.055
Variance	1.72	282632.53	0.00071	4.22
Skewness	3.20	-2.52	4.33	3.88
Kurtosis	27.60	15.35	23.99	20.58

Table 19.8 Parameters for Uncertainty Analysis - PSFαSMC (α=1.0, FLC)

Statistics	y_1	y_2	y_3	s
Max	11.06	4534.98	0.00	14.81
Min	-6.032	-513.78	-0.22	-6.55
Mean	0.0027	2659.48	-0.16	0.23
Median	-0.076	2640.38	-0.16	0.11
Mode	-6.032	-513.78	-0.22	-6.55
STD	1.93	560.63	0.028	2.77
Variance	3.74	314312.74	0.00079	7.71
Skewness	1.25	-0.73	3.10	2.66
Kurtosis	9.53	8.98	18.93	13.79

Table 19.9 Parameters for Uncertainty Analysis - PSFαSMC (α=0.0, SkyhookSMC)

Statistics	y_1	y_2	y_3	s
Max	11.06	3890.82	0	12.99
Min	-5.11	-1089.93	-0.18	-2.15
Mean	0.010	2647.74	-0.16	0.24
Median	-0.0045	2714.45	-0.16	-0.031
Mode	-5.11	-1089.93	-0.18	-2.15
STD	1.26	577.50	0.027	1.83
Variance	1.61	333507.072	0.00073	3.37
Skewness	3.39	-3.07	4.49	4.17
Kurtosis	31.05	16.65	23.55	23.84

Variance), asymmetry (*Skewness*) and flatness (*Kurtosis*) [Rocquigny, et al. (2008), International Organization for Standardization (2008), Choi, et al. (2007)] can be obtained by estimating the corresponding values of the variable of interest Y and s. In addition, the uncertainty measures are matched to the figures for the system outputs as discussed above.

19.10 Conclusion and Future Work

In conclusion, the architecture of the proposed PSFαSMC control method has been discussed; the MOμGA is utilized to optimize the parameters for the PFs in the offline step. Then, the PFs are applied to supervise the parameter generation in the online step for the SA suspension ride comfort control. The simulation results demonstrate that the PSFαSMC can adjust the control effects between the FLC and SkyhookSMC methods by the factor α, which provides flexible control effects for comfortable ride under road uncertainty. Generally, the framework of the PSFαSMC, which can provide multi-state ($\alpha=0,0.5,1$) control effects, has been verified through the simulation results; other control methods, such as H_∞ and *PID* control, can also be used as one of the control states. In this paper, the controlled state response comparison of the FLC, SkyhookSMC and PSFαSMC methods validates the effectiveness of the last control method with uncertainty analysis. In the offline step, the MOμGA is applied as an optimizer for the parameters of the PF; the optimized results confirm the usability of the MOμGA as a practical approach. In the online step, the optimized PFs supervise the control effects for ride comfort and provide an efficient way to adjust the key parameters of the PSFαSMC controller.

Given the flexibility of the parameter selection, the PSFαSMC method can be applied to a wide range of engineering applications, such as space tether deployment and vehicle navigation systems. According to the requirements of each type of engineering application, other new methods can also be modulized as derivatives under the design framework of this proposed PSFαSMC method, for example, as a central pattern generator (CPG) for the locomotion control of a legged robot or a propelled robotic fish optimized by swarm intelligence methods.

Part IV
CIAD for Social Sciences

Part IV introduces some CIAD's applications in social science, for example, exchange rate investigation, electricity consumption, spatial analysis for urban study, etc.

20

Exchange Rate Modeling and Decision Support

20.1 Introduction

The modeling of exchange rate movements is a challenging task in international finance. A strong consensus in academic research is that macroeconomic fundamentals have no explanatory power for exchange rate fluctuations in the short term [Obstfeld and Rogoff (2000), Rogoff (1996)]. In contrast, microstructure approaches focus on information concerning macro fundamentals and both non-fundamentals and their transfers in the foreign exchange market as well as their impact on the movement of exchange rates. Empirical evidence demonstrates the significant positive link between exchange rates and their corresponding contemporaneous order flow, which is defined as the net value between buyer-initiated trades and seller-initiated trades [Lyons (2001), Killeen, et al.(2001), Payne (2003)].

Some evolutionary computation (EC) methods have been utilized for exchange rate analysis and other financial studies. In 1996, Hann and Steurer [Hann and Steurer(1996)] analyzed the influences of data frequency on American Dollar/Deutsch Mark forecasting using ANNs; the studies reported that ANN does not greatly improve the forecasting accuracy when monthly data are utilized. In 2003, Qi and Wu [Qi, et al. (2003)] proposed a multi-layer feed-forward network to forecast exchange rates, the numerical results of which concluded that the ANN was not efficient in terms of out-of-sample forecast accuracy. In 2007, Yadav et al. [Yadav, et al. (2007)] applied a standard multilayer neural network (SMN) to predict a set of time-series data for exchange rate prediction from 2002 to 2004. In 2005, Rimcharoen et al. [Rimcharoen, et al. (2005)] proposed a method of adaptive evolution strategies (ES) for the prediction of the stock exchange of Thailand, in which the GA method was combined with the ES method. No further studies on ES for the prediction of exchange rates were conducted since then. A differential evolution algorithm, combining the strengths of multiple strategies, was proposed by Worasucheep and Chongstitvatana [Worasucheep and Chongstitvatana (2009)] in 2009; however, there have been no further studies on exchange rate determination using DE or DE-related methods. The PSO method is a swarm intelligence algorithm;

this method uses a population-based search algorithm mimicking the social behavior of individuals (particles) moving among a multi-dimensional search space. The PSO method was applied to stock market forecasting, using an ANN, by Nenortaite and Simutis [Nenortaite and Simutis (2004)] in 2004 and by Zhao and Yang [Zhao and Yang (2009)] in 2009; however, neither reported on PSO applications for exchange rate prediction.

Exchange rate determination has been regarded as one of the most challenging applications of high-frequency time-series trading [Lyons (2001), Killeen, et al.(2001), Payne (2003), Tenti (1996), Gujarati (2004)]. Thus, to provide investors and researchers with more precise predictions, various models have been presented in which prices follow a random walk phenomenon. This is suitable for GAs with stochastic and nonlinear searching abilities.

In this chapter, the Mendel-GA will be applied to the studies of empirical analysis on exchange rate determination, which can provide an evolutionary and computational method for the exchange rate determination problem. Specifically, this work attempts to compare the performance of the Mendel-GA and traditional estimation methods, for instance, ordinary least square (OLS) and linear least squares (LS) estimation. OLS and LS are methods for estimating the unknown parameters in a linear regression model. These methods minimize the sum of squared distances between the observed responses in the data set and the responses predicted by the linear approximation. Compared with OLS and LS, the Mendel-GA, through the evolutionary process, can address linear and nonlinear models with higher complexity and can be used as an active optimization solver for switching from one prediction model to another.

20.2 Exchange Rate Determination Model

In 2001, Kileen et al. [Killeen, et al.(2001)] found the co-integration relationship between exchange rates and cumulative order flow (COF), which is the proximate determinant of price in all microstructure models. In 2003, Payne [Payne (2003)] used non-standard vector autocorrelation to examine the causality of exchange rates from the order flow.

The order flow is defined as the net of buyer- and seller-initiated currency transactions and is taken as a measure of net buying pressure [Lyons (2001)]. According to previous reports [Obstfeld and Rogoff (2000), Killeen, et al.(2001), Payne (2003), Reitz, et al. (2007), Rime, et al. (2008)], the order flow is intimately related to a broad set of current and expected macroeconomic fundamentals and as such, the order flow is regarded as a useful predictor of the movements in exchange rates.

Microstructure approaches use order flow to proxy the information reflecting the movement of exchange rates. An intensive study by Lyons [Obstfeld and Rogoff (2000)] in 2000 demonstrated that order flow contains information concerning the movements of exchange rates.

By starting from conventional exchange rate theories, the exchange rates can be expressed as the discounted present value of current and expected fundamentals, as given by Equation (20.1) [Obstfeld and Rogoff (2000), Killeen, et al.(2001), Payne (2003), Rime, et al. (2008), Andersen, et al. (2003)].

$$s_t = (1-b) \sum_{q=0}^{\infty} b^q E_t^m f(x_{t+q}) \tag{20.1}$$

where
- s_t is the spot exchange rate, which is defined as the domestic price of foreign currency;
- b is the discount factor;
- q is the future periodic factor;
- $f(x_t)$ denotes the fundamentals at time t;
- x_t is the contemporaneous order flow;
- $E_t^m f(x_{t+q})$ is the market-makers's expectations about future fundamentals at q-periods after time t, conditional on information available at time t.

Specifically, by iterating Equation (20.1) forward and rearranging terms as given by Equation (20.2), the relation between the spot exchange rates s_t and the contemporaneous order flow x_t can be specified as follows:

$$\Delta s_{t+1} = \frac{1-b}{b}(s_t - E_t^m f(x_t)) + \varepsilon_{t+1} \tag{20.2}$$

where
- Δ denotes the first difference of the series;
- ε_t is the disturbance term, as stated in Equation (20.3) [Bacchetta and Wincoop (2006)], which shows that the future exchange rate change is a function of the gap between the current exchange rates and the expected current fundamentals. It is also a term that captures changes in expectations about fundamentals;

$$\varepsilon_{t+1} = (1-b) \sum_{q=0}^{\infty} b^q \left(E_{t+1}^m f(x_{t+q+1}) - E_t^m f(x_{t+q+1}) \right) \tag{20.3}$$

Purchasing foreign currency increases the demand of the foreign currency, leading to appreciation of the foreign currency and depreciation of the domestic currency, i.e., the spot exchange rates s_t increase. We expect a positive correlation between the spot exchange rates and the corresponding order flow.

20.3 Fitness Function of Regression Modeling

To evaluate the performance of the Mendel-GA comparison process with the exchange rates and COF feature, the R^2 metric, which is a statistical measurement of the agreement between the observed data sequence y_i and the Mendel-GA-generated specimen data sequence $\hat{y}_i$, has been borrowed for the fitness function definition.

Using the basic idea of the R^2 metric, the association between exchange rates and COF is specified by Equation 20.4. We define the fitness function as Equation 20.5, which attempts to use the Mendel-GA to find the best parameters that maximize R^2.

$$\Delta \hat{y}_i(t) = \beta_1 + \beta_2 \Delta x_i(t) + \beta_3 \Delta y_i(t-1) \qquad (20.4)$$

$$J = fitness(\beta_1, \beta_2, \beta_3) \to R^2(\Delta \hat{y}_i, \Delta y_i) \qquad (20.5)$$

where

○ $\Delta \hat{y}_i(t)$ is the estimated (modeled) value from the regression model at time t

○ β_1, β_2 and β_3 are known as the intercept coefficient, slope coefficient and historical data coefficient, respectively

○ x_i is the observed COF of a foreign currency

○ y_i defines the ratio of a foreign currency vs. the US dollar in $\log(*)$ space. Equations (20.6) and (20.7) are the data scaling operations for the cases of the Deutsche Mark vs. the US Dollar and the Japanese Yen vs. the US Dollar, respectively

$$y_i = \log\left(\frac{DM}{USD}\right) \times 10000 \qquad (20.6)$$

$$y_i = \log\left(\frac{JPY}{USD}\right) \times 10000 \qquad (20.7)$$

○ $\Delta x_i(t)$ and $\Delta y_i(t-1)$ are the difference data sequence obtained from the pre-data-handling process at times t and $t-1$, as stated in Equations (20.8) and (20.9)

$$\Delta x_i(t) = x_i(t) - x_i(t-1) \qquad (20.8)$$

$$\Delta y_i(t-1) = y_i(t-1) - y_i(t-2) \qquad (20.9)$$

As shown in Equation (10.6), the numerical simulation generates minimal errors by $(y_i - \hat{y}_i)$ and $(y_i - \bar{y}_i)$, which could result in a loss of precision. Given the scaling factor, the numerical errors can be scaled to an acceptable numerical range to ensure computational precision. The scaling factor can be selected on a case-by-case basis. As shown in Figure 20.1, to reduce the possibility of numerical simulation errors, the data sequence y_i has been scaled by multiplying by a scaling factor (a constant) of 10000 for the DM vs. USD and JPY vs. USD cases. Then, the difference data sequences of Δx_i and Δy_i are generated. The next step is to generate the fitness function using Equation (20.5).

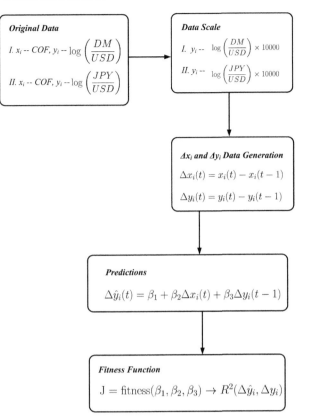

Fig. 20.1 Original data pre-handling process and fitness function generation

20.4 Empirical Results and Discussion

The original exchange rate data came from Evans and Lyons [Evans and Lyons (2002)] in 2002, where there are two groups of trading data:

    ○ Data-I: First trading data pair of Deutsche Mark (DM) vs. US Dollar (USD);

    ○ Data-II: Second trading data pair of Japanese Yen (JPY) vs. US Dollar (USD);

Empirical results are obtained using a specially devised simulation toolkit for the Mendel-GA in *MATLAB*®, known henceforth as *SGALAB* [SGALAB (2009)]. Unless stated otherwise, all the results are generated using the following parameters for the GAs, as listed in Table 20.1. Binary encoding/decoding, tournament selection, single-point crossover and mutation are utilized by the Mendel-GA evolutionary process. The Mendel-GA's results are also compared to the results of standard GA and OLS methods.

As listed in Table 20.1, the total number of simulations is 1000. For the fitness results, the max, min and mean values of the fitness are calculated to represent the

Table 20.1 Empirical Parameters for Mendel-GA

Max generations	100
Crossover probability	0.8
Mutation probability	0.001
Population	50
Selection operator	tournament
Crossover operator	single point
Mutation operator	single point
Encoding method	binary
β_1 Regression parameter	$[-10, 10]$
β_2 Regression parameter	$[0, 5]$
β_3 Regression parameter	$[-1, 1]$
Mendel percentage	1 (full chromosome length)
Number of simulations	1000

Mendel-GA's single-run performance ($fitmax_i, fitmin_i$ and $fitmean_i$) and for the Mendel-GA's 1000 experiments in total, as shown in Figures 20.2 and 20.3. The plots for the average data of all $fitmax_i$, $fitmin_i$ and $fitmean_i$ over the 1000 experiments are expressed.

In addition, for the Mendel-GA's 1000 experiments in total, the mean E(*) and variance VAR(*) of the max fitness values of all experiments are evaluated as the Mendel-GA's overall performance indices, as listed in Tables 20.2, 20.3, 20.4 and 20.5.

Figures 20.2 and 20.3 describe the evaluation process of the Data-I (DM vs. USD) fitness values and the Data-II (JPY vs. USD) fitness values, respectively. Over the full evolution time defined by max generation, the fitness values increase quickly to a steady state with the initial parameters in Table 20.1. The solid lines in Figures 20.2 and 20.3 are the mean values of the average of the total fitness, the fitness data indicated by '+' at the top are the max values of the average of the total fitness, and the fitness data indicated by '+' in the bottom are the min values of the average of the total fitness.

Tables 20.2 and 20.3 present the mean and variance results of Data-I (DM vs. USD). Specifically, in Table 20.2, the $E[\beta_1]$, $E[\beta_2]$ and $E[\beta_3]$ of the Mendel-GA, standard GA and OLS, respectively, are similar. Among the $E[R^2]$ of the Mendel-GA, standard GA and OLS, the Mendel-GA's $E[R^2]$ is the largest, the standard GA's $E[R^2]$ is the second largest and OLS's $E[R^2]$ is the smallest, which indicates that the Mendel-GA outperforms the two other methods.

Table 20.3 shows that the $VAR[\beta_1]$, $VAR[\beta_2]$ and $VAR[\beta_3]$ of the Mendel-GA, standard GA and OLS are slightly different from each other and the Mendel-GA's $VAR[R^2]$ is smaller than the standard GA's $VAR[R^2]$. This means that the Mendel-GA's results remain within a smaller scattered range compared to those of the standard GA. The OLS's VAR[*] values are zero because traditional numerical methods are not used and the results are same for every experiment.

Similarly, Tables 20.4 and 20.5 are the mean and variance results of Data-II (JPY vs. USD) and provide evidence of the Mendel-GA's better searching ability compared to the standard GA and OLS.

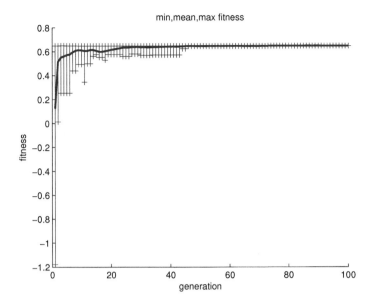

Fig. 20.2 Data-I (DM vs. USD) evolutionary fitness data

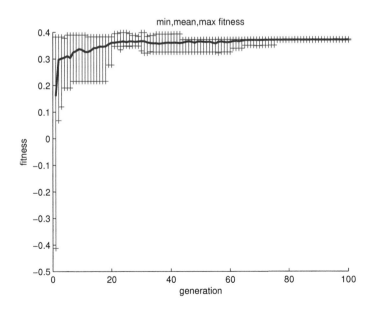

Fig. 20.3 Data-II (JPY vs. USD) evolutionary fitness data

Table 20.2 Mendel-GA Results of Mean for Data-I (DM vs. USD)

	Mendel-GA	SGA	OLS
$E[\beta_1]$	-4.9297	-4.9251	-4.9143
$E[\beta_2]$	0.2211	0.2203	0.2194
$E[\beta_3]$	0.1490	0.1478	0.1399
$E[R^2]$	0.6520	0.6415	0.6400

Table 20.3 Mendel-GA Results of Variance for Data-I (DM vs. USD)

	Mendel-GA	SGA	OLS
$VAR[\beta_1]$	0.2341	0.2572	-
$VAR[\beta_2]$	0.00233	0.00242	-
$VAR[\beta_3]$	0.0013	0.0018	-
$VAR[R^2]$	0.012	0.016	-

Table 20.4 Mendel-GA Results of Mean for Data-II (JPY vs. USD)

	Mendel-GA	SGA	OLS
$E[\beta_1]$	-4.9625	-4.9012	-4.8334
$E[\beta_2]$	0.3185	0.3035	0.2952
$E[\beta_3]$	-0.1411	-0.1402	-0.1396
$E[R^2]$	0.4073	0.4064	0.40

Table 20.5 Mendel-GA Results of Variance for Data-II (JPY vs. USD)

	Mendel-GA	SGA	OLS
$VAR[\beta_1]$	0.2343	0.2532	-
$VAR[\beta_2]$	0.00112	0.00127	-
$VAR[\beta_3]$	0.0025	0.0031	-
$VAR[R^2]$	0.0025	0.0037	-

As shown in Figures 20.4 and 20.5, the 'o' data points indicate the sampling data, and the '*' are the data points from regression modeling using Equation 20.4. Both types of data show how the regression model is estimating the exchange rate trading behavior. The estimations of β_1, β_2 and β_3 are given in Tables 20.2 and 20.4.

According to the simulation results obtained using the Mendel-GA method, β_1 and β_2 are in a relatively steady state and β_3 shows how the historical data of $\Delta y_i(t-1)$ can produce the estimation data $\Delta \hat{y}(t)$. Figures 20.6 and 20.7 present how β_3 affects R^2, which measures the performance of the regression modeling.

As shown in Tables 20.2 and 20.4, compared with the study of Evans and Lyons [Evans and Lyons (2002)], the coefficient of determination is improved: for the DM/dollar, the coefficient of determination is improved from 60 per cent $\rightarrow$ 64 per cent; for the yen/dollar, the coefficient of determination is slightly improved, from 40 per cent $\rightarrow$ 40.73 per cent.

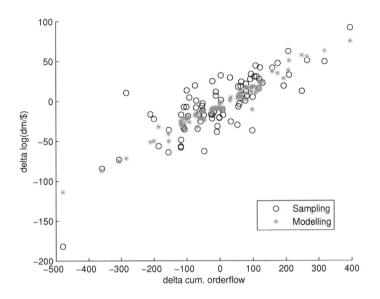

Fig. 20.4 Sampling and modeling Δx_i vs. Δy_i for Data-I (DM vs. USD)

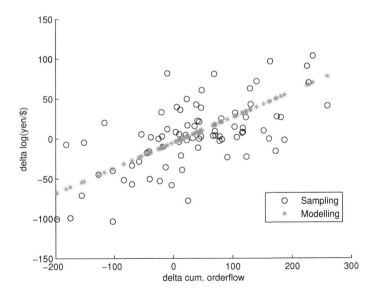

Fig. 20.5 Sampling and modeling Δx_i vs. Δy_i for Data-II (JPY vs. USD)

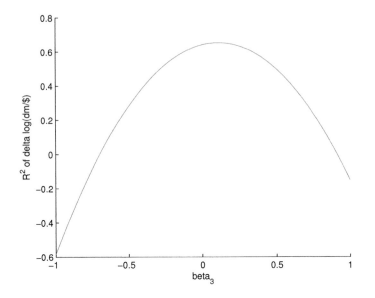

Fig. 20.6 β_3 vs. $\Delta \log(DM/USD)$ when $(\beta_1 = -4.9297, \beta_2 = 0.2211)$

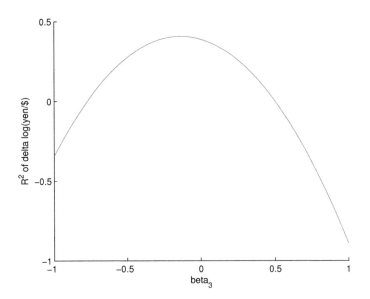

Fig. 20.7 β_3 vs. $\Delta \log(JPY/USD)$ when $(\beta_1 = -4.9625, \beta_2 = 0.3185)$

20.5 Conclusions and Future Work

Interestingly, we find the variations characterizing the association between β_3 and the coefficient of determination. In the case of the DM/dollar, β_3 being positive makes the coefficient of determination obtain a minimum, whereas in the case of the yen/dollar, β_3 being negative makes the coefficient of determination obtain a minimum. This reflects the different feedback trading behaviors in the different cases of the foreign exchange market.

Data-I (DM vs. USD) indicates a positive trend tracing behavior, which means that over the sample period, positive exchange rate returns induce buyer-dominant trading behaviors. However, in the case of Data-II (JPY vs. USD), the results suggest that a positive exchange rate return induces a reverse trading behavior.

According to the simulation results for Data-I (DM vs. USD) and Data-II (JPY vs. USD), a significant positive association between exchanger rate return and the corresponding order flow change is observed, which is consistent with the theoretical hypothesis.

A comparison of the results obtained using the Mendel-GA, standard GA and OLS indicates that the Mendel-GA outperformed the standard GA and OLS methods, which add to the research efforts to bridge the divide between macro and micro approaches to exchange rate economics by examining the linkages between exchange rate movement, cumulative order flow and expectations of macroeconomic variables.

For trading decisions, technical analysis is sometimes utilized to assist traders in making buying and selling decisions, therein attempting to properly utilize available time-series data. This paper provides an evolutionary alternative to exchange rate determination, which differs from conventional econometrical methods; we also would like to apply this proposed technique to other trading data sets in practice.

In addition, the integration of experts' skills in this model is considered to be essential for trading decision-making purposes. In further studies, a fuzzy logic system will be used for developing decision models in which the experience of a trader can be incorporated in a natural or artificial manner.

This paper's results suggest some interesting issues for further investigations in real-time trading policy making, in which the R^2 metric will be one of the trading decision criteria, for example, Value-at-Risk (VaR) with volatility predictions.

21

Quantitative Modeling of Electricity Consumption

21.1 Introduction

The causality relationship between energy consumption (EEC) and social or economic factors is one of the central themes in energy research. In the last decade, there has been increasing research interest in the interactions among energy consumption, economic growth, and sustainable energy policies. The UK, for example, strives to reduce its greenhouse gas emissions by at least 80 per cent (from the 1990 baseline) by 2050, which reflects a reorientation away from specific technological solutions and moves toward structural transformation. The rapid development in China has contributed to global prosperity and electricity is the lifeblood of its economic sectors. However, economic growth also creates social-environmental problems such as pollution, global warming, and the depletion of natural resources. This is because the majority of today's electricity is generated from fossil fuel-fired power plants that release large amounts of greenhouse gases and pollutants. In addition to China and the UK, other countries (USA, New Zealand, Australia, India, Indonesia, the Philippines, Thailand, etc.) will also be included in the following literature review section.

In the literature, four major areas have been investigated pertaining to energy consumption: area (1) the causal relationship between economic growth and EEC; area (2) the relationship between microeconomic factors (e.g., income) and EEC; area (3) the correlation between carbon dioxide (CO_2) emissions and EEC; and area (4) the modeling of EEC and related techniques. These issues are reviewed in the following paragraphs and are listed in Tables 21.1 to 21.4.

The first research area concerns the study of the causal relationship between economic growth and EEC, as given in Table 21.1. Stern investigated the causal relationship between gross domestic product (GDP) and energy consumption for the period of 1947–1990 in the USA [Stern (1993)]. Masih et al. illustrated the long-term equilibrium relationship between economic growth rate and energy consumption [Masih and Masih (1996)]. Wesseh and Zoumara investigated the causal dependency between energy consumption and economic growth in Liberia using a boot-

Table 21.1 Summary of Literature on Area(1): the Causal Relationship Between Economic Growth and Energy Consumption (EEC)

Author(s)	Countries	Methodology	Relationship*
Stern [Stern (1993)]	USA (1947–1990)	Multivariate VAR model	EEC→GDP
Ghosh [Ghosh (2002)]	India (1950–1997)	Granger causality	EEC→GDP
Fatai et al. [Fatai, et al. (2004)]	New Zealand (1960–1999), Australia, India, Indonesia, Philippines, Thailand	Granger causality	EEC↔GDP
Oh and Lee [Oh and Lee (2004)]	South Korea (1981–2000)	Granger causality, error correction model	EEC→GDP
Lee [Lee (2005)]	18 developing countries (1975–2001)	Granger causality, FMOLS	EEC→GDP
Chen [Chen, et at. (2007)]	Hongkong (2004), Korea; Indonesia; Singapore, Thailand, Taiwan	Granger causality	GDP→EEC; EEC→GDP; EEC≠GDP
Zhang and Cheng [Zhang and Cheng (2009)]	China (1960–2007)	Granger causality	GDP→EEC
Wesseh and Zoumara [Wesseh and Zoumara(2012)]	Liberia (1980–2008)	granger causality, bootstrap	EEC→GDP
Abbas and Chaudhary [Abbas and Choudhury (2013)]	India (1972–2008)	Granger causality	GDP→EEC

* Definitions of notation: →, ↔ and ≠ represent unidirectional causality, bidirectional causality, and no causality, respectively.

Table 21.2 Summary of Literature on Area(2): the Relationship Between Microeconomic Factors and EEC

Author(s)	Countries	Methodology	Relationship
Masih and Masih [Masih and Masih (1996)]	India; Pakistan; Malaysia; Singapore; Indonesia; Philippines	Granger causality, unit root test, non-stationarity test	EEC→income; EEC↔income; EEC≠income; EEC≠income; income→EEC; EEC≠income
Asafu-Adjaye [Asafu-Adjaye (2000)]	India (1973–1995); Indonesia (1973–1995); Philippines (1971–1995); Thailand (1971–1995)	Granger causality, error-correction	EEC→income; EEC→income; EEC↔income; EEC↔income
Soytas and Sari [Soytas and Sari (2003)]	Argentina, Italy, Korea, Turkey, France, Germany, Japan	Granger causality	EEC↔income
Narayan and Smyth [Narayan and Smyth (2005)]	Australia	Granger causality	income→EEC, employment →EEC
Sari and Soytas [Sari and Soytas (2007)]	6 developing countries (1971–2002)	GVD, GIR	EEC→income
Ajmi et al. [Ajmi, et al. (2013)]	G7 countries (1960–2010)	nonlinear causality	EEC↔income

strap methodology [Wesseh and Zoumara(2012)]. Fatai et al. [Fatai, et al. (2004)] examined the causal relationship between GDP and various types of energy, in-

Table 21.3 Summary of Literature on Area(3): the Correlation Between CO_2 Emission and EEC

Author(s)	Countries	Methodology	Relationship
Ramanathan [Ramanathan (2005), Ramanathan (2006)]	17 countries (1996), International Energy Annual (1980–2001)	DEA	EEC↔CO_2
Martiskainen [Martiskainen (2007)]	UK	behavioural models	EEC↔CO_2
Soytas et al. [Soytas and Sari (2007)]	US (1960–2004)	Granger causality	EEC↔CO_2
Soytas et al. [Soytas and Sari (2009)]	Turkey (1960–2004)	Granger causality	CO_2→EEC
Chang [Chang (2010)]	China (1981–2006)	Granger causality, OLS, VECM	EEC↔CO_2
Bian et al. [Bian, et al. (2013)]	China (1978–2009)	non-radial DEA	EEC→CO_2

Table 21.4 Summary of Literature on Area(4): the Modelling of EEC

Author(s)	Countries	Methodology	EEC Models
Saab et al. [Saab, et al. (2001)]	Lebanon (1970–1999)	AR, ARIMA, AR(1)/highpass	EEC forecasting
Tso and Yau [Tso and Yau (2007)]	Hong Kong	MRA, DT, ANN	EEC forecasting
Yohanis et al. [Yohanis, et al. (2008)]	Northern Ireland (2003–2004)	correlation analysis	type of dwelling, location, ownership, size, household appliances, number of occupants, income, age, occupancy patterns → EEC forecasting
Swan and Ugursal [Swan and Ugursal (2009)]	Canada	top-down, bottom-up models	EEC forecasting
Bianco et al. [Bianco, et al. (2009)]	Italy (1970–2007)	regression	EEC forecasting
Kankal et al. [Kankal, et al. (2011)]	Turkey	ANN, regression	EEC forecasting
Kiran et al. [Kiran, et al. (2012)]	Turkey	ABC, PSO	EEC forecasting

cluding coal, natural gas, electricity and oil, and they concluded that energy conservation policies do not have significant impacts on GDP growth in industrialized countries. Oh and Lee studied two models—a demand-side model consisting of energy consumption, GDP and energy price and a production-side model consisting of GDP, energy generation, capital, and labor [Oh and Lee (2004)]. Shiu and Lam applied an error-correction model to examine the causal relationship between electricity consumption and real GDP in China from 1971–2000 [Shiu and Lam (2004)]. Lee provided a study on a long-run co-integration relationship by considering the heterogeneity effects of nations and the causality relationship between energy consumption and GDP in 18 developing countries [Lee (2005)]. Mohamed and Bodger studied the influence of selected economic and demographic variables on annual electricity consumption in New Zealand [Mohamed and Bodger (2005)]. Zhang and Cheng investigated the existence and direction of Granger causality between economic growth, energy consumption and carbon emissions in China from 1960 to 2007 [Zhang and Cheng (2009)]. Abbas et al. performed an empirical study to determine the causality between electricity consumption and economic growth at aggregated and disaggregated levels in the agricultural sector [Abbas and Choudhury (2013)].

The second research stream explores the relationship between microeconomic factors and EEC, as given in Table 21.2. Leveraging the Granger causality test, Masih and Masih [Masih and Masih (1996)] studied the co-integration between total energy consumption and real incomes in six Asian countries—India, Pakistan, Malaysia, Singapore, Indonesia and the Philippines. By combining error-correction techniques, Asafu-Adjaye [Asafu-Adjaye (2000)] estimated the causal relationships between energy consumption and incomes in India, Indonesia, the Philippines and Thailand. Soytas and Sari [Soytas and Sari (2003)] studied the causality relationship between energy consumption and income based on time-series models in the top 10 emerging markets and the G7 countries—USA, UK, France, Germany, Italy, Canada and Japan. Narayan and Smyth [Narayan and Smyth (2005)] examined the relationship between EEC, employment and real income in Australia, which indicated a long-run employment and real income Granger causality relationship with EEC. Sari and Soytas [Sari and Soytas (2007)] investigated the inter-temporal link between EEC and income in six developing countries using generalized variance decompositions (GVDs) and generalized impulse response (GIR). Ajmi et al. [Ajmi, et al. (2013)] tested the real causal links between EEC and national income of G7 countries for policy decision making.

In the third research area, the goal is to investigate how EEC influences CO_2 emissions and renewable energy policy, as given in Table 21.3. Ramanathan [Ramanathan (2005), Ramanathan (2006)] investigated the relationship between energy consumption and CO_2 emissions in 17 countries in the Middle East and North Africa using data envelopment analysis (DEA). Martiskainen [Martiskainen (2007)] reported different household consumption behaviors and the goal of reducing energy use and CO_2 emissions in the UK. Soytas et al. [Soytas and Sari (2007)] and Soytas and Sari [Soytas and Sari (2009)] studied the effect of energy consumption and the release of CO_2 in the US and Turkey. Chang [Chang (2010)] utilized multivariate co-integration Granger causality tests to investigate the correlations between carbon emissions, energy consumption and economic growth in China using ordinary least squares (OLS) and the vector error correction model (VECM). Bian et al. [Bian, et al. (2013)] estimated the potential energy saving and potential CO_2 emission reduction implications in China.

The fourth research thrust focuses on the modeling of energy consumption by accommodating different factors using a wide range of techniques, as given in Table 21.4. Saab et al. [Saab, et al. (2001)] investigated three univariate models—the autoregressive (AR), autoregressive integrated moving average (ARIMA), and autoregressive with a high-pass filter (AR(1)/highpass) models. Then, the case of electricity in Lebanon was studied for electrical EEC forecasting. Tso and Yau [Tso and Yau (2007)] presented three modeling techniques for predicting ELC in Hong Kong, including multiple regression analysis (MRA), decision tree (DT) and ANNs. Yohanis et al. [Yohanis, et al. (2008)] studied the patterns of ELC with dwelling characteristics in ELC data for Northern Ireland. Swan and Ugursal [Swan and Ugursal (2009)] conducted a literature review of various modeling techniques for residential energy consumption. Bianco et al. [Bianco, et al. (2009)] investigated the influence of economic and demographic variables on annual electricity consumption in Italy with the intent to develop a long-term forecasting model. Payne [Payne (2010)] published a survey on empirical prediction models and discussed various hypotheses on the causal relationship between electricity consumption and economic growth. Kankal et al. [Kankal, et al. (2011)] proposed a multivariate approach to modeling energy consumption in Turkey with socio-economic and demographic variables (GDP, population, import and export amounts, and employment) using an ANN and regression analyses. Kiran et al. [Kiran, et al. (2012)] proposed two electrical energy estimation models based on artificial bee colony (ABC) and PSO techniques to estimate electricity demand in Turkey.

Despite the large number of publications, the literature on energy consumption models fails to address the following concerns: (1) These studies do not capture the coupling effects on energy consumption resulting from the interactions among social, economic, and environmental factors. (2) Few effective methods have been developed that allow for quantitative analysis, model verification, and parameter optimization. (3) A general quantitative modeling framework or platform has yet to be presented. In response to the three above-mentioned concerns, this chapter employs the CIAD framework to construct and optimize the electricity consumption model. The model has been validated based on a 33-year dataset collected from China.

21.2 Quantitative Modeling of National Electricity Consumption

$$E\hat{E}C(C_0, \Theta, \Omega) = C_0 + \theta_1 X_1^{\omega_1} + \theta_2 X_2^{\omega_2} + \theta_3 X_3^{\omega_3} + \theta_4 X_4^{\omega_4} + \theta_5 X_5^{\omega_5} \quad (21.1)$$

To estimate and predict national electricity consumption, a quantitative model based on three design variables, C_0, Θ, and Ω, and five impact factors, X_i, is embedded in Equation (21.1), where

○ $E\hat{E}C$ is the estimation of annual national electricity consumption, $GWh/year$

○ C_0 is a design variable of data shifting

○ $\Theta = [\theta_1, \theta_2, ..., \theta_i, ..., \theta_{n_1}]$ is a design variable of the coefficient vector of X_i, n_1 is the the the number of design variables

○ $\Omega = [\omega_1, \omega_2, ..., \omega_i, ..., \omega_{n_1}]$ is a design variable of the exponent vector of X_i;

○ $X_i = [X_1, X_2, X_3, X_4, X_5]$ is the vector of the impact factors, where X_1 is the GDP; X_2 is the electricity price; X_3 is the efficiency, which is the ratio of national output over electricity consumption; X_4 is the economic structure, which is the ratio of residential consumption over industrial production; and X_5 is the CO_2 emissions in billion of metric tons

As stated by Equation (21.1), X_1 represents macroeconomic performance, and X_2 is obtained using producer production prices. It is critical to include the two impact factors X_3 and X_4 in this conceptual model and data analysis due to the dynamic nature of the economic structure and consumption efficiency. X_5 are the environmental factors concerning the original data collection.

The '+' operator is applied to refine this model and identifies the contribution of each impact factor to the dynamic behaviors of $E\hat{E}C$ and the interaction among the five impact factors within the same quantitative scale (normalization).

21.3 Fitness Function

The fitness function represents the approximation to the national electricity consumption. As shown in Figure 21.1, four steps are involved in defining the fitness function. Step 1 collects raw data from specific data sources, including yearbooks and research reports. Step 2 screens, filters, and pre-processes the source data. Step 3 estimates the national electricity consumption, as described in Section 21.2. Step 4 creates the fitness function given by Equation (21.2), where the fitness function is defined as the root mean square (RMS) errors of $E\hat{E}C$ and EC_0. Here, mmAP is taken as an index for the quantitative analysis driven by the CIAD approach, and $E\hat{E}C$ is stated in Equation (21.1). The goal of this paper is to find the optimal combination of C_0, Θ, and Ω that simultaneously maximizes the electricity consumption $E\hat{E}C$ based on historical data EC_0.

$$F = Maximize : \left\{ mmAP\left(-RMS\left(E\hat{E}C\left(C_0, \Theta, \Omega \right) - EC_0 \right) \right) \right\}$$ (21.2)

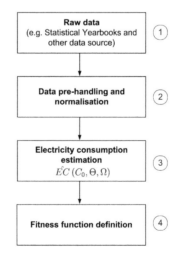

Fig. 21.1 Fitness function definition work-flow

21.4 Numerical Results

Maximizing the fitness function yields the minimum RMS of $E\hat{E}C$ and EC_0, which is obtained using the specially designed toolboxes *SwarmFirefly* [SwarmFirefly (2017)] and *SECFLAB* [SECFLAB (2012)]. The computer used for the simulations is equipped with a 2.1 GHz Intel dual-core processor, Windows XP Professional v 5.01 Build 2600 service pack 3, 2.0 GB of 800 MHz dual-channel DDR2 SDRAM, and MATLAB® R2008a. The initial parameters are listed in Table 21.5, where the max generation number is 100, which serves as the termination condition of each test; the test number is also 100; the randomness is set to 0.2; the randomness reduction is 0.98; the population is 50, in which the non-replaceable population P_N and replaceable population P_R are 40 and 10, respectively; the absorption coefficient is 1; and the ranges of C_0, Θ and Ω are [-100,100], [-10,10] and [0,3], respectively. X is collected from both the China yearbooks and reference [Guo (2010)]. In this setting, data between 1980 and 2009 are utilized for the EEC approximation, and data between 2010 and 2012 are applied for EEC prediction. Table 21.6 gives the optimal combinations (MEAN±STD) of C_0, $\Theta = [\theta_1, \theta_2, \theta_3, \theta_4, \theta_5]$ and $\Omega = [\omega_1, \omega_2, \omega_3, \omega_4, \omega_5]$.

Figure 21.2 gives the mmAP curves, with their upper and lower mmSTD boundaries, where the fitness increases very quickly and reaches a plateau from generations 1 to 20. Note that all lines converge by generation 100. Figure 21.3 depicts the fitness mmSTD over the entire simulation. The curve decreases quickly within the first 20 generations and finally reaches 0 by generation 100. These two figures indicate that the proposed optimization algorithm is efficient and accurate.

Table 21.5 Initial Parameters of the Swarmfirefly

Max generations	100
Test number	100
Randomness	0.2
Randomness reduction	0.98
Population	50
Non-replaceable population	40
Replaceable population	10
Absorption coefficient	1
C_0	[-100, 100]
Θ	[-10, 10]
Ω	[0, 3]

Figure 21.4 shows the histograms of the fitness function, from which it can be observed that most fitness data are located at the very right (larger fitness) end and their values are approximately 0. The distribution of fitness values across the full fitness range indicates that FAVP is an efficient tool for optimization, as introduced in Section 7.4.

Table 21.6 The Normalized Optimal Parameters by FAVP

Variable	MEAN±STD
C_0	0.6548 ± 5.0385
θ_1	5.2501 ± 3.0927
θ_2	0.2838 ± 1.9646
θ_3	-0.2952 ± 1.4105
θ_4	-0.6391 ± 1.3201
θ_5	-0.2527 ± 1.7784
ω_1	1.3093 ± 0.9530
ω_2	1.2381 ± 0.9525
ω_3	1.2671 ± 1.4878
ω_4	1.7667 ± 1.4061
ω_5	1.4285 ± 1.3181

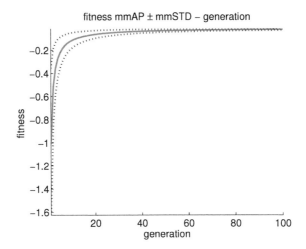

Fig. 21.2 Fitness curves of mmAP ± mmSTD over the simulation

Figure 21.5 shows the approximation of the electricity consumption using the statistical data from 1980 to 2009, where the solid line is the MEAN value of the estimated $E\hat{E}C$ (theory) and the circles are the statistical EC_0 (real). For the period from 1980 to 2009, Figure 21.5 demonstrates the dynamic behaviors of China's electricity consumption, which increases each year in accordance with the acceleration of GDP growth. These optimal results should be restored to the original scale for practical use and then utilized for predicting the consumptions in the years 2010 to 2012 in this case.

As shown in Figure 21.6, the EEC approximation error (for the years 2010 to 2012) is plotted with '*' (history), and the EEC prediction error (for the years 2010 to 2012) is plotted with 'o' (prediction). Note that the errors of the estimated and real EEC between 1980 and 2009 are within the range of ±15 per cent. Similarly,

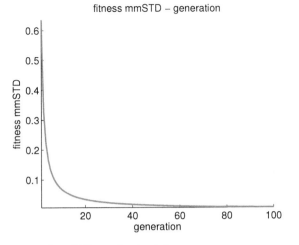

Fig. 21.3 Fitness curve of mmSTD over the simulation

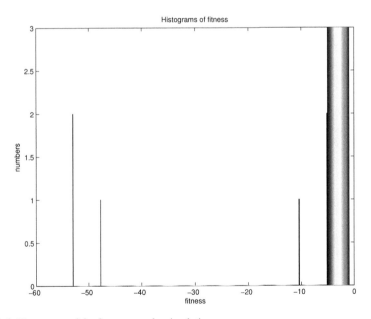

Fig. 21.4 Histograms of the fitness over the simulation

the errors for the years 2010 to 2012 are in the range of [-8 per cent, -5 per cent], as shown in Figure 21.6. The errors are also presented in Table 21.7, which lists the error values associated with the approximations between 1980 and 2009 and the predictions between 2010 and 2012, respectively.

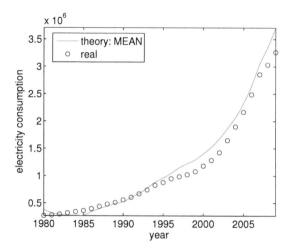

Fig. 21.5 EEC approximation, 1980 to 2009

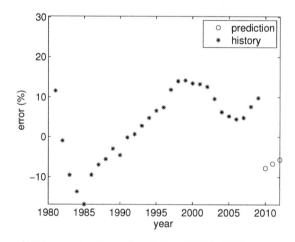

Fig. 21.6 Error of EEC approximation and prediction, 1980 to 2012

Table 21.7 Errors of EEC Approximation and Prediction

	Year	Error (%)
Approximation	1980 to 2009	5.9250 ± 13.9429
Prediction	2010	-7.7111 ± 5.45
Prediction	2011	-6.6781 ± 6.78
Prediction	2012	-5.6106 ± 5.49

21.5 Social, Economic and Environmental Impacts

To demonstrate the social, economic and environmental impacts using the proposed model, Figures 21.7 to 21.11 present the '1-to-1' relationship between various pairs of quantitative variables in the '2D' graphs, and Figures 21.12 to 21.15 show the '1-to-2' relationships among the design variables in the '3D' graphs, which are generated under the condition that all the remaining variables (aside from the x-y-z variables) remain constant in these figures with the corresponding mean values.

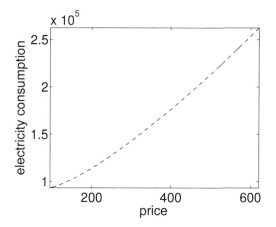

Fig. 21.7 EEC vs. price

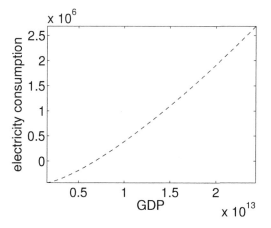

Fig. 21.8 EEC vs. GDP

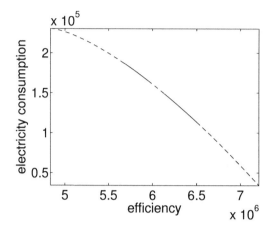

Fig. 21.9 EEC vs. efficiency

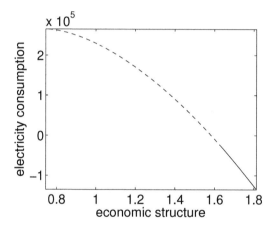

Fig. 21.10 EEC vs. economic structure

From an economic perspective, the increase in price will drive the EEC growth, as shown in Figure 21.7. Figure 21.8 shows that GDP growth requires greater EEC, that is, under the current economic structure and social environment, the pursuit of GDP leads to greater demands for electricity. As can be observed in Figure 21.9, the increase in the efficiency leads to a decrease in EEC, which implies that efficiency improvements can reduce the EEC. Figure 21.10 shows that, when the economic structure increases, the EEC decreases, suggesting that industrial consumption dominates residential consumption, and a reduction in industrial production's EEC causes a reduction in total EEC. Figure 21.11 shows how the environmental factor indicates the EEC behavior, where the increasing values of CO_2 correlate with the decreasing EEC under the current industrial conditions of the electricity generation and consumption structure.

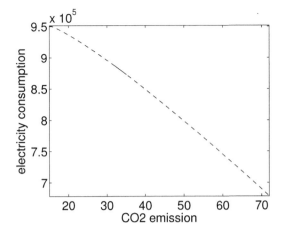

Fig. 21.11 EEC vs. CO_2

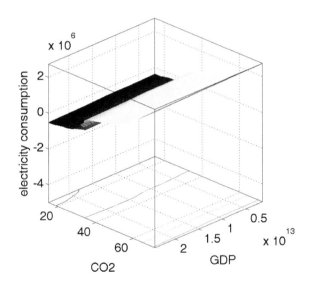

Fig. 21.12 EEC - GDP - CO_2

To further demonstrate the social-economic-environmental coupled impacts, Figures 21.12 to 21.15 present four '3D' figures for evaluating these impacts: Figure 21.12 shows that CO_2 emissions are strongly affected by EEC, an increase in EEC leads to an increase in CO_2 emissions, and EEC is not sensitive to changes in GDP, which indicates that a variation in EEC could cause greater environmental impacts (CO_2 emissions) and smaller economic impacts (GDP). Figures 21.13 and 21.14 show that EEC is sensitive to changes in utility prices. Figure 21.15 shows that

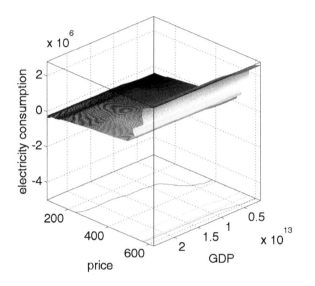

Fig. 21.13 EEC - price - GDP

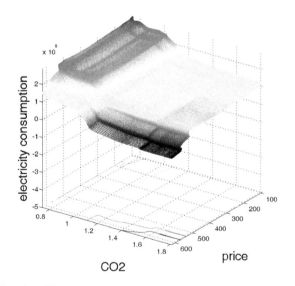

Fig. 21.14 EEC - price - CO$_2$

industrial production consumes more electricity than do residential users, which justifies the social-economic coupled impacts on EEC.

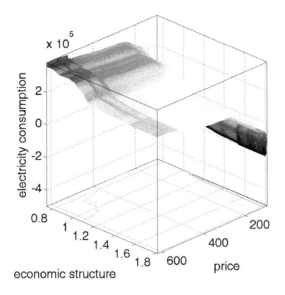

Fig. 21.15 EEC - price - ecostruc

21.6 Conclusions and Future Work

In this study, a quantitative model incorporating social, economic and environmental coupled impacts was proposed to analyze the long-term relationship between electricity consumption and its major macroeconomic variables. The proposed model was verified and validated on China's national electricity consumption data between 1980 and 2012. Based on the trend indices mmAP and mmSTD, a fitness function was constructed with three variables, i.e., the data shifting variable, the coefficient vector of the impact factors, and the exponent vector of the impact factors. Under the CIAD framework, the model parameters are optimized with a variable-population-based firefly algorithm. The research findings can be summarized in three aspects.

- The proposed prediction model for the national electricity consumption achieves good agreement with the actual consumption data. We also found that EEC is sensitive to the price and the economic structure, thereby providing a strong argument for implementing sustainable energy policy
- Under the current social-economic structure, EEC has a direct impact on the environment due to the heavy dependency on conventional generation technology. Hence, it is imperative to migrate from fossil fuel-based technology to wind and solar power
- Energy consumption may not be able to directly accelerate GDP growth under the current social-economic conditions and industry infrastructure, although GDP

growth always implies a higher demand for electricity. Deploying energy conservation policies could be a good vehicle to sustain and accelerate GDP growth

Our further research will focus on developing new types of CI algorithms, such as the swarm bat algorithm, the swarm fish algorithm and the MO GA, to optimize energy consumption and on validating the models based on sustainable and renewable energy data. To achieve long-term energy savings and carbon reductions, additional experimental research is expected to establish social, economic and environmental coupled behaviors.

22

CIAD Gaming Support for Electricity Trading Decisions

22.1 Introduction

This chapter illustrates how CIAD may be used to evolve decision support in the form of game theory, which is intractable via conventional optimization means. It is generally believed that opening the power industry to competition would benefit the trading participants and improve economic efficiency. Since the 1980's, much effort has been made to restructure the traditional monopoly electricity industry. Whilst the details differ, the core of this reform involves the introduction of competition among electricity generators and suppliers through the creation of a deregulated electricity market. However, the California electricity trading market, which is regarded as a benchmark example for worldwide energy industries, shockingly collapsed after an energy crisis and, meanwhile, some generation companies in California made speculatively high profits [Martha McNeil Hamilton and Greg Schneider (2001)]. Further, the cause to the biggest blackout in eastern America recently is still unclear [Richard Perez-pena (2003), John M. Broder (2002)].

Ideally, the market structure and management rules in an electricity market are expected to be well designed. However, the emergence of market power and collusion among energy companies has been drawing more attention to strategic gaming behavior and market systems on global electricity trading.

In March 2001, the New Electricity Trading Arrangement (NETA) was implemented to operate power markets in England and Wales. The trading management mechanisms are still on trial operation for improvement. The improvements so far do not however alter the fact that there exist loopholes which can be exploited by market participants and could have led to collapse [The New Electricity Trading Arrangements (1999a)] of the Californian electricity crisis of 2001. System models that reflect human intelligence in trading, market power and gaming strategies need to be developed and enhanced so as to avoid or significantly reduce disruptive trading operations and distorted market prices in NETA [The New Electricity Trading Arrangements (1999a)].

An intelligent decision-making and support technique, 'game theory', is often used in market practice. Game theory is a discipline concerned with how individuals make decisions when they are partly aware of how their action might affect each other and when each individual might take this into account. In general, there are three ways to develop intelligent and optimal trading strategies. The first one relays estimations of the market clearing price (MCP) in the next trading period, which is relatively simple in principle. Based on estimates of the MCP, it is straightforward for a power supplier to determine its strategy by simply offering a price a little cheaper than the MCP. The second utilizes estimates of bidding behaviors of rival participants, which is more challenging. The third is game theory which is most sophisticated and involves market simulation and empirical methods [Paule Stephenson (2001)].

A good market model therefore requires taking gaming behaviors into account, since gaming strategies are widely practiced in trading systems for both decision-making and decision support. Modeling such a system is, however, extremely challenging, since conventional optimization breaks down in dealing with non-numerical inferences. However, evolutionary computation (EC) with its a-posterior and coded search power can make such model building realizable.

In this chapter, following the analysis of intelligent market behaviors in Section 2, models are developed in Section 3 primarily to analyze the process of how power generators attempt to employ gaming strategies on NETA. Then these models are validated in Section 4 and their consequences are studied and possible market equilibria under such situations are searched for in Section 5, again using the power of EC. The research is formulated as both a decision-support problem and an extra-numerical global optimization problem. Finally, conclusions are drawn in Section 6.

22.2 Modelling Intelligent Market Behaviors

22.2.1 NETA Market Price Formulation

NETA is an electricity trading arrangement for bilateral trading between generators and suppliers. On NETA, there is a high level of over-capacity in the market and hence market prices often gradually drop down. An objective behind generators' present gaming strategies is, therefore, to manipulate the market prices through reaching coalition among main generators to drive the marketplace to an oligopoly situation for a higher profit margin. In order to optimize the efficiency of these strategies, all sides involved need to improve in real-time and constantly search for optimal solutions during the trading procedure.

A basic structure of these models was published in the documentation of Office of Gas and Electricity Markets [Paule Stephenson (2001), Brian Saunders (2001), The New Electricity Trading Arrangements (1999b), NGC Incentives Under NETA (2000)] and [Settlement Administration Agent User Requirements Specification (2000)]. Figure 22.1 demonstrates a representative structure of these models, based upon two sequential markets [Brian Saunders (2001)].

To explain this market mechanism, let PX stand for forward and spot markets, which evolve in response to the requirements of participants. This will allow bilateral contracts for electricity to be struck over time-scales ranging from long term to on-the-day markets. The PX market clearing price is defined as:

$$PXP = f(Q_{SPX}^{1},.., Q_{SPX}^{i}, P_{SPX}^{1},.., P_{SPX}^{i};$$

$$P_{BPX}^{1},.., P_{BPX}^{j}, Q_{BPX}^{1},.., Q_{BPX}^{j}) \quad (22.1)$$

where $i = 1, 2, ... n$ represents the number of generators involved; $j = 1, 2, ... M$ represents the number of suppliers; Q_{SPX}^{i} and P_{SPX}^{i} are the quantity and price generator i wants to sell at PX, Q_{BPX}^{j} and P_{BPX}^{i} are the quantity and price supplier j wants to buy at PX.

After the PX, market participants submit a set of offer-bid pairs to a Balancing Mechanism (BM) to indicate the willingness of participants to operate at a level above or below their final bilateral contracts. The quantities contracted will be compared with the quantities generated or consumed to calculate the imbalances. If a plant is generating more than it has contracted or if a supplier is consuming less than it has contracted, transactions will be made at the System Sell Price (SSP), which is a weighted average of accepted bids. If a plant is generating less than it has contracted or if a supplier is consuming more than it has contracted, one pays the bid prices, i.e., System Buy Price (SBP), which is a weighted average of accepted offers.

The spread between the two prices is intended to provide a penalty for being out of balance. The SSP (SBP) is expected to be considerably lower (higher) than forward market price PXP [The New Electricity Trading Arrangements (1999b), NGC Incentives Under NETA (2000), Settlement Administration Agent User Requirements Specification (2000)] and [Balancing Mechanism Reporting System].

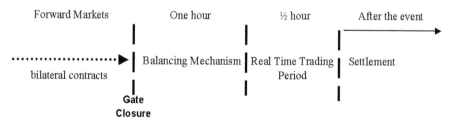

Fig. 22.1 Trading stream on NETA

22.2.2 Generator Gaming Strategies

The objective of generation companies' gaming strategy is to set up a collusion agreement among some main generators, at which these particular players keep withholding power output volumes during some specific periods, i.e., a number of weeks in the winter or summer, hence leading the whole marketplace to an oligopoly situation and drive up the market prices.

There are some uncertainties involved with this strategy that need be addressed by the agreement members, i.e.,

- As NETA consists of two separate markets, i.e., PX and BM, how do they arrange output volumes between them to make the most profits on this strategy?
- For each generator, what should be the optimal withheld output capacity and selling prices?
- How do they keep coalition generators loyal to carry out the agreement?
- Is it likely that there exists equilibrium that collusive generators can make best profits meanwhile the markets trading can be kept in balance, i.e., does it not collapse?

The market clearing prices, i.e., PXP, SSP and SBP are results of interactions among all market players' bid/offer prices and output volumes. Because most of power volume trading is carried out on PX, imbalance penalties in BM are much higher than PX clearing prices. Therefore, the core of gaming generators' strategies is to withhold their output volumes on PX and to make the supply/demand unbalanced in this market. As a consequence of maintaining this strategy, power suppliers will be driven to BM and purchase the shortfall with imbalance charges.

This sort of collusion among gaming generators is termed as 'cooperative strategy' in game theory. As its legality might be doubted, the agreement exists only in verbal form, which is actually tolerated and even encouraged by a number of European countries [Charles L. Cole (1995)].

When evolving an intelligent model, the following features of the market need to be considered:

- Trading participants include n electricity generators (sellers), as some generation companies that sell energy in the market, m power suppliers (buyers), as energy service companies, i.e., power transmission companies that buy electricity to serve end-users, and the System Operator (SO) who operates the markets
- Double-side bidding mechanism is adopted following the fact on NETA
- Colluded generators are concerned about the expected payoff in the long run rather than the pay-off in a particular round of auction
- System Operator broadcasts 2–14 day-ahead demand forecast and provides real-time information and offers made and accepted, as is the case on NETA
- For simplicity, the demand elasticity, transmission constraints and loss are ignored when SO matches selling and buying quantity

The following cooperative behaviors also need to be taken into account:

- Each member of the agreement withholds a portion of its total capacity, as variable X, expressed as a percentage of its total generation capacity. The range of X is assumed to be between 10 per cent to 40 per cent. Then the remaining volume $Q_{smax}{}^i.(1–X)$, is traded into the PX
- After the suppliers are driven to BM and have to submit bids for getting extra supply with paying SBP, the gaming generators need to provide offers to BM to meet the shortfall demand and determine how much volume should be taken from the withheld volume $Q_{smax}{}^i. X$ to trade in BM. Given the part taken from $Q_{smax}{}^i. X$ is Y, expressed as a percentage from 0–100 per cent

- The last part of the cooperative strategies is to optimize the trading on forward markets. Because the state of suppliers is no longer superior as before when the market is under an oligopolistic condition, generators can improve their selling curves to drive up the market prices as high as the suppliers could accept under PX

In building the market model, each generator is characterized by three sets of parameters:

- Fixed electricity generation parameters, i.e., maximal generation capacity $Q_{smax}{}^i$, marginal cost $P_{ma}{}^i$, etc.
- Strategic variables: X being generator i's portionparameter on PX, $P_{SPX}{}^i$ being the price that generator i wants to sell on PX, $Q_{SPX}{}^i$ being the quantity that generator i wants to sell on PX, portfolio instrument l expressed as a percentage of its total generation capacity, BM offer price $P_o{}^i$ and $Q_{SBM}{}^i$ being the quantity generator i wants to sell at BM. Their relationship is formulated as:

$$Q_{smax}{}^i \cdot 1 = Q_{SPX}{}^i + Q_{SBM}{}^i \qquad (22.2)$$

- Collusion parameters: PTR, T, Qcomp and Qcoop

There are two types of gaming generators—one follows a strategy called 'opportunistic collusion', whereby generators withhold capacity from the market only when an 'opportunity' to raise profits by doing so exists. Opportunistic collusion might result in a generator setting aside a portion of their capacity and deciding for each hour whether or not to offer that capacity to the market depending on the expectations of raising profits. This is different from the other type, suggesting that generators should 'always' withhold a portion in anticipation of an agreement. The second kind is named 'loyal co-operator'.

For making the agreement more efficient, a more extreme management-enforcement is utilized to constrain the agreement members by 'loyal co-operator'. In this application, a well-known game theory strategy, 'trigger price strategies' [Bierman, et al. (1998)], is employed to enhance this agreement. On a trigger price strategy, collusive generators make inferences about any member in this agreement from the observation of market price P_{PX}. If the price remains above some critical value—the trigger value—then the generators will infer no cheating on the coalition and will maintain a cooperative output level. If the price falls below the trigger, then some punishment must be imposed on the cheater(s).

Trigger price strategy depends on four parameters, P_{TR}, T, Q_{comp} and Q_{coop}, where P_{TR} is the trigger price, T is the number of time periods the punishment will last, Q_{comp} is the competitive output given 100 per cent generation volume $Q_{smax}{}^i$ and Q_{coop} is the cooperative output given $Q_{smax}{}^i \cdot (1{-}X)$.

There exist other generators who do not join the collusion and each of them independently trades all of its generation volume on PX. As a result, the state of such generators is inferior to suppliers because the latter have enough choices to select generators with low selling prices to make contracts, and hence all suppliers' demands are theoretically satisfied. The contracted prices, as forward markets prices, could be as low as what suppliers could accept. Consequently, generators can only sell out parts of their total volumes at P_{PX} level.

22.2.3 Supplier Gaming Strategies

For suppliers, the dual cash out prices of BM are intended to discourage market participants from being out of balance because the penalty for contracting at less than actual demand can be extremely high. Suppliers have, therefore, responded to NETA imbalance prices by over-contracting to reduce exposure to SBP [The Review of the First Year of NETA (2002)]. The cost of over-contracting can be viewed as an insurance premium that reduces exposure to the potentially high risks of being short.

Each supplier's objective is to optimize its contract position, as well as trading prices, to minimize the cost of contracting in order to maximize total daily profits. The strategy of each supplier j, is characterized as following [McClay (2002)]:

$$C_L = \sum_r^{48} PXP^r . Q_D^{\ r} \tag{22.3}$$

where C_L is the marginal cost of supplier j, r is the settlement period number, PXP is the PX clearing price and $Q_D^{\ j}$ is the actual demand at settlement period r:

$$C_S = \sum_r^{48} (PXP^r . Q_C^{\ r} - Max[0, Q_C^{\ r} - Q_D^{\ r}] . SSP^r + Max[0, Q_D^{\ r} - Q_C^{\ r}] . SBP^r) \tag{22.4}$$

where C_S is the contracted cost of supplier j, $Q_C^{\ j}$ is the contracted volume at settlement period r on PX. A percentage premium for supplier's strategy is derived from (C_S/C_L); the lower the premium the more efficient the strategy.

22.3 Intelligent Agents and Modeling

Based on the analysis in the previous section, a basic model structure involving thermal, nuclear, combined cycle gas turbine (CCGT) and renewable plants is presented in Figure 22.2. Cooperative generators have many strategic variables, i.e., $P_{SPX}^{\ i}, Q_{SPX}^{\ i}, l, X, P_O^{\ i}, Q_{SBM}^{\ i}$, that need to be optimized. Further competitive generators face a dilemma. On the one hand, they have to make their selling prices appropriately low to win contracts; on the other, they need to offer selling prices higher than individual marginal cost $P_{ma}^{\ i}$ to cover the production cost. On the other side of this competition, supply companies also face the evaluation and optimization problems expressed in Equations (22.3) and (22.4).

The major task here is to model generators and suppliers as decision-making participants. Many performance variables in the power market trading strategy development do not present accurate measurements. Conventional mathematical models are hence inadequate here. Many incommensurable and competing objectives require meeting before any solution is considered adequate. However, these can be handled by a genetic algorithm (GA), which is a representative of search, machine learning and optimization techniques that are non-deterministic and aposterior. GA employs coding and hence deals with non-numerical variables. An organism's

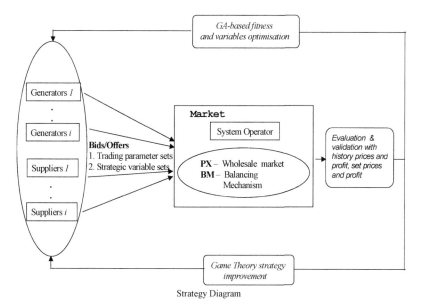

Fig. 22.2 Structure of a market model

genetic code is its position in solution space while its survival in its environment and its number of offspring indicates probabilistically the degree to which it meets its objectives. It is a powerful technique in the search for global optimal solutions and hence, a GA is employed to solve these game theory activated extra-numerical search and optimization problems.

Strategic variables and parameters of market players are mapped into GA chromosomes. Each auction round represents a generation. The GA population is divided into sellers and buyers. Information is exchanged solely within each type of trader. There is no information exchange between buyers and sellers other than the amount of profit they made known. The fitness of each trader is proportional to the profit made in the auction round and is recalculated in every round. Once a population of individuals with assigned fitness values arises, the next step is to preferentially select a subset of individuals that should survive into the next generation. Figure 22.3 shows the search and evaluation process and Table 22.1 lists GA parameters used.

Here, the tournament selection scheme is employed, based on group competitions. The population is divided into subgroups, which can be of any size; or members with the best fitness among the subgroups get selected. The uniform crossover method is used in which offspring individuals are created from a randomly generated uniform bit mask. An elitism technique is also implemented.

It is worth noting that bounded rational agents are built to learn about their environment and improve their trading behavior with experience. However, at the same time the agents' behavior should not be too non-rational (some lower bounds were imposed on the agents' level of rationality). It has to be made sure that the agents do not choose completely unreasonable actions, be it in the early stages of the learning process.

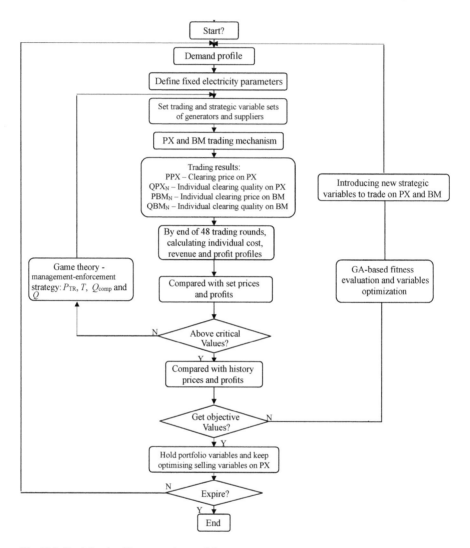

Fig. 22.3 Evolving intelligent gaming models

In order to avoid inconsistent behavior during the learning process, some lower bounds of rationality are imposed through operational rules, including:

- *Adaptive expectation*: Never bid (or offer) above (or below) the previous SBP (or SSP).
- *Avoidable cost*: Never pay more than the marginal cost for 'speculative' decrement.

Table 22.1 GA Parameters in Evolutionary Model Building

Input Specifications	Values
Number of generations (Suppliers)	480
Population size (Suppliers)	20
Chromosome length (Suppliers)	9
Number of generations (Generators)	480
Populations size (Generators)	20
Chromosome length (Generators)	15
Selection mechanism	Tournament
Crossover	70% uniform
Mutation	0.1%

22.4 Model Analysis and Verification

Once NETA market is successfully modeled, the model confirms that generators have an incentive to withhold capacity from the market. The model also reveals that there are two types of gaming generators. The first is the classical 'tacit collusion' that occurs in static repeated withholding output capacity, where the object is for all players to learn that they can always make excess profits if they withhold the amount of capacity from the market. This kind is referred to 'loyal co-operators', suggesting that these generators should 'always' withhold a portion in anticipation of an agreement. The second type asserts, however, that it is not always profitable to withhold capacity from the market, since the opportunity for raising profits does not always exist due to internal factors and external, such as collaborative generators breaking the agreement, the demand bid, imbalance prices, etc. We refer this phenomenon as 'opportunistic tacit collusion' to distinguish it from the classical 'tacit collusion'. These generators follow an 'opportunistic collusion' strategy whereby generators withhold capacity from the market only when they perceive an 'opportunity' to raise profits by doing so exists. Opportunistic collusion might result in a generator setting aside a portion of the capacity and deciding on each trading round whether or not to offer that capacity to the market, depending on expectations of raising profits. Once this is learned, suppliers 'tacitly collude' to sustain high market prices.

For the 'opportunistic collusive' generators, it is difficult to judge an 'opportunity' to earn more profit by estimating possible profit with cooperative strategy. Since in a certain market environment where a wide number of market participants are trading interactively, there are uncertainties and it is unlikely to precisely predict all the participants' future moves and trading consequences. Nevertheless, the market clearing price in PX, *PXP* and individual generators' capacity used in both PX and BM, are introduced as the reference for the 'opportunistic collusive' generators to decide whether or not to join the coalition agreement and withhold capacity from the market.

22.4.1 Small-scale Model Simulation

In order to gain a better view of the effects of gaming trading strategies on NETA, the first application experiment is carried out based on a small-scale model. The total available generation capacity is assumed at 33.3 GW in this experiment. The number of generators, m, is assumed to be 5, and suppliers', n, is assumed to be 4 in this experiment. The total demand is set as 25 GW; therefore, the individual maximal demand $Q_{dmax}{}^i$, of each supplier is set at 6250 MW. These experiments are based on the standard daily demand profile of November 2004, published by NETA and shown in Figure 22.4.

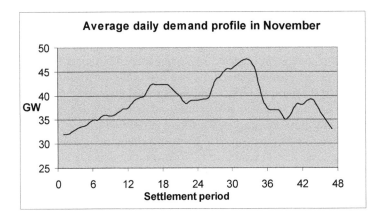

Fig. 22.4 Standard daily demand profile in November [Charles L. Cole (1983)]

The generation system self-parameter of this model is demonstrated in Table 22.2.

Table 22.2 Generators' System Self-parameters

Type of Generation Plants	Nuclear	Combined Cycle Gas Turbine (CCGT)	Large Coal	Gas Turbines	Oil
Marginal generation cost (£/MWH)	24.50	9.72	33.23	66.95	87.91
Maximal generation capacity (MW)	6600	6600	6600	6600	6600

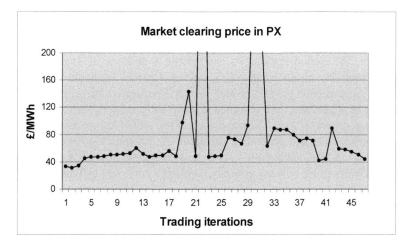

Fig. 22.5 Mean daily market clearing price in PX

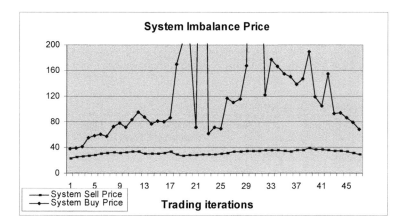

Fig. 22.6 Mean daily System Imbalance Price

The wholesale market clearing price PXP, imbalanced settlement prices—System Buy Price (SBP) and System Sell Price (SSP) are major model outputs to assess the performance of market players' different strategies. They are presented and evaluated as follows. Figure 22.5 demonstrates the mean daily power exchange market prices in a week.

Figure 22.6 demonstrates the mean daily price curves of imbalanced settlement prices SSP and SBP in a week.

22.4.2 Large-scale Model Simulation

In order to compare the effects of market player strategies under different market circumstance, the second application experiment is taken on a large-scale model which is comparably similar to the NETA market. The total available capacity is set same as the experimental model at which the non-cooperative strategy is employed. The number of generators, m, is assumed to be 15 and suppliers, n, is assumed to be 10 in this experiment as well. The generators of the same type are assumed to have similar marginal costs and generation capacity. The generators' system parameters, i.e., estimated marginal generation costs $P_{mc}{}^i$, assumed maximal generation capacity $Q_{smax}{}^i$, of each generator on each generation type are presented in Table 22.3 below. The total available generation capacity is set as 66.7 GW. Oppositely, the maximal market demand is 50 GW, which is same as the market scale of the NETA, so that the average maximal demand $Q_{dmax}{}^i$, of each supplier is 1125 MW. The ratio of maximal market demand to total available generation capacity is set as 0.75, following the real situation in NETA. Large-scale experiments are based on the same winter daily demand profile as introduced in the previous experiment. The generation system self-parameters are shown in Table 22.3.

The model major outputs market clearing price PXP, imbalanced settlement prices—System Buy Price SBP and System Sell Price SSP—are presented and evaluated. Figure 22.7 shows the mean daily market clearing price in PX in a week, while Figure 22.8 gives the mean daily System Imbalance Price in a week.

Table 22.3 Generators' System Self-parameters

Type of Generation Plants	Nuclear	Combined Cycle Gas Turbine (CCGT)	Large Coal	Gas Turbines	Oil
Marginal generation cost (£/MWH)	24.50	9.72	33.23	66.95	87.91
Maximal generation capacity (MW)	4450	4450	4450	4450	4450

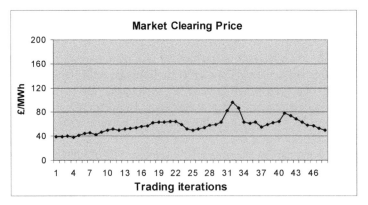

Fig. 22.7 Mean daily market clearing price in PX

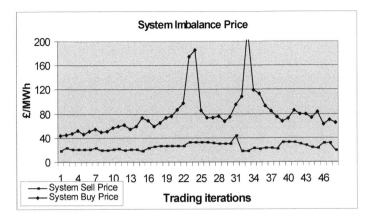

Fig. 22.8 Mean daily System Imbalance Price

22.5 Applications of the Model

A set of tests are carried out in developing different gaming strategies using the model. The estimates used are consistent with those used in published studies on the UK electricity market, i.e., actual demand profiles, generation and supply parameters. There are different sorts of power generation concerned in this model. After the computing is finished, output sets are evaluated to find out the market states and consequences of the strategic generators that maintain the gaming behaviors.

22.5.1 Competitive Strategy

Test results show that market prices on PX are kept on a quite low level and an equilibrium state is achieved when all generation companies are independently trade without any collusion. Figure 22.10 displays this tendency based on published data.

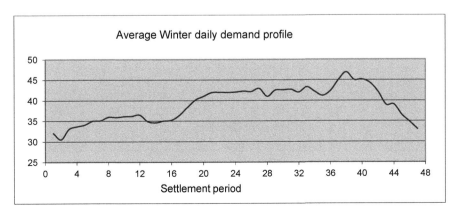

Fig. 22.9 Average winter demand profile

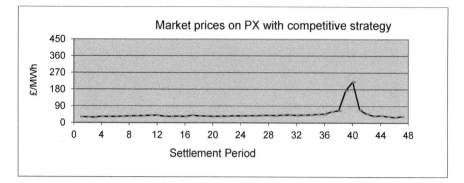

Fig. 22.10 Trend of PX without gaming strategy

22.5.2 Cooperative Strategy

On the second set generators are divided into two groups: some adopt competitive strategy with never joining coalition agreement; the others are gaming players. Figure 22.11 presents that the market prices on PX are pushed up to a significantly high level and a serious price spike is caused about 20 times than average marginal cost. This result is still far lower than the spike happened in California energy crisis that was 40 times than average marginal cost. Programming results demonstrate that when half of main generators, who use 'opportunistic collusion' strategy at trading, quit this collusive agreement sometimes when they find that they can make more profits if they are outside the coalition, market price cannot be pushed over £500/MWh.

The difference between the profit of CCGT (combined cycle gas turbine) generator with gaming strategy and the one without strategy is shown in Figure 22.12.

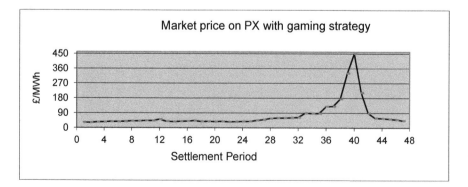

Fig. 22.11 Trend of PX with gaming strategy

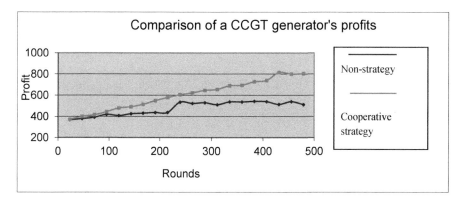

Fig. 22.12 Comparison of a CCGT generator's profits

Another method to evaluate the two strategies is the 'percentage premium'.

$$\text{Percentage premium} = \text{Imbalance penalty/Profit} \qquad (22.5)$$

To avoid the risk of being penalized at the imbalance settlement stage, generators offer over-contract as an insurance against plant outage, running their sets below their optimum efficiency, whilst suppliers ensure that their contracts exceed their highest estimate of demand with insurance against the symmetric penalties. This over-contracting strategy is expressed as over-contracting premium on functions (22.3) and (22.4). Figures 22.13 and 22.14 present the comparison between the over-contracting premium for competitive strategy and gaming strategy, respectively for a CCGT generator and a supplier. Figures 22.15 and 22.16 illustrate the comparison of the unbalancing volumes and prices for this particular generator. Figure 22.18 shows that the CCGT generator pays much less penalty when gaming strategy is employed.

To evaluate efficiency of this strategy, model outputs are compared with the results of another strategy [Bunn (2001)]. Figures 22.17 and 22.18 demonstrate the difference.

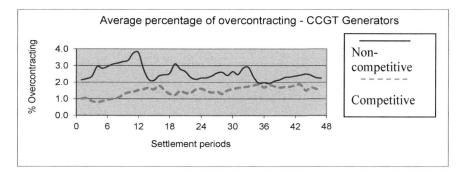

Fig. 22.13 Comparison of CCGT's percentage over contraction

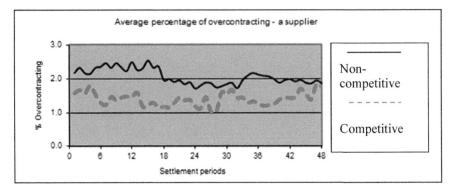

Fig. 22.14 Comparison of a supplier's percentage over contraction

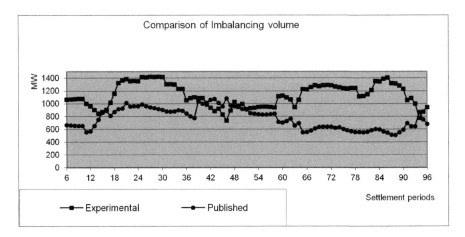

Fig. 22.15 Comparison of the unbalancing volume

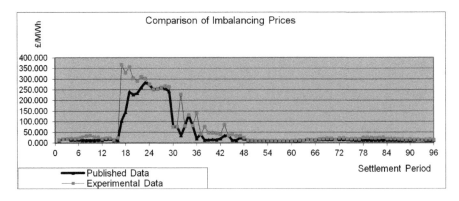

Fig. 22.16 Comparison of the unbalancing prices

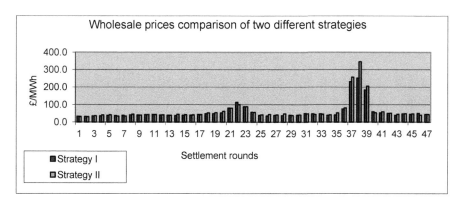

Fig. 22.17 Comparison of wholesale prices

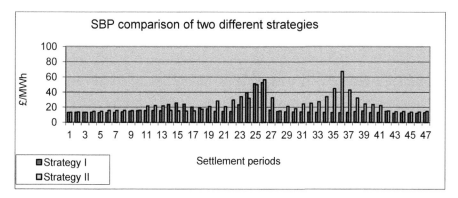

Fig. 22.18 Comparison of SBP prices

22.6 Conclusions

A GA and its genotype and phonotype parameters have enabled intelligent modeling of the NETA electricity market. The ability of EC in including game theory allows gaming strategies of power generation firms to be well simulated and built into the model. The market manipulation and strategic trading behaviors made by market players can hence be analyzed and predicted to a certain degree. Therefore, a hybrid approach combining the advantage of both the game theory and evolutionary computing has been developed as intelligent support techniques for decision-making on electricity trading strategies under NETA.

Test results show that the model evolved is a helpful tool in developing competent strategies and that profits of generators with a gaming strategy are much greater than generators who trade independently. The model also reveals that

- When some generation companies adopt a cooperative strategy, the market prices are driven up to a level significantly higher than when no gaming is played

- Fitness, percentage premium and overcontracting premium have shown that the profits of generators with a gaming strategy are much greater than generators who trade without
- 'Trigger price strategies' prove to be a useful method of executing the agreement as the market clearing prices and management-enforcement are kept considerably high
- On a non-natural oligopoly electricity market, where the total supply exceeds total demand, the effort of finding a coalition among main generators to drive the market to an oligopoly setting cannot achieve its original targets because some main generators are always on the agreement
- Manipulation on market prices can be accomplished when at least half of generation-volume holders loyally carry out the agreement. An equilibrium trading state is kept on a balanced price level.

23

Dynamic Behavior of Rural Regions with CO_2 Emission Estimation

23.1 Introduction

In the past decade, China's fast economic growth has been contributing to global economic growth, although this economic growth has created serious environmental problems [Liu and Diamond (2005)]. In particular, the effects of China's large population, rapid urbanization, the increase in householder's wealth demands and local economic reconstruction have influenced regional development and caused low-efficiency environmental recycling. As a representative research issue, the impacts of various air pollutants, such as carbon monoxide (CO), nitrogen oxides (NO_x), silicon dioxide (SO_2), on regional economic development have been discussed.

This line of study is part of the discussion on global warming, which is one of the most serious environmental issues that humans currently face. The mechanism of global warming is not yet fully understood. However, it has been considered significantly related to the emission trends of anthropogenic greenhouse gases, such as carbon dioxide (CO_2) and methane (CH_4), which has aroused researchers' interest and resulted in various counter measures. Meanwhile, technological development and abatement policies based on economic planning and international negotiations have been discussed and partially applied [Peters, et al. (2006), Akimoto, et al. (2006)].

As has been widely accepted, CO_2 emissions are connected with most human activities mainly through fossil-fuel-based energy consumption, for example, from fossil fuel power stations. To evaluate the dynamic behavior of particular regions concerning energy consumption and CO_2 emissions, various interdisciplinary studies have been proposed, which are reviewed as follows. Berling-Wolff and Wu [BERLING-WOLFF and Wu (2004)] reviewed the historical development of urban growth models and summarized that various disciplines and diverse theories, such as fuzzy logic theory and neural network theory, could be combined to generate new models. Nejadkoorki et al. [Nejadkoorki, et al. (2008)] proposed a model to estimate the road traffic CO_2 emissions of an urban area with three components. Weiss et al. [Weiss, et al. (2009)] presented a bottom-up model for estimating the non-energy use of fossil fuels and CO_2 emissions. Briefly, these studies suggest that work on dynamic models needs to be considered for integration with interdisciplinary factors, especially the interaction between economic activities and environmental factors. In this context, the studies attempt to examine the dynamic behavior of rural areas, with particular focus on the spatial features of CO_2 emission and economic growth.

Specifically, in this paper, we propose a GA method applying Mendel's principles (Mendel-GA) for a spatial analysis that incorporates three economic components and a carbon dioxide (CO_2) emission factor. Our empirical analysis evaluates the shaped dynamic behaviors of the rural regions of Chongqing at the administrative county level.

23.2 CO_2 Emission Estimation of Productive Activity

According to the 'Revised 1996 IPCC Guidelines (Reference Manual, Volume 3)' [IPCC (1997)], the estimation of CO_2 emissions, which relates to productive activity, is based on three major factors for each fuel type—the heat conversion factors (HCFs), the carbon emission factor (CEF) and the fraction of carbon oxidized (FOC) in each sector. As listed in Table 23.1, the three major series are obtained from different data sources [IPCC (1997), Tsinghua (2000), Qu, et al. (2010)] and include three parts: liquid (row A, $i = 1$ to 10), solid (row B, $i = 11$ to 16) and gaseous (row C, $i = 17$).

To establish specific values of the FOC, Qu [Qu, et al. (2010)] suggests that the sector-specific values of the FOC for coal vary between 80–95 per cent, which are smaller than the IPCC standard of 98 per cent.

Equation (23.1) is the CO_2 emission estimation (E) approach provided by the IPCC [IPCC (1997)], which includes main fuel combustion, as listed in Table 23.1. China typically converts all its energy statistics into "metric tons of standard coal equivalent" (TCE), a unit that bears little relation to the heating value of the coals actually in use in China [National Research Council (2000)], whose HCFs, CEF and FOC are given in row $i = 0$ in Table 23.1.

$$E = \sum_{i=1}^{n_1} \tau \gamma_i \left(\alpha_i \beta_i A_i \times 10^{-3} - S_i \right) \tag{23.1}$$

where
  - E are the CO_2 emissions, $10^3 t$ (ton)
  - $\tau = 44/12$ is the molecular weight ratio of CO_2 to C [1]
  - γ_i is the fraction of carbon oxidized ($i = 1, 2, ..., n_1$), $n_1 = 17$
  - α_i is the heat conversion factor, $TJ/10^3 t$

---

[1] Conversion between C and CO_2 can be performed using the relative atomic weights of the carbon and oxygen atoms, $C = 12$ and $O = 16$. The atomic weight of carbon is 12, and that of CO_2 is 12 + 16 + 16 = 44 (i.e., one carbon and two oxygen atoms). To convert from C to CO_2, multiply by 44/12 or 3.67 [IPCC (1997)]

Table 23.1 Heat Conversion Factors, Carbon Emission Factor, and Fraction of Carbon Oxidized of Various Fossil Fuels [IPCC (1997), Tsinghua (2000), Qu, et al. (2010)]

i	Fuel	Conversion Factor $\alpha_i(TJ/10^3\ t)$	Emission Factor $\beta_i(t\ C/TJ)$	Fraction of Carbon Oxidized γ_i
0	TCE	29.27	24.74	0.90
	(A) Liquid			
	Primary Fuels:			
1	Crude Oil	42.62	20.00	0.98
	Secondary Fuels:			
2	Gasoline	44.80	18.90	0.98
3	Kerosene	44.67	19.55	0.98
4	Diesel Oil	43.33	20.20	0.98
5	Residual Fuel Oil	40.19	21.10	0.98
6	LPG	47.31	17.20	0.98
7	Naphtha	45.01	20.00	0.80
8	Bitumen	40.19	22.00	1.00
9	Lubricants	40.19	20.00	0.50
10	Other oil	40.19	20.00	0.98
	(B) Solid			
	Primary Fuels:			
11	Raw Coal	20.52	24.74	0.90
	Secondary Fuels:			
12	Cleaned Coal	20.52	24.74	0.90
13	Washed Coal	20.52	24.74	0.90
14	House Coal	20.52	24.74	0.90
15	Coking Coal	28.20	29.50	0.97
16	Coal tar	28.00	22.00	0.75
	(C) Gaseous			
17	CNG	48.00	15.30	0.99

○ β_i is the carbon emission factor, which is considered as the carbon content per unit of energy due to its close link between the carbon content and energy value of the fuel, tC/TJ

○ S_i is the non-energy use, $10^3 t$

○ A_i is the apparent consumption, as given by Equation (23.2), t

$$A_i = O_i + I_i - X_i - B_i - R_i \qquad (23.2)$$

○ O_i is the energy production, t

○ I_i is the import energy, t

○ X_i is the export energy, t

○ B_i is the international bunkers, t

○ R_i is the stock change, t

Subject to the requirements outlined above and with the intention of ensuring the comparability of country inventories, the IPCC approach to the calculation of emissions encourages the use of fuel statistics collected by an officially recognized national body.

However, in practice, this IPCC approach can hardly be applied to the CO_2 emissions of a local area for two reasons. First, recent satellite data have shown that the decrease in coal consumption is most likely unrealistic, and the coal consumption data should not be used [Peters, et al. (2006), Akimoto, et al. (2006)]. Second, the CO_2 emission data by fuel combustion for each local district are not directly available.

To ensure comparability for the local districts, an alternative to TCE/GDP-based CO_2 emission estimation ($\hat{E}$) is proposed in this context, as introduced by Equation (23.3).

$$\hat{E} = \sum_{j=1}^{n_2} \kappa_j \tag{23.3}$$

$$\kappa_j = \tau \gamma_0 \alpha_0 \beta_0 \left(\hat{A}_j + v_0 \hat{B}_j \right) \times 10^{-3} \tag{23.4}$$

where
- $\hat{E}$ is the total TCE CO_2 emission estimation of the CQ rural area, $10^3 t$
- As defined by Equation (23.4), κ_j is the CO_2 emission estimation of each district j based on GDP data, as listed in Table 23.2
- g_j is the gross domestic product (GDP) of district j, $10^4 CNY$
- e_j is the TCE energy consumption per GDP unit of district j, TCE/$10^4 CNY$
- f_j is the electricity consumption per GDP unit of district j, $kWh/10^4 CNY$
- v_0 is the ratio of electricity to TCE, which is 0.01182 in this case[2]
- $\hat{A}_j$ is the CO_2 estimation of the equivalent *energy* consumption of district j, given by Equation (23.5), t

$$\hat{A}_j = e_j g_j \tag{23.5}$$

- $\hat{B}_j$ is the CO_2 estimation of the equivalent *electricity* consumption of district j, given by Equation (23.6), t

$$\hat{B}_j = f_j g_j \tag{23.6}$$

- τ, γ_0, α_0 and β_0 are the molecular weight ratio of CO_2 to C, the TCE fraction of carbon oxidized, the TCE heat conversion factor and the TCE carbon emission factor, as given in Table 23.2, $j = 1,...,n_2$

As a study case, the TCE/GDP energy statistics of the rural area of Chongqing (CQ) are listed in Table 23.2; $j = 1,2,...n_2$, $n_2 = 31$. The energy consumption data are constructed based on the gross domestic product (GDP) g_j, the GDP-related energy

---

[2] Electricity is converted to TCE through the equation, 104 kWh = 1.229 TCE, that is, v_0 = 1.229/104 = 0.01182 [Chinese Energy Statistical Yearbook (2009)]

consumption e_j and the electricity consumption f_j, which are officially provided by the Chinese Energy Statistical Yearbook and the Chongqing Statistical Yearbook.

Table 23.2 Chongqing Energy Statistical Data [Chongqing Statistical Yearbook (2009)]

j	District	e_j	f_j	g_j	$\hat{A}_j$	$\hat{B}_j$	κ_j
1	Wansheng	3.662	2578.51	323630	1185133.06	9442.50362	2,832.3
2	Shuangqiao	1.440	2214.30	300008	432011.52	3188.592	1,032.5
3	Fuling	1.610	1136.81	2534758	4080960.38	1830.2641	9,752.2
4	Changshou	2.669	1795.25	1438173	3838483.737	4791.52225	9,172.8
5	Jiangjin	2.523	1574.31	2192439	5531523.597	3971.98413	13,218.6
6	Hechuan	1.278	774.00	2034734	2600390.052	989.172	6,214.1
7	Yongchuan	1.093	604.35	1920729	2099356.797	660.55455	5,016.8
8	Nanchuan	1.560	941.99	1004876	1567606.56	1469.5044	3,746.1
9	Qijiang	1.061	1030.67	1252081	1328457.941	1093.54087	3,174.6
10	Tongnan	0.832	513.83	863084	718085.888	427.50656	1,716.0
11	Tongliang	0.849	656.26	1109987	942378.963	557.16474	2,252.0
12	Dazu	0.856	555.19	1018873	872155.288	475.24264	2,084.2
13	Rongchang	2.138	900.14	1100317	2352477.746	1924.49932	5,621.7
14	Bishan	1.170	1304.30	1134928	1327865.76	1526.031	3,173.2
15	Wanzhou	1.272	967.26	2560553	3257023.416	1230.35472	7,783.2
16	Liangping	0.967	684.99	754187	729298.829	662.38533	1,742.8
17	Chengkou	2.756	4384.73	195382	538472.792	12084.31588	1,287.1
18	Fengdu	0.766	623.26	574684	440207.944	477.41716	1,052.0
19	Dianjiang	0.853	669.16	815592	695699.976	570.79348	1,662.5
20	Zhongxian	0.848	415.53	778005	659748.24	352.36944	1,576.6
21	Kaixian	2.102	883.94	1106848	2326594.496	1858.04188	5,559.8
22	Yunyang	0.890	599.86	664771	591646.19	533.8754	1,413.8
23	Fengjie	0.875	525.84	753320	659155	460.11	1,575.2
24	Wushan	1.108	888.53	337123	373532.284	984.49124	892.6
25	Wuxi	1.172	828.84	235560	276076.32	971.40048	659.8
26	Qianjiang	0.963	876.09	604842	582462.846	843.67467	1,391.9
27	Wulong	1.159	839.29	498120	577321.08	972.73711	1,379.6
28	Shizhu	0.989	704.96	439487	434652.643	697.20544	1,038.7
29	Xiushan	2.464	4241.37	500364	1232896.896	10450.73568	2,946.5
30	Youyang	1.280	1748.30	329180	421350.4	2237.824	1,006.9
31	Pengshui	1.026	936.46	500474	513486.324	960.80796	1,227.1
	$\hat{E}$						103,203.1

23.3 Hybrid Modeling of the Functional Region

The city of Chongqing (CQ) is one of the largest of the four direct-controlled municipalities of China and is the only such municipality in south-west China. The rural area of CQ spans over 80,000 km^2, covering 31 out of 40 district-level divisions [Chongqing Statistical Yearbook (2009)]. Figure 23.1 shows the location of CQ and the rural districts. To estimate the rural-urban (R-U) spatial interactions, a functional region affecting index (FRAI) with a 'law-of-gravity' interpretation [Hua and Porell (1979), Rooij (2008)] can be characterized by Equation (23.7) using the basic form of a Cobb-Douglas (C-D) production function [Gujarati (2004)]. This index can be used as a qualitative tool for performing a distance-related interactive analysis and as the fitness function for the Mendel-GA optimization.

Taking the district Wulong as an example, as shown in Figure 23.1, O is the CQ city center, r_1 and r_2 are the shortest and longest distances[3] from O to *Wulong*, A and B are the closest and farthest locating-points of Wulong.

Fig. 23.1 Functional distances of Chongqing districts

$$FRAI = \sum_{j=1}^{n_2} \left(\alpha_j \frac{y_j}{\kappa_j} \frac{z_j}{x_j r_j^{\beta} + \varepsilon} \right)^{\eta_j} \tag{23.7}$$

---

[3] The distances r_{1j} and r_{0j} are provided by the GIS system, *China Map*, www.51ditu.com

where

○ $FRAI$ is the functional region affecting index

○ x_j is the number of enterprises in the district j, where n_2 is the number of rural districts

○ y_j are the total profits of district j, $10^4 CNY$

○ z_j is the non-agricultural employment of district j (10^4 persons)

○ $\alpha_j = \alpha_{1j}/\alpha_{0j}$ is the population effect coefficient of district j, where α_{1j} is the local non-agricultural population and α_{0j} is the population of district j (10^4 persons)

○ κ_j is the CO_2 emissions of district j, as listed in Table 23.2

○ η_j is the output elasticity of district j, as given by Equation (23.8), in which $\bar{g}$ is the mean of all g_j in Equation (23.9)

$$\eta_j = \begin{cases} 1 & \text{if } g_j \geq \bar{g} \\ -1 & \text{if } g_j < \bar{g} \end{cases} \tag{23.8}$$

$$\bar{g} = \frac{\sum_{j=1}^{n_2} g_j}{n_2} \tag{23.9}$$

○ β is the factor of the functional distance r_j, $\beta \in [0.5, 3]$

○ ε is the floating-point relative accuracy, which prevents a singularity in the case where r_j^β is approaching 0 and $FRAI$ is approaching ∞ numerically

○ r_j is the mean value of the point-to-point distance from the CQ city center to the town center of district j, as defined by Equation (23.10), kilometer (km)

$$r_j = \frac{r_{j1} + r_{j2}}{2} \tag{23.10}$$

23.4 Fitness Function

To evaluate the dynamic behavior relating to the functional distance, CO_2 emissions and population for the economic production of the CQ rural area, the $FRAI$ index has been adopted as the fitness function, as defined by Equation (23.11) in the form of Equation (23.7). The $FRAI$ index is a hybrid measurement for the functional region and the CO_2 emission factor, which is a spatial determination of the agreement between the statistical data and the Mendel-GA-driven data pool. All the statistical data for the parameters of Equation (23.11) are listed in Table 23.3. The data sources are the Chongqing Statistical Yearbook 2009 [Chongqing Statistical Yearbook (2009)] and the China Statistical Yearbook 2009 [China statistical yearbook (2009)].

$$Fitness = F(X,Y,Z) = \sum_{j=1}^{n_2} \left(\alpha_j \frac{y_j}{\kappa_j} \frac{z_j}{x_j r_j^\beta + \varepsilon} \right)^{\eta_j} \tag{23.11}$$

Table 23.3 Chongqing Rural Area Statistical Data [Chongqing Statistical Yearbook (2009), China statistical yearbook (2009)]

No. j	District	r_{j1}	r_{j2}	α_{0j}	α_{1j}	μ_{x_j}	μ_{y_j}	μ_{z_j}
1	Wansheng	58.3	97.4	26.82	13.23	578	-7220	2.90
2	Shuangqiao	70.5	79.5	5.01	2.82	299	5636	1.01
3	Fuling	38.4	108.3	113.80	33.04	3094	121250	10.42
4	Changshou	45.0	99.0	89.87	21.46	2467	117276	7.21
5	Jiangjin	27.2	117.0	148.65	40.02	2488	78214	10.87
6	Hechuan	41.1	104.7	153.89	32.04	2950	70417	6.36
7	Yongchuan	45.8	101.2	110.18	29.72	3309	130936	6.02
8	Nanchuan	38.8	111.9	66.09	11.71	2376	66422	2.33
9	Qijiang	44.6	124.9	95.00	22.18	1360	18777	4.72
10	Tongnan	69.4	119.4	93.26	11.99	1216	9372	2.09
11	Tongliang	69.3	84.7	82.42	14.95	1857	31717	2.33
12	Dazu	53.8	105.9	95.02	17.72	1890	40290	3.62
13	Rongchang	81.6	122.5	83.07	18.51	1585	60004	3.29
14	Bishan	29.8	56.1	62.14	15.92	1736	83208	3.39
15	Wanzhou	181.3	257.6	172.54	51.06	4123	96974	11.64
16	Liangping	133.8	194.3	91.07	11.81	559	47499	3.12
17	Chengkou	301.3	359.2	24.15	2.95	331	14344	0.96
18	Fengdu	100.8	162.0	82.44	15.05	883	25533	2.34
19	Dianjiang	86.6	146.7	93.94	14.43	1938	25829	3.65
20	Zhongxian	119.6	190.9	99.22	16.01	2272	17161	1.93
21	Kaixian	202.4	327.2	159.72	21.29	1590	6861	4.71
22	Yunyang	240.9	312.8	133.58	18.40	732	528	3.14
23	Fengjie	268.2	356.7	104.76	13.22	898	7961	2.49
24	Wushan	325.6	399.6	62.35	10.29	453	2902	1.71
25	Wuxi	294.4	393.5	53.52	6.29	571	3111	1.73
26	Qianjiang	186.5	235.9	52.17	10.36	576	52422	2.45
27	Wulong	68.7	144.1	41.01	5.92	650	3362	1.84
28	Shizhu	145.8	209.4	53.34	8.83	720	11305	2.27
29	Xiushan	241.5	291.0	64.35	8.62	760	32654	1.70
30	Youyang	182.1	274.8	80.81	8.56	958	1712	2.15
31	Pengshui	138.3	200.4	67.38	6.67	614	36854	1.97

The following assumptions concerning the parameter uncertainties are made:

∘ $X = [x_1, x_2, ..., x_j, ..., x_{n_2}]$ is the vector of the enterprise number, $X \sim N(\mu_X, \sigma_X^2)$. Let $\mu_X = [\mu_{x_1}, \mu_{x_2}, ..., \mu_{x_j}, ..., \mu_{x_{n_2}}]$ and $\sigma_X = [\sigma_{x_1}, \sigma_{x_2}, ..., \sigma_{x_j},$

..., $\sigma_{x_{n_2}}$] be the mean and standard deviation (STD) vectors of X, respectively, that is, $x_j \sim N(\mu_{x_j}, \sigma_{x_j}^2)$, μ_{x_j} and σ_{x_j} are the mean and standard deviation of x_j

○ $Y = [y_1, y_2, ..., y_j, ..., y_{n_2}]$ is the vector of the total profits, $Y \sim N(\mu_Y, \sigma_Y^2)$. Let $\mu_Y = [\mu_{y_1}, \mu_{y_2}, ..., \mu_{y_j}, ..., \mu_{y_{n_2}}]$ and $\sigma_Y = [\sigma_{y_1}, \sigma_{y_2}, ..., \sigma_{y_j}, ..., \sigma_{y_{n_2}}]$ be the mean and standard deviation vectors of Y, respectively, such that $y_j \sim N(\mu_{y_j}, \sigma_{y_j}^2)$, where μ_{y_j} and σ_{y_j} are the mean and standard deviation of y_j

○ $Z = [z_1, z_2, ..., z_j, ..., z_{n_2}]$ is the vector of the non-agricultural employment, $Z \sim N(\mu_Z, \sigma_Z^2)$. Let $\mu_Z = [\mu_{z_1}, \mu_{z_2}, ..., \mu_{z_j}, ..., \mu_{z_{n_2}}]$ and $\sigma_Z = [\sigma_{z_1}, \sigma_{z_2}, ..., \sigma_{z_j}, ..., \sigma_{z_{n_2}}]$ be the mean and standard deviation vectors of Z, respectively, such that $z_j \sim N(\mu_{z_j}, \sigma_{z_j}^2)$, μ_{z_j} and σ_{z_j} are the mean and standard deviation of z_j

○ δ_{x_j}, δ_{y_j} and δ_{z_j} are the coefficients of variation (CV) of x_j, y_j and z_j, respectively, with the definition given by Equation (23.12) [Lyman and Longnecker (2001)]. The remaining parameters, α_j, β, r_j, κ_j, ε and η_j, are as introduced in Section 23.3

$$\delta_j = \frac{\sigma_{x_j}}{\mu_{x_j}} \tag{23.12}$$

23.5 Empirical Results and Discussion

The empirical results for the Mendel-GA-driven hybrid modeling of the CQ functional region are obtained with *SGALAB* [SGALAB (2009)] and the parameters of the Mendel-GA calculation are reported in Table 23.4. Here, the selection operator is the tournament method; the crossover and mutation operators are the single-point method; the encoding is the binary method; the Mendel percentage is with the full chromosome length; the CVs δ_j for all x_j, y_j and z_j are set to 15 per cent; and the experiment number is 100.

Table 23.4 Parameters for the Simulations

Max generations	1000/10000/100000
Crossover probability	0.8
Mutation probability	0.001
Population	50
Selection operator	tournament
Crossover operator	single point
Mutation operator	single point
Encoding method	binary
Experiment number	100
CV (δ_j)	15%
Mendel percentage	1 (full chromosome length)

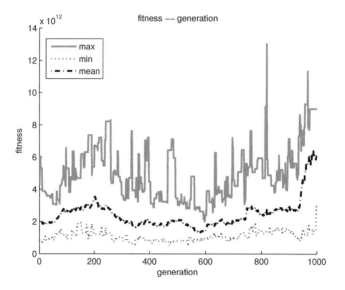

Fig. 23.2 FRAI fitness plot @ generation = 1000

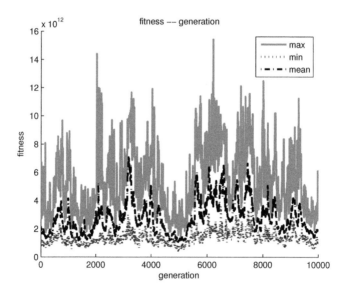

Fig. 23.3 FRAI fitness plot @ generation = 10000

Figures 23.2, 23.3 and 23.4 are the fitness plots for generations g = 1000, 10000 and 10000, whose 'x-axis' is the evolution generation of the Mendel-GA and the 'y-axis' is the fitness value. These figures demonstrate the short-term (g = 1000), middle-term (g = 10000) and long-term (g = 100000) dynamic behaviors of the functional regions of the CQ rural area, in which the solid line, dotted line and dash-dot line are the 'max', 'min' and 'mean' values, respectively, of the fitness evolution process. In contrast to the fitness shape for engineering applications, the fluctuations of the fitness curves indicate the non-monotonic behaviors of the CQ rural area.

As shown in Table 23.5, nine measures of the dynamic behaviors of fitness plots are devised to describe the uncertainties of the 'max' ($\mu_{F(X,Y,Z)}|_{max}$), 'mean' ($\mu_{F(X,Y,Z)}|_{mean}$) and 'min' ($\mu_{F(X,Y,Z)}|_{min}$) fitness plots: *Max*, *Min*, *Mean*, *Median*, *Mode*, *STD*, 25th percentile (Q_1), 50th percentile (Q_2) and 75th percentile (Q_3).

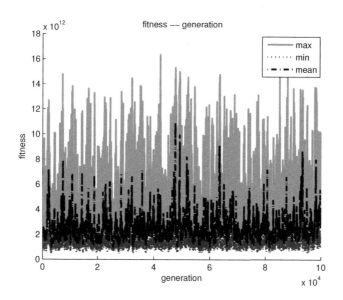

Fig. 23.4 FRAI fitness plot @ generation = 100000

Table 23.5 Fitness Uncertainty Analysis

Measures	$\mu_{F(X,Y,Z)}|_{max}$	$\mu_{F(X,Y,Z)}|_{mean}$	$\mu_{F(X,Y,Z)}|_{min}$
Max	1.6332	1.1008	0.45981
Min	0.1395	0.0974	0.04727
Mean	0.4994	0.2582	0.12872
Median	0.4404	0.2345	0.12222
Mode	0.5695	0.1193	0.09740
STD	0.2267	0.1034	0.03863
Q_1	0.3340	0.1892	0.10253
Q_2	0.4404	0.2345	0.12222
Q_3	0.6089	0.2979	0.14697

Table 23.6 The Estimated $\hat{X}$, $\hat{Y}$ and $\hat{Z}$ Optimized by the Mendel-GA

No. j	District	$\hat{x}_j$	$\Delta x_j(\%)$	$\hat{y}_j$	$\Delta y_j(\%)$	$\hat{z}_j$	$\Delta z_j(\%)$
1	Wansheng	454.1	-21.4	-5873.9	-18.6	1.0	-65.3
2	Shuangqiao	183.2	-38.7	5411.9	-4.0	0.7	-28.1
3	Fuling	527.6	-82.9	73545.4	-39.3	8.9	-14.6
4	Changshou	1678.9	-31.9	97562.5	-16.8	7.2	-0.7
5	Jiangjin	2251.1	-9.5	38926.7	-50.2	4.1	-62.2
6	Hechuan	1214.2	-58.8	58146.4	-17.4	2.5	-60.9
7	Yongchuan	1442.9	-56.4	101450.4	-22.5	3.6	-39.6
8	Nanchuan	1723.4	-27.5	35494.6	-46.6	2.2	-4.0
9	Qijiang	1209.3	-11.1	8935.8	-52.4	4.4	-6.2
10	Tongnan	785.2	-35.4	2930.0	-68.7	1.7	-17.3
11	Tongliang	946.9	-49.0	21328.0	-32.8	1.6	-30.3
12	Dazu	988.1	-47.7	8948.8	-77.8	3.4	-5.5
13	Rongchang	1149.3	-27.5	9618.9	-84.0	1.3	-60.2
14	Bishan	1608.0	-7.4	42294.6	-49.2	1.2	-65.7
15	Wanzhou	1276.9	-69.0	36627.3	-62.2	9.6	-17.4
16	Liangping	518.2	-7.3	8379.1	-82.4	1.3	-59.5
17	Chengkou	188.7	-43.0	4284.6	-70.1	0.2	-80.3
18	Fengdu	655.1	-25.8	5916.4	-76.8	2.3	-3.3
19	Dianjiang	1541.9	-20.4	9170.7	-64.8	0.6	-82.3
20	Zhongxian	1301.1	-42.7	6333.3	-63.1	0.3	-84.7
21	Kaixian	746.1	-53.1	2586.9	-62.3	2.4	-49.9
22	Yunyang	638.6	-12.8	521.9	-1.1	3.0	-5.1
23	Fengjie	366.1	-59.2	7509.2	-5.7	2.3	-7.0
24	Wushan	173.0	-61.8	1324.4	-54.4	1.1	-35.3
25	Wuxi	397.2	-30.4	2603.2	-16.3	1.6	-9.3
26	Qianjiang	100.5	-82.6	19219.0	-63.3	1.7	-30.7
27	Wulong	597.5	-8.1	1630.7	-51.5	1.1	-39.3
28	Shizhu	200.6	-72.1	8374.4	-25.9	0.5	-76.3
29	Xiushan	230.4	-69.7	28798.1	-11.8	1.7	-2.0
30	Youyang	779.2	-18.7	484.5	-71.7	0.8	-64.0
31	Pengshui	347.31	-43.4	16500.8	-55.2	0.8	-60.7

Table 23.6 presents one of the sets of estimation $\hat{X} = [\hat{x}_1, \hat{x}_2, ..., \hat{x}_j, ..., \hat{x}_{n_2}]$, $\hat{Y} = [\hat{y}_1, \hat{y}_2, ..., \hat{y}_j, ..., \hat{y}_{n_2}]$ and $\hat{Z} = [\hat{z}_1, \hat{z}_2, ..., \hat{z}_j, ..., \hat{z}_{n_2}]$ for each district in the CQ rural area. Compared with the yearbook data μ_{x_j}, μ_{y_j} and μ_{z_j} in Table 23.3, the negative percentage errors of $\Delta x_j(\%)$, $\Delta y_j(\%)$ and $\Delta z_j(\%)$ indicate that the scale of the non-agricultural activities in the CQ rural area present a decreasing trend, which matches the practical situation, that is, the rural area of CQ is designed as an agriculture-oriented functional region, administratively. On the other hand, the urban area of CQ is an industry-oriented functional region [Chongqing Statistical Yearbook (2009)].

23.6 Conclusions and Future Work

Technical analysis is frequently adopted to assist in region planning and attempts to fully utilize available statistical data. This paper provides an evolutionary alternative to the dynamic behavior modeling of functional regions using the distance factor and the CO_2 emission factor. In this paper, first, the Mendel-GA is introduced; then, the GDP-based TCE method is applied to the CO_2 emission estimation. For the hybrid modeling of the functional region of the CQ rural area, the FRAI index describes the interaction between the production inputs and the outputs with the source data uncertainties. This index indicates a type of distance-based relationship between an FRAI combination of the X, Y and Z inputs and the possible maximum output under current production capabilities, with which the experience of policy makers can be incorporated in a natural or artificial manner. Clearly, CO_2 emissions are related to the complicated dynamic behavior of the functional region. The GDP/TCE method for CO_2 emission estimation represent a practical method for addressing the lack of direct fossil fuel data sources.

Our empirical simulation demonstrated the practical behavior of functional regions of the CQ rural area and suggested some interesting issues for further investigation in real-time public sector policy making. Here, the consumer price index (CPI) will be one of the dynamic decision criteria for the reliability analysis of a multi-state FRAI modeling or the analysis of regional system maintenance and sustainability.

24

Spatial Analysis of Functional Region of Suburban-Rural Areas

24.1 Introduction

As a result of the spatial development associated with the large population of China, concerns have been raised that economic integration at the functional region level may enhance the connections of suburban-rural economic interactions, e.g., the spatial concentration of districts and industrial specialization of regions. Spatial analytical techniques, geographical analysis and modeling methods are therefore required to analyze data and to facilitate the decision process at all levels (macro, meso and micro). Mostly, dynamic industries concentrate in the city's core regions (city center), which cause the peripheral regions (rural areas) to grow slower in terms of income, employment, etc.

The interdependence between research, available data and statistical measures has largely been neglected, which allows for a systematic assessment of the properties of the measures based on the background of the available data and the research purpose has not been determined [Coricelli and Dosi (1988), Hanson and Pratt (1988), Hanson and Pratt (1995)]. Evolutionary methods have been applied since the 2000s to a few aspects, such as entrepreneurship, industrial dynamics [Arthur (1994), Swann and Prevezer (1996), Klepper (2001), Klepper (2002), Boschma and Frenken (2003)], network analysis [Boschma and Frenken (2006), Giuliani (2006)] and spatial systems [Freeman and Perez (1988), Boschma (2004), Saviotti and Pyka (2004)]. The problem here is that most well-known economic theories do not actually concern rural areas, not even with China's available source data.

Beijing (BJ) is one of the four direct-controlled municipalities of China and is also its capital. Studies of its rural area represent a fast-growing interest because of the socioeconomic effects of its spatial spreading. In this chapter, we present an alternative technique for the spatial analysis of BJ's suburban-rural area, where the VPμGA, as introduced in Section 2.4, explores the economic significance of complexity, interpreted in terms of the qualitative as well as quantitative adaptations of a model system.

24.2 Spatial Modeling of the Functional Regions

To estimate the urban-rural spatial interactions, a functional region affecting index (Θ) with a 'law-of-gravity' interpretation [Hua and Porell (1979), Rooij (2008)] can be characterized by Equation (24.1) using the basic form of a Cobb-Douglas (C-D) production function [Gujarati (2004)]. This can be used as a quantitative tool to perform a distance-related interactive analysis and it can be utilized as the fitness function for VPμGA optimization.

Beijing is the capital of China and is an independently-administered municipal district. As the nation's political and cultural center, there are 16 districts and two counties; 13 suburban districts are discussed in this context [China statistical yearbook (2009)]. As shown in Figure 24.1, taking *Yanqing* county ($i = 13$) as an example, which is northwest of the BJ city center (core region), *A* and *B* are the closest and farthest measuring points to the city center *O*, respectively, where the distance from *O* to *A* is defined as the a shortest radius r_1 of *Yanqing*. Similarly, the distance from *O* to *B* is defined as the largest radius r_2 of *Yanqing*. Generally, the functional distance r_i is defined as the mean value of the measured distances[1] r_{i1} and r_{i2}, which are the closest and farthest distances of a given district *i* to the city center of Beijing.[2]

$$\Theta = \sum_{i=1}^{n_1} \left(\frac{\alpha_i y_i z_i}{x_i r_i^{\beta} + \varepsilon} \right)^{\eta_i} \tag{24.1}$$

where
- Θ is the functional region affecting index
- x_i is the number of enterprises in the district *i*, n_1 is the number of districts
- y_i are the total profits of district *i*, $10^4 CNY$
- z_i is the industrial employment of district *i*, 10^4 persons
- α_i is the density of the local population of district *i*, people/m^2
- η_i is the output elasticity factor, which is designed using a two-state representation, as given by Equation (24.2). ζ_i is the personal income of district *i*, and ζ_0 is the average personal income of the 13 districts

$$\eta_i = \begin{cases} -1 \text{ if } \zeta_i > \zeta_0 \\ \\ 1 \;\; \text{if } \zeta_i < \zeta_0 \end{cases} \tag{24.2}$$

- β is the factor of the functional distance r_i, $\beta \in [0.5,3]$
- r_i is the functional distance from Beijing's central area to the center of suburban town *i*, as defined by Equation (24.3), where r_{i1} is the near-end radius and r_{i2} is the far-end radius

---

[1] *China Map*, www.51ditu.com, is an online GIS system
[2] Usually, the city center of Beijing is cited as the area circled by the second city ring motorway

Fig. 24.1 The functional distance of Beijing

$$r_i = \frac{r_{i1} + r_{i2}}{2} \tag{24.3}$$

∘ ε is the floating-point relative accuracy, which prevents a singularity in the cases of r_i^{β} approaching 0 or Θ approaching ∞

24.3 Sensitive Analysis to Functional Distance

For investigating the sensitivity of the functional-distance-related spatial interactions, the quantized connection to the functional distance is given by Equation (24.4), which provides a distance-based sensitive description of the functional distance r_i to the region affecting index Θ.

$$\frac{\partial \Theta}{\partial r_i} = -\frac{\beta r_i^{\beta-1} \eta_i x_i y_i z_i \alpha_i}{\left(x_i r_i^{\beta} + \varepsilon\right)^2} \left(\frac{\alpha_i y_i z_i}{x_i r_i^{\beta} + \varepsilon}\right)^{\eta_i - 1} \tag{24.4}$$

24.4 Fitness Function

Borrowing the idea of the functional region affecting index Θ, as stated in Equation (24.1), the fitness function F_1 for the VPμGA evaluational process is defined by Equation (24.5).

$$F_1 = fitness(X, Y, Z) = \sum_{i=1}^{n_1} \left(\frac{\alpha_i y_i z_i}{x_i r_i^{\beta} + \varepsilon}\right)^{\eta_i} \tag{24.5}$$

where

○ $X = [\mu_{x_1}, \mu_{x_2}, ..., \mu_{x_i}, ..., \mu_{x_{n_1}}]$ is the mean vector of the number of enterprises; each element μ_{x_i} is the mean of the number of enterprises of district i, which is assumed to follow a normal distribution $N[\mu_{x_i}, \sigma_{x_i}^2]$; and $\sigma_{x_i}^2$ is the variance of x_i

○ Similarly, $Y = [\mu_{y_1}, \mu_{y_2}, ..., \mu_{y_i}, ..., \mu_{y_{n_1}}]$ and $Z = [\mu_{z_1}, \mu_{z_2}, ..., \mu_{z_i}, ..., \mu_{z_{n_1}}]$ are the mean vectors of the total profits y_i and employment z_i, which are assumed to follow the normal distributions $N[\mu_{y_i}, \sigma_{y_i}^2]$ and $N[\mu_{z_i}, \sigma_{z_i}^2]$, where μ_{y_i} and μ_{z_i} are the mean parameters and $\sigma_{y_i}^2$ and $\sigma_{z_i}^2$ are the variance parameters

○ The coefficients of variation (CV) are given by Equation (24.6) [Lyman and Longnecker (2001)], and α_i, η_i, β, r_i and ε are the same as discussed in Section 24.2, as listed in Table 24.1. The relative errors $\Delta\mu_i$ (%) between the original data and the optimized data are defined by Equation (24.7), in which $\hat{\mu}_i$ is the VPμGA evaluated output and μ_i is the mean parameter of the statistics

$$CV_i = \frac{\sigma_i}{\mu_i} \tag{24.6}$$

$$\Delta\mu_i = \frac{\hat{\mu}_i - \mu_i}{\mu_i} \times 100 \tag{24.7}$$

Table 24.1 Non-agricultural Parameters and Functional Distance [China statistical yearbook (2009), Beijing statistical yearbook (2009)]

No. i	District i	ζ_i	r_{i1}	r_{i2}	μ_{x_i}	μ_{y_i}	μ_{z_i}	α_i	CV_i
1	Chaoyang	15090	2.5	22.6	680	53273	51967	6775	15
2	Fengtai	11584	3.7	30.5	895	33429	59252	5733	15
3	Haidian	14319	5.7	36.8	3208	48710	62206	6802	15
4	Mentougou	10282	20.6	84.6	9891	30046	30122	190	5
5	Fangshan	10073	17.8	85.6	39589	31931	285884	455	10
6	Tongzhou	10213	13.5	49.7	23826	25793	211286	1146	10
7	Shunyi	10402	20.4	57.8	25368	40154	213558	711	10
8	Changping	10121	14.3	59.8	2661	30914	80785	701	10
9	Daxing	10103	8.5	51.4	27930	40978	174984	1059	10
10	Huairou	9871	43.4	130.9	8939	33211	49800	169	5
11	Pinggu	9790	50.4	90.7	4050	25157	62278	448	5
12	Miyun	9529	50.5	125.8	12660	25409	80874	205	5
13	Yanqing	9385	54.5	99.6	7214	25826	33606	144	5

24.5 Empirical Results and Discussion

The functional-distance-based dynamical behaviors are essential to urban-rural interaction studies and need to be maintained in a reliable manner under practical boundary conditions to assess the socioeconomic performance of the BJ suburban-rural area. The real-world data-driven simulations are performed using the functional region affecting index, and the statistics of each district i are listed in Table 24.1 [China statistical yearbook (2009), Beijing statistical yearbook (2009)].

Numerical results are obtained using a specially devised simulation toolkit, known as 'Simple Genetic Algorithm Laboratory (*SGALAB*)' [SGALAB (2009)]. Unless stated otherwise, all the results are obtained using the parameters for the GA calculation in Table 24.2.

Figures 24.2 to 24.5 show the fitness function of the VPμGA evolution process at 100, 1000, 10000 and 100000 generations, respectively. Figures 24.2 to 24.5 demonstrate the Θ dynamic behaviors of the suburban-rural area of BJ for relatively 'short'-term, 'middle'-term, 'middle-long'-term and 'long'-term simulation periods.

Table 24.2 Parameters for the VPμGA Calculation

Max generations	100000
Number of experiments	100
Crossover probability	0.8
Max internal generation	4
Initial internal population	4
Initial external population	100
Selection operator	tournament
Crossover operator	single point
Mutation operator	no
Encoding method	binary
Reproduction rate	[-1,1]

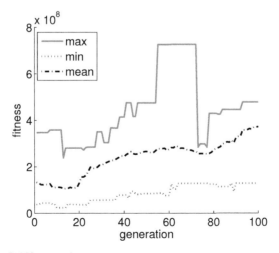

Fig. 24.2 Fitness @ 100 generations

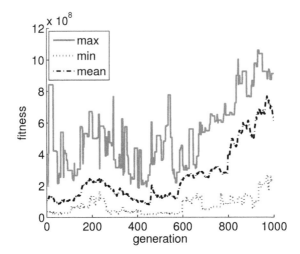

Fig. 24.3 Fitness @ 1000 generations

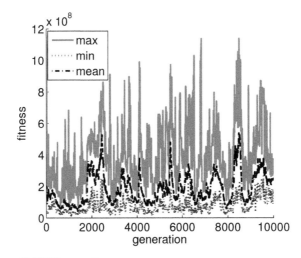

Fig. 24.4 Fitness @ 10000 generations

Over the full evolution length, the fitness's stochastic activity indicates that the performance of the functional-distance-based regional area is based on the combination of the 13 districts' distributed suburban-rural interactions. Table 24.3 lists the functional distance r_i, the VPμGA-evaluated outputs $\hat{\mu}_i$ and the relative error $\Delta\mu_i$ for the 13 districts and shows that the economic activities of the non-agricultural industries (e.g., township enterprises) increase ($\Delta\mu_i > 0$) closer ($r_i < 55$ km) to the city center O and decrease ($\Delta\mu_i < 0$) in districts further ($r_i > 55$ km) from the city center O. This functional-distanced-based phenomenon indicates that the package of policies for BJ urbanization [Hutchison (2009)] is actively affecting the

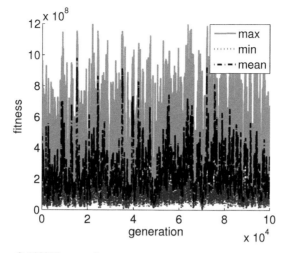

Fig. 24.5 Fitness @ 100000 generations

Table 24.3 Simulation Results for 13 Districts of Town and Township Enterprises

No. i	District i	r_i	$\hat{\mu}_{x_i}$	$\Delta\mu_{x_i}$	$\hat{\mu}_{y_i}$	$\Delta\mu_{y_i}$	$\hat{\mu}_{z_i}$	$\Delta\mu_{z_i}$
1	Chaoyang	12.55	772.87	13.66	54866.32	2.99	89107.28	71.47
2	Fengtai	17.10	1413.30	57.91	33582.26	0.46	79623.10	34.38
3	Haidian	21.25	2730.15	-14.90	50038.15	2.73	68426.30	10.00
4	Mentougou	52.60	12492.78	26.30	54426.75	81.14	18971.82	-37.02
5	Fangshan	51.70	44856.66	13.31	25460.33	-20.26	423118.55	48.00
6	Tongzhou	31.60	26562.99	11.49	26074.15	1.09	205610.95	-2.69
7	Shunyi	39.10	22310.69	-12.05	45442.06	13.17	231730.98	8.51
8	Changping	37.05	2208.40	-17.01	61146.60	97.80	98404.74	21.81
9	Daxing	29.95	28592.36	2.37	39204.69	-4.33	78761.56	-54.99
10	Huairou	87.15	10291.10	15.13	2545.79	-92.33	23900.90	-52.01
11	Pinggu	70.55	3046.33	-24.78	24449.39	-2.81	2799.39	-47.33
12	Miyun	88.15	14785.55	16.79	25273.16	-0.53	31908.20	-60.55
13	Yanqing	77.05	14895.64	106.48	42198.90	63.40	14628.11	-56.47

movement of non-agricultural industries from further BJ suburban-rural areas to closer BJ suburban areas.

Figure 24.6 is the phase portrait (F_1 vs. F_1') of the 'max' fitness of Figure 24.5, which shows that the VPμGA-driven evolution process is a stable dynamical system converging to a limit cycle at approximately $[(2,0) \to (6,2) \to (10,0) \to (6,-2)] \times 10^8$.

Figure 24.7 and Table 24.4 show that the urbanized districts ($i = 1,2,3$) are highly sensitive to their functional distances to the city center O, that is, the locations of the business branches are closely related to their operational outputs such as transportation and catering. As can also be observed in Table 24.4, districts

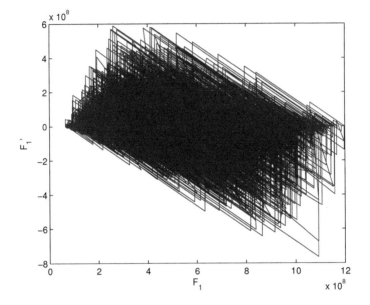

Fig. 24.6 The phase portrait of the fitness behavior

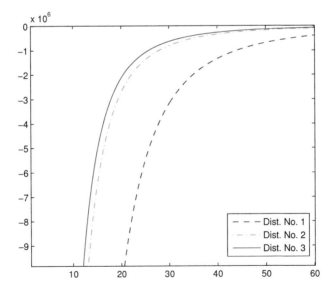

Fig. 24.7 The sensitivity portrait of district $i = 1,2,3$

4 to 13 are with small sensitivities ([e-09, e-07] to [e-08, e-04]) to the functional distance r_i.

Table 24.4 The Sensitivity Comparison of the Functional Distances r_i to Θ

	$\dfrac{\partial \Theta}{\partial r_i}$			
Distance (km)	District 1	District 2	District 3	Districts 4 to 13
1.0	-6.4e+10	-1.6e+10	-1.3e+10	[e-09, e-07]
1.2	-3.9e+10	-9.9e+09	-7.7e+09	...
1.3	-3.1e+10	-7.9e+09	-6.2e+09	
2.0	-9.2e+09	-2.3e+09	-1.8e+09	
2.6	-4.3e+09	-1.1e+09	-8.6e+08	
2.8	-3.5e+09	-8.8e+08	-7.0e+08	
4.4	-9.4e+08	-2.4e+08	-1.9e+08	
5.6	-4.6e+08	-1.2e+08	-9.2e+07	
6.1	-3.6e+08	-9.1e+07	-7.2e+07	
9.6	-9.4e+07	-2.4e+07	-1.9e+07	
12.1	-4.7e+07	-1.2e+07	-9.4e+06	
13.1	-3.7e+07	-9.4e+06	-7.4e+06	
20.8	-9.4e+06	-2.4e+06	-1.9e+06	
26.1	-4.8e+06	-1.2e+06	-9.5e+05	
28.3	-3.7e+06	-9.5e+05	-7.4e+05	
44.8	-9.5e+05	-2.4e+05	-1.9e+05	
55.5	-5.0e+05	-1.3e+05	-9.9e+04	...
60.0	-4.0e+05	-1.0e+05	-7.9e+04	[e-08, e-04]

24.6 Conclusions and Future Work

In this paper, the VPμGA was proposed and utilized for the spatial analysis of the functional region of the BJ suburban-rural area; here, the functional region affecting index Θ was borrowed as the fitness function for the VPμGA evolutionary process. Given the statistics of the 13 districts of the BJ suburban-rural area, the simulation results demonstrate a stable non-linear dynamic behavior that is in agreement with the tendency of the functional-distanced-measured spread of cities and city-state processes. Through the sensitivity analysis, non-agricultural industries have been shown to be transferring from rural areas to suburban/urban areas. Central to this endeavor is the development of functional-distanced-based sustainable suburban-rural area modeling and decision support tools.

The functional distance can be defined in other forms, e.g., the weighted mean value of the farthest and closest distances. These definitions can all be used to perform the same function but may be subject to different economic loads and environmental conditions. In addition, the computational technique of the GA has potential applications in public sector resource reconfiguration, health care policy making, top management team behavioral integration, etc.

To integrate various activities, such as climate impacts, the rural economy, land use, transport and the ecological conditions, future work will provide new insights into the direct and indirect impacts of active factors in suburban-rural areas. These insights will form the basis for the development of strategies for adaptation in rural areas.

CIAD for Industry 4.0 Predictive Customization

25.1 Introduction

A new era of industrialization is dawning upon us in the digital age. Until recently, Industry 4.0 has been not much more than a concept [Kull (2015a)]. With i4, design for manufacture is shifting to a new paradigm, targeting innovation, lower costs, better responses to customer needs, optimal solutions, intelligent systems, and alternatives towards on-demand production. All these trends have in common the integration of several features in the same place as a response to challenges of computerized decision making and big data, which are proliferated by the Internet and cloud computing.

As companies are already making a good effort 'to network machinery, control systems and sensors together, such that all the data from the production process can be used to make decisions on manufacturing' for i4 (Siemens), this chapter aims to model and develop the link that is currently missing in the i4 value chain—smart design with market informatics for smart manufacturing. This can be achieved by using the CIAD approach. At the technological heart of CIAD is CAutoD, which elevates the traditional CAD-resultant 'rapid prototyping' and '3D printing' to target smart prototyping for direct manufacture. Utilization of computational intelligence, cloud computing and big data, i4-CAutoD links design and manufacture with customer needs and wants have been subconsciously hidden in the cloud and big data, but which can now be fed back automatically to design and manufacture. With automated value creation chain and cyber-physical integration, the i4-CAutoD approach will therefore help revolutionize the way that designs are created and machines are built.

Historic industrial revolutions had led to a paradigm shift, starting with the steam-motor improvement in the 18th century, then mass production systems in the early 19th century because of electricity commercialization, and to the advancement of information and communication technology (ICT) and introduction of automation systems in the late 20th century. Innovations in manufacturing have revolutionized the way products were manufactured, services were given and business was made.

Advances in ICT technologies have continually progressed in numerous fields, including both software and hardware, which might bring a revolution or evolution in the manufacturing industry. For this revolution, Smart Manufacturing could be the driving force. Integration of various technologies can promote a strategic innovation of the existing industry through the convergence of technology, humans, and information. Contrary to lean manufacturing targeted cost saving by focusing on waste elimination in the 1980's and 1990's, Smart Manufacturing represents a future growth engine that aims at sustainable growth through management and improvement of the major existing factors, like quality, flexibility, productivity and delivery based on technology convergence and numerous elements over societies, environment and humans [Kang, et al. (2016)].

The main idea of i4 is the combination of several technologies and concepts, such as Smart Factory, CPS, industrial Internet of Things (IoT), and Internet of Services (IoS) interact with one another to form a closed-loop production value chain [Flores Saldivar, et al. (2016a)]. Not many industries can produce individual goods in a completely automated fashion for the beneficial dilemma inside the production line: variety vs. productivity, which is different from similar ambitious strategies like advanced manufacturing partnership in the US [Flores Saldivar, et al. (2016a)] and the 'Manufacturing 2025' plan in China. For this to become a reality, not only the machines but occasionally even the parts themselves need to become smart [Kull (2015b)].

The prime aim of this chapter is to address the integration of several technologies inside a closed-loop cycle to retrieve information from the existing inputs before obtaining the prediction for decision making and customize the intelligent design of products. This framework is proposed under the i4 principles due to the capacity of integration with cloud computing, big data analytics, ICT, CPS and business informatics inside the manufacturing production systems. Prediction and selection of best attributes and customers' needs and wants, utilization of fuzzy c-means and Genetic Algorithm (GA) selection for customized designs for smart manufacture are some of the objectives of this research.

In Section 2 of this chapter, challenges and trends of i4 are discussed together with the issues about mass customization. In Section 3, we tackle the smart design issue for mass customization and present a self-organizing tool for predicting customer needs and wants. We demonstrate the effectiveness of the proposed methodology through a case study in Section 4. Lastly, Section 5 draws conclusions with discussions on future work.

25.2 Customization in Industry 4.0

Coined in the late 1980's, the term 'mass-customized production' has become a subject of research along with the proliferation of information throughout the IoT in the 21st century, affecting business strategies and acquiring goods and services [Möller (2016)]. This implies that mass customization in the manufacture supply chain, material flow, information and a connection between product types had a direct effect on customer satisfaction [Yang and Burns (2003)].

Customized manufacture describes a manufacturing process for which all involved elements of the manufacturing system are designed in a certain way to enable high levels of product variety at mass production costs [Möller (2016)]. This is why today companies are facing challenges as a result of customers' increasing demands for individualized goods and services. With the development and introduction of CPS into the manufacturing process, manual adjustments and variations in product quality can be minimized by connecting the virtual part of the process through computer-aided design (CAD) and comparing the desired information to target optimal features. Finally, all the data that intervene alongside the process help to monitor the manufacturing process and apply changes if necessary. It results in more informed processes and leads to reliable decisions by having a closed loop to constantly retrieve information inside the customized design and customer satisfaction [Flores Saldivar, et al. (2016b)].

The next section describes how data and CPS can be integrated into a framework for manufacturing application.

25.2.1 CPS with Data Analytics Framework for Smart Manufacturing

The use of sensors and networked machines has increased recently, resulting in persistent generation of high volume data known as Big Data [Lee, et al. (2013)]. In this way, CPS can be developed for managing Big Data and exploiting the interconnectivity of machines to reach the goal of resilient, intelligent and self-adaptable machines. Boosting efficiency in production lines for meeting customers' needs and wants is the key in i4 principles. Since CPS is still in the experimental stage, a proposed methodology and architecture described in [Lee, et al. (2015)] consists of two main components: (1) advanced connectivity that guarantees real-time data procurement from the physical world and information feedback from the digital space; and (2) intelligent data analytics, management and computational capability that constructs the cyber space. Presented in 0, the value is created when combining CPS from an earlier data acquisition and analytics.

From the above framework, one can see that the smart connection plays an important role. It acquires reliable and accurate data from machines and components, and customers' feedback which tells the insides of the design and the best approaches to meet their needs and wants. Data also flows in enterprise manufacturing systems as enterprise resource planning (ERP), manufacturing execution system (MES) and supply chain management (SCM) through smart connection. Thus all the collected data can be transformed into action through updated information in real time and obtain a reliable inside on the product [Lee, et al. (2015)].

Industry 4.0 also describes the overlap of multiple technological developments that comprise products and processes. The purpose of this chapter is to provide a well-founded methodology to give possible solutions to the missing gap of how big data offer opportunities and solutions for individualistic manufacture (customized production). The next section discusses the relation between smart products and machine learning for i4 environments.

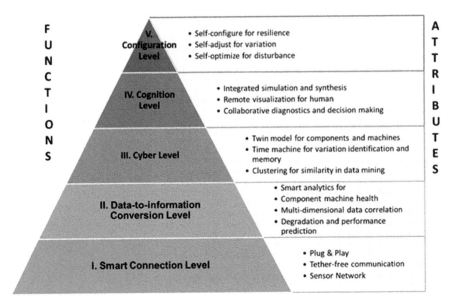

Fig. 25.1 Architecture for implementing CPS [Lee, et al. (2015)]

25.2.2 Smart Products and Product Lifecycle for Industry 4.0

Defined by [Mühlhäuser (2008)], a smart product is an entity (software, tangible object, or service) made and designed for self-organized embedding into different (smart) environments in the direction of its lifecycle, providing boosted simplicity and openness through upgraded product-to-user and product-to-product interaction by means of proactive behavior, context-awareness, semantic self-description, CI planning, multimodal natural interfaces and machine learning.

The interaction with products' environment is what makes a product smart and under the i4 principles, Radio Frequency Identifiers (RFID) give each product identity. In this sense, the increase in volume, variety and velocity of data creation represents a challenge which is identifying the best attributes in smart product designs and detecting exactly what customers really want for an individual product. Today with the IoT, data is collected constantly, thereby creating a continuous stream of data, leading to an evolving data base that comprises videos, sounds and images that can trigger best design for products, best quality, meeting customer needs and wants and process operations [Schmidt, et al. (2015)].

The digitalization of the value chain, how to optimize a process and bring flexibility lead to a whole value chain fully integrated. Customers and suppliers are included in the innovation of the product through social software [Nurcan and Schmidt (2009)]. Then cloud services connect to the networked product in the use phase. During its entire lifecycle, the product stays connected and maintains data collection. Hence Big Data can be used to create a feedback loop into the production phase, using algorithms and models that are able to process data in an unprecedented velocity, volume and variety [LaValle, et al. (2011)].

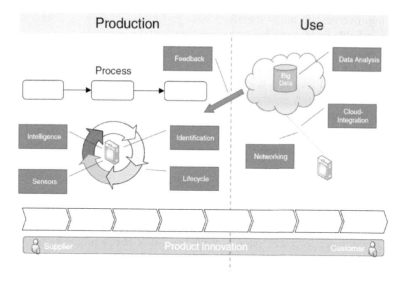

Fig. 25.2 Framework for smart product's innovation [Schmidt, et al. (2015), Schmidt (2013)]

Creating smart products for i4 technologies also determines the necessary base technologies, which can be named as follows: mobile computing, Big Data and cloud computing [Schmidt, et al. (2015)]. More than providing scalable compute capacity, i4 aims to provide services that can be accessed globally via the Internet, which emphasises the importance of cloud computing and mobile computing [Schmidt, et al. (2013)]. The [Schmidt, et al. (2015)] framework of smart product's innovation is proposed as depicted in 0.

How data is managed and analyzed is the key to this work. CPS will only implement mass production, but mass customization needs to be designed beforehand, and it is often found that the customer is not clear what his needs and wants are [Isaacson (2012)]. Eventually, how data is managed will lead to evolution for the innovation floor by this constant communication and linkage that IoT enables.

Next section reviews the machine learning techniques together with CI for addressing prediction in customized production.

25.2.3 Computational Intelligence for Customized Production

As discussed previously, the main components of the i4 or factory of the future vision are: CPS with the ability to connect everything through the IoT and IoS in digitalized environment, comprising decentralized architectures and real-time capability to analyse huge quantities of data (Big Data analytics) in a modular way.

In this context, classical and novel machine learning and CI techniques, among which is Artificial Neural Networks (ANN), are found in application which have been developed exactly to extract (hidden) information from data for pattern recognition, prediction issues and classification. Such techniques have a huge potential to provide a clear improvement in many transformation processes, as well as to services by providing reliable insides to what customers' really need and want.

Addressing prediction in larger datasets can be but one application of machine learning techniques, but first it's necessary to understand the characteristics of the data in order to find the most suitable method according to data inputs [Ji-Hyeong and Su-Young (2016)]. A good understanding of the dataset is crucial to the choice and to the eventual outcome of the analysis. Within the context of i4, there are two main sources of data: human-generated data and machine-generated data, both of which present huge challenges for data processing. Many of the algorithms developed so far are iterative, designed to learn continually and seek optimized outcomes. These algorithms iterate in milliseconds, enabling manufacturers to seek optimized outcomes in minutes versus months.

Facing the era of the IoT in [Shewchuk (2014)] is discussed the integration of machine-learning databases, applications and algorithms into cloud platforms and most of all automate processes in terms of the feasibility of controlling high-complex processes. An architecture is proposed by [Shewchuk (2014)] and presented in 0.

This presented framework encloses four key components: customer relationships, design and engineering, manufacturing and supply chain, and service and maintenance. The enterprise and business process are connected inside the cloud that retrieves information already processed from the industrial equipment. Here intelligence is used in the form of system's service agent. Local technicians report events, status or alarms if necessary for remote experts to evaluate each event; in this process, business intelligence takes part when accessing all the data that the platform processed to generate prediction models. Finally, a cloud-based machine learning platform facilitates the analysis and new knowledge is obtained, which experts need to verify to check the reliability of the obtained prediction.

Machine learning can also be implemented inside business intelligence where prediction must be achieved by using descriptive statistics that give insight into customer relations. In [Nauck, et al. (2008)] is suggested the following approaches for identifying customer relations:

- Use linear models for data analysis, which is regularly performed in simple ways. Since numerous implicit assumptions from linear statistics are about mutual independence between variables and normally distributed values, it can be helpful in the initial stage of exploration
- Dealing with stochastic distributions, the Hidden Markov Models (HMM) [Rabiner and Juang (1986)] focus on the analysis of temporal sequences of separate (discrete) states, as well as those that are used for creating predictions on time-stamped events
- When analyzing customer satisfaction, the use of Bayesian networks are suggested in [Heckerman and Wellman (1995)], which are based on a graphical model representing inputs as nodes with directed associations among them. Nevertheless, because these are developed for academic level and do not provide the needed levels of intuition, automation and integration into corporate environments, accessible Bayesian network software is not suitable. Enabling this can create an accessible approach for business users

IoT Services Architecture & Platform Components

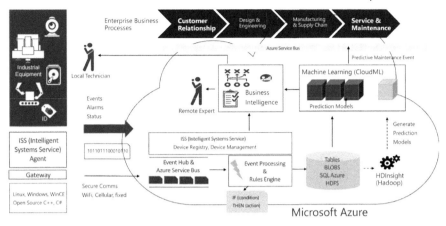

Fig. 25.3 Architecture for IoT services proposed by Microsoft [Shewchuk (2014)]

As discussed in [Nauck, et al. (2008)], customers play a significant role in Smart Manufacturing environments because of the improvement of customer-business relations and the responsiveness of business to take actions in real-time when based on customer lifecycle. This is not a trivial task that can be implemented overnight by using existing business informatics models. Two main factors can be attributed to this [Flores Saldivar, et al. (2016a)]: (i) lack of an automated closed-loop feedback system that can intelligently inform business processes to respond to changes in real time based on the inputs (for example, data trends, user experience, etc.) received, and (ii) existing analytical tools cannot accurately capture and predict consumer patterns.

The use of digital models is a possible way forward for (i), which a digital model is able to achieve automation in a closed-loop. A solution for (ii) when analyzing business contained in data using intelligence should be considered as the use of gathered information into data and finally into action. Intelligence in this sense comes from the expert knowledge that can be integrated in the analysis process, the knowledge-based methods used for analysis and the new knowledge created and communicated by the analysis process.

The next section presents the methodology used for addressing prediction in customer relations, determining customers' needs and wants and selecting the best attributes.

25.3 Methodology and CIAD Approaches

With all the revised methods and tools available from different researches, it was determined to use machine learning as unsupervised learning. It used fuzzy c-means for clustering and genetic algorithms for selection of best attributes once the fuzzy clustering finished classifying. Following the next sections, the fuzzy c-means is described together with the Genetic Algorithm (GA) selection. After the tools used, a proposed framework is shown, which integrates the i4 principles for design and manufacture, data analytics, machine learning and Computational Intelligence Aided Design (CIAD). With this, the closed-loop for automation can finally close the missing gap in determining customers' needs and wants in order to achieve customized design and processes.

25.3.1 Fuzzy c-Means Approach

Cluster approaches can be applied to datasets that are qualitative (categorical), quantitative (numerical), or a mixture of both. Usually the data (inputs) are observations of some physical processes. Each observation consists of n measured variables (features), grouped into an *n-dimensional* column vector $z_k = \lfloor z_{1k}, \ldots, z_{nk} \rfloor^T$, $z_k \in R^n$ [Ludwig (2015)].

N Observations set is denoted by $Z = \{z_k | k = 1,2, \ldots N\}$, and is represented as a $n \times N$ matrix:

$$Z = \begin{pmatrix} z_{11} & z_{12} & \cdots & z_{1N} \\ z_{21} & z_{22} & \cdots & z_{2N} \\ \cdots & \cdots & \cdots & \cdots \\ z_{n1} & z_{n2} & \cdots & z_{nN} \end{pmatrix}$$

Many clustering algorithms have been introduced and clustering techniques can be categorized depending on whether the subsets of the resulting classification are fuzzy or crisp (hard). Hard clustering methods are based on classical set theory and require that an object either belongs or not to a cluster. Hard clustering means that the data is partitioned into a specified number of mutually exclusive subsets. Fuzzy clustering methods, however, allow the objects to belong to several clusters simultaneously with different degrees of membership [Ludwig (2015)]. Fuzzy clustering assigns membership degrees between 0 and 1, thus indicating their partial membership. Cluster partition is vital for cluster analysis, as well as for identification techniques that are based on fuzzy clustering.

Most analytical fuzzy clustering algorithms are based on optimization of the basic c-means objective function, or some modification of the objective function. Optimization of the c-means function represents a nonlinear minimization problem, which can be solved by using a variety of methods including iterative minimization [Bezdek (1981a)]. The most popular method is to use the simple Picard iteration through the first-order condition for stationary points, known as the FCM algorithm. Bezdek [Bezdek (1981b)] has proven the convergence of the FCM algorithm. An optimal c partition is produced iteratively by minimizing the weight within group sum of squared error objective function:

$$J = \sum_{i=1}^{n} \sum_{j=1}^{c} (u_{ij})^m d^2(y_i, c_j)$$

where $Y = [y_1, y_2, \ldots ,y_n]$ is the dataset in a d-dimensional vector space, n is the number of data items, c, $2 \leq c \leq n$, is the number of clusters, which is defined by the user, u_{ij} is the degree of membership of y_i in the *j*th cluster, m is a weighted exponent on each fuzzy membership, c_j is the center of the cluster *j* and $d^2(y_i, c_j)$ is a square distance measure between object y_i and cluster c_j.

The following steps were used inside MATLAB® for the fuzzy c-means algorithm:

1) Input → c = centroid matrix, m = weighted exponent of fuzzy membership, ϵ = threshold value used as stopping criterion, $Y = [y_1, y_2, ... , y_n]$: data
   Output → c = update centroid matrix

2) Randomly start the fuzzy partition matrix $U = \lceil u_{ij}^k \rceil$

3) Repeat 1) and 2)

4) Calculate the cluster centres with U^k:

$$c_j = \sum_{i=1}^{n} (u_{ij}^k)^m y_i / \sum_{i=1}^{n} (u_{ij}^k)^m$$

Update the membership matrix U^{k+1} using:

$$u_{ij}^{k+1} = 1/\sum_{k=1}^{c} \left(\frac{d_{ij}}{d_{kj}}\right)^{\frac{2}{(m-1)}}$$

where

$$d_{ij} = \|y_i - c_j\|^2$$

until $max_{ij} \|u_{ij}^k - u_{ij}^{k+1}\| < \epsilon$.

5) Return c

After that, the selection of the best attributes is done inside MATLAB using the code for GA; the process is described in 0.

25.3.2 *Framework for Predicting Potential Customer Needs and Wants*

In 0 is depicted the framework proposed to solve several of the aforementioned challenges in i4. Based on i4 and Smart Manufacturing, there are some key objectives, i.e., achieve self-prediction, and self-configuration in order to manufacture products and provide services tailor-made at mass production rates.

In the first block of the proposed framework, shown in 0, the customer needs and wants are first captured and processed to extract key design characteristics. The information is then fed into a CIAD engine [Yun Li, et al. (2004)] where the design requirements, features and performance objectives are mapped into 'genotypes' for further analyses. This process, which is commonly known as rapid virtual prototyping, uses intelligent search algorithms, such as the GA or PSO, to explore the design search space for optimal solutions. In the proposed framework, this process takes place over the Cloud and produces a set of optimized virtual prototypes at the end of the search.

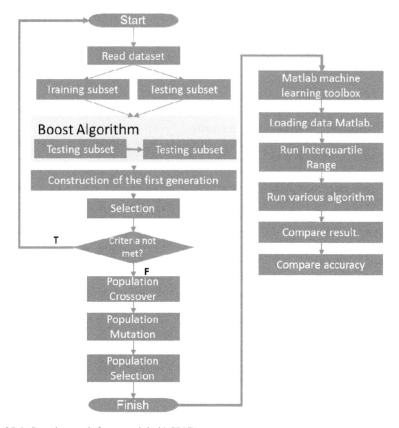

Fig. 25.4 Genetic search framework in i4 CIAD

The second block of the closed loop in 0 shows the virtual prototype, which is obtained from the selection and design process in CIAD. Through the integration of CPS or Cyber-Physical Integration (CPI), the virtual prototype in the second block is transformed into a physical product, i.e., the Smart Product as shown in 0.

The next part of the framework refers to Business Informatics and how the Smart Products are connected to IoT. Here is where the big data takes part. Through the performance of the product and the feedback from the customer, more features can be considered, which cover the necessary attributes that make the product to be manufactured in optimal ways.

Following this, the response obtained from the customer is automatically fed back to the system for further analysis and to fine-tune the virtual prototype. It is necessary to perform the analysis, which is related to prediction, by using node or dynamic analysis that can perform clustering, selection and detection of patterns and visualization. After that, the fuzzy c-means clustering completes the update of selected attributes by comparing the latest input to the existing cluster and tries to identify one cluster that is similar to the input sample. Then several features are fed back into the Cloud again.

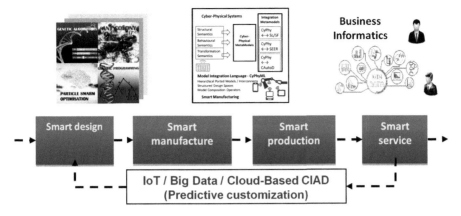

Fig. 25.5 Industry 4.0 value chain with predictive customer needs and wants fed back for automated customization

The analysis can result in two outcomes [Flores Saldivar, et al. (2016a)]: (i) Similar clusters are found and, this will be reflected as an existent attribute and the algorithm will update the existing cluster using information from the latest sample. (ii) Non-similar clusters are found. The algorithm will hold its operation with the current sample until it sees enough count-of-cluster samples.

When the number of out-of-cluster samples exceeds a certain amount, it means that there exists a new behavior in the data that has not been modeled. Then the algorithm will create a new cluster to represent such a new behavior.

The following section shows that a clustering problem is solved by using a fuzzy c-means network which is designed in MATLAB toolbox for machine learning. A widely used concise representation of a system's behavior is produced when fuzzy c-means is grouped into n clusters with every data-point in the dataset belonging to one cluster to a certain degree [Bezdek (1981a)]. Industry 4.0 value chain with predictive customer needs and wants is fed back to automated customization [Flores Saldivar, et al. (2016a)].

25.4 Case Study

Cluster analysis was performed with fuzzy c-means to the data set found in [Schlimmer (1985)], where the data set consists of three types of entities: (a) specification of an auto in terms of various characteristics, (b) its assigned insurance risk rating, (c) its normalized losses in use as compared to other cars. The second rating corresponds to the degree to which the auto is riskier than its price indicates. Cars are initially assigned a risk factor symbol associated with their price. If it is riskier (or less), this symbol is adjusted by moving it up (or down) the scale. Actuaries call this process 'symbolling'. A value of +3 indicates that the auto is risky; –3, that it is probably safer.

The third factor is the relative average loss payment per insured vehicle year.

Table 25.1 Automobile Data

Attribute	Attribute Range	Attribute	Attribute Range
symbolling	-3, -2, -1, 0, 1, 2, 3.	curb-weight	Continuous from 1488 to 4066.
normalized-losses:	Continuous from 65 to 256.	engine-type	dohc, dohcv, l, ohc, ohcf, ohcv, rotor.
make	alfa-romero, audi, bmw, chevrolet, dodge, honda, isuzu, jaguar, mazda, mercedes-benz, mercury, mitsubishi, nissan, peugot, plymouth, porsche, renault, saab, subaru, toyota, volkswagen, volvo.	num-of-cylinders	Eight, five, four, six, three, twelve, two.
fuel-type	Diesel, gas.	engine-size	Continuous from 61 to 326.
Aspiration	Std, turbo.	fuel-system	1bbl, 2bbl, 4bbl, idi, mfi, mpfi, spdi, spfi.
num-of-doors	Four, two.	bore	Continuous from 2.54 to 3.94.
body-style	Hardtop, wagon, sedan, hatchback, convertible.	stroke	Continuous from 2.07 to 4.17.
drive-wheels	4wd, fwd, rwd.	compression-ratio	Continuous from 7 to 23.
engine-location	Front, rear.	horsepower	Continuous from 48 to 288.
wheel-base	Continuous from 86.6 to 120.9.	peak-rpm	Continuous from 4150 to 6600.
Length	Continuous from 141.1 to 208.1.	city-mpg	Continuous from 13 to 49.
Width	Continuous from 60.3 to 72.3.	highway-mpg	Continuous from 16 to 54.
height	Continuous from 47.8 to 59.8.	price	Continuous from 5118 to 45400.

This value is normalized for all autos within a particular size classification (two-door small, station wagons, sports/specialty, etc...), and represents the average loss per car per year.

Database contents are shown in 0

This dataset comprises 205 instances, 26 attributes as shown in 0. The results of the fuzzy c-means are shown in 0. The partition of 3 clusters in which the scatter plot shows the connections between all the instances. From here, MATLAB function for fuzzy c-means updates the cluster centers and membership grades of each data point and clusters are iteratively moved from the center to the right location inside the dataset. The selected parameters for the fuzzy c-means were 3 clusters, exponent = 3, the maximum of iterations = 100, and minimum improvement = 1e-05. Iterations are based on minimizing an objective function that represents the distance from any given data point to a cluster center weighted by that data point's membership grade. Membership function plots are presented in 0. For each cluster, the plots are given when maximum of iterations is reached or when the objective function improvement between two consecutive iterations is less than the minimum amount of improvement specified. Once the clustering was done, the training data was processed to obtain the attribute classification inside the MATLAB toolbox for machine learning and it was embedded parallel routine for speeding up the whole process. Testing with several classifier algorithms, the results are presented in 0.

All these values in green show the corrected classified instances, based on the attribute that best reflected the desired selection: manufacturer or make. The red slots represent the incorrect instances. Here the manufacturer (make) was selected as the predictive variable in order to provide which of the observed brands are more attractive to customers based on all the considered variables.

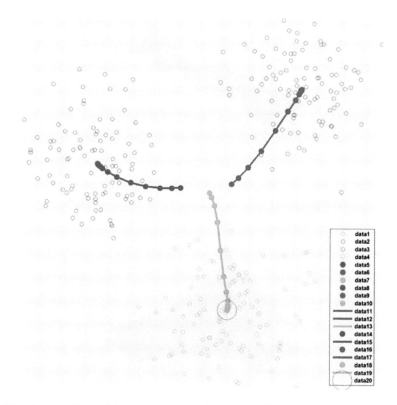

Fig. 25.6 Results of tested data. Fuzzy c-means with 3 clusters found

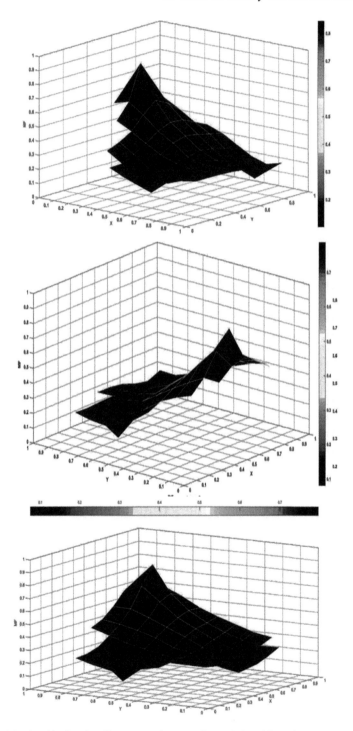

Fig. 25.7 Membership function. From top to bottom: cluster 1, 2 and 3 results

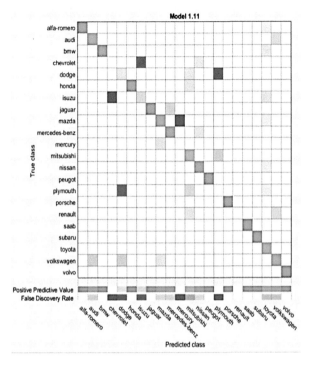

Fig. 25.8 Confusion matrix obtained for positive predictive values

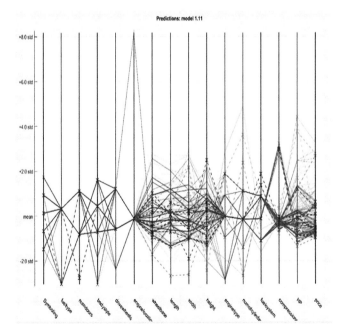

Fig. 25.9 Parallel coordinates plot for membership functions

From this plot in 0 is inferred what type of attributes represent the most correctly classified instances to the predictive model. The selected response variable was the manufacturer and each color represents the brand related to the predictors (fuel-type, number of doors, body style, engine locations, HP, etc.). The strongest relation is found with the engine location, number of cylinders and the HP variables. Moreover, once the attribute selection was performed using the GA selection, the following instances were selected: number-of-doors, drive-wheels, height, engine-type, number of cylinders. These were performed with a crossover probability of 0.6, a max of generations of 20, mutation probability of 0.033, initial population size of 20 and an initial seed.

25.5 Discussion and Conclusion

The use of fuzzy c-means to identify clustering, classify attributes and select instances using GA search delivered a promising performance. It was found that visualization of results facilitates the analysis in real time. Identification of values for customer's acquisition of a car based on categorical and numerical inputs can be achieved with fuzzy clustering.

Through the development of a predictive tool for mining customers' subconscious needs and wants, selection of best designs can thus be achieved in a smart way. The following features are summarized through the development of this work:

- In the case study, the results reveal that customer behavior is based on 5 attributes: number of doors, drive-wheels, height, engine-type, number of cylinders
- Fuzzy c-means performed a good partition on the dataset and identified 3 clusters for classification
- A feedback design process is suitable for automation with CIAD
- Intelligent search within the design process allows needs and wants to be predictively covered, with virtual prototypes further tuneable by the customer
- A CPS interconnected to the designed virtual prototypes would implement customisation efficiently
- A smart product may be gauged with business informatics and reliable data constantly, which can be fed back to smart design with IoT in the loop of the i4 value chain
- Since the 'Internet of Everything (IoE)' facilitates connection through the Cloud, it could make it faster to satisfy customer needs and wants
- Customer-oriented decision by the manufacturer becomes easier to make with customer-driven informatics, design and automation
- Big data analytics help visualize the influence of product characteristics, clustering and interpretation of subconscious customer needs and wants.

References

Aaronson, S. 2004. Multilinear formulas and skepticism of quantum computing. pp. 118–127. *In*: Proceedings of 36th Annual ACM Sympotium on Theory of Computing (STOC'04). ACM Press.

Abbas, F. and Choudhury, N. 2013. Electricity consumption-economic growth Nexus: An aggregated and disaggregated causality analysis in India and Pakistan. Journal of Policy Modeling 35: 538–553.

Abdelnour, G.M. et al. 1991. Design of a fuzzy controller using input and output mapping factors. IEEE Trans. on Systems, Man and Cybernetics 21(5): 952–960.

Aberdeen Group. 2006. The Transition from 2D Drafting to 3D Modeling Benchmark Report.

ABET. 2017. Criteria for Accrediting Engineering Programs, 2016–2017.

Abraham, A., Jain, L. and Goldberg, R. (eds.). 2005. Evolutionary Multiobjective Optimization-Theoretical Advances and Applications. Springer, London.

Adleman, L. 1994. Molecular computation of solutions to combinatorial problems. Science 266: 1021–1024.

Adleman, L. 1998. Computing with DNA. Scientific American 279: 54–61.

Ajmi, A.N., Montasser, G.E. and Nguyen, D.K. 2013. Testing the relationships between energy consumption and income in G7 countries with nonlinear causality tests. Economic Modelling 35: 126–133.

Akimoto, H., Ohara, T., Kurokawa, J.-I. and Horiid, N. 2006. Verification of energy consumption in china during 1996–2003 by using satellite observational data. Atmospheric Environment 40(40): 7663–7667 (1999).

Al-Enezi, J.R., Abbod, M.F. and Alsharhan, S. 2010. Artificial immune systems-models, algorithms and applications. International Journal of Research and Reviews in Applied Sciences 3: 118–131.

AL-Hussaini, E.K. and ABD-EL-Hakim, N.S. 1989. Failure rate of the inverse Gaussian Weibull mixture model. Ann. Inst. Statist. Math. 41(3): 617–622.

Allen, M.P. 2007. Understanding Regression Analysis, Springer.

Altman, N.S. 1992. An introduction to kernel and nearest-neighbor nonparametric regression. The American Statistician 46:3, 175–185.

Amphlett, J.C., Baumert, R.M., Mann, R.F., Peppley, B.A., Roberge, P.R. and Harris, T.J. 1995. Performance modeling of the ballard mark IV solid polymer electrolyte fuel cell. J. Electrochem. Soc. 142(1): 1–8.

Andersen, T., Bollerslev, T., Diebold, F.X. and Vega, C. 2003. Micro effects of macro announcements: real-time price discovery in foreign exchange. American Economic Review 93: 38–62.

Andresen, S.L. 2002. John McCarthy: Father of AI. IEEE Intelligent Systems 17: 84–85.

Andrews, P.S. and Timmis, J. 2006. A computational model of degeneracy in a lymph node. Artificial immune systems. Lecture Notes in Computer Science 4163: 164–177.

Ang, K.H., Chong, G. and Li, Y. 2005. PID control system analysis, design, and technology. IEEE Trans. Control Systems Technology 13(4): 559–576.

Annunziata, M. and Evans, P.C. 2012. Industrial Internet: Pushing the Boundaries of Minds and Machine, General Electric.

Anthony Kadrovach, B., Michaud, S.R., Zydallis, J.B., Lamont, G.B., Secrest, B. and Strong, D. 2001. Extending the simple genetic algorithm into multi-objective problems via mendelian pressure. Genetic and Evolutionary Computation Conference, San Francisco, California, 7–11 July 1: 181–188.

Antorg. 2015. Ant Colony Optimization. www.aco-metaheuristic.org.

Arthur, W.B. 1994. Increasing returns and path dependence in the economy. Ann, Arbor, MI: University of Michigan Press.

Asafu-Adjaye, J. 2000. The relationship between energy consumption, energy prices and economic growth: time series evidence from Asian developing countries. Energy Economics 22: 615–625.

Aström, K.J. and Wittenmark, B. 1984. Computer Controlled System: Theory and Design, Prentice-Hall Information and System Science Series. Kailath, T. (ed.). Englewood Cliffs, N.J.

Bäck, T., Hoffmeister, F. and Schwefel, H.-P. 1991. A survey of evolution strategies. pp. 2–9. In: Belew, R.K. and Booker, L.B. (eds.). Proceedings of the Fourth International Conference on Genetic Algorithms. Morgan Kaufmann, San Mateo, CA, USA.

Bacchetta, P. and Wincoop, E. 2006. Can information heterogeneity explain the exchange rate determination puzzle? American Economic Review 96: 552–576.

Back, T. 1996. Evolutionary Algorithms in Theory and Practice. Oxford University Press, New York.

Bakar, R. and Watada, J. 2008. DNA computing and its applications: Survey. CIC Express Letters 2: 101–108.

Balancing Mechanism Reporting System. OFGEM, UK. http://www.bmreports.com/bwx_reporting.htm.

Banks, A. Vincent and Anyakoha, J.C. 2007. A review of particle swarm optimization. Part I: Background and Development 6: 467–484.

Barker, J.S.F. 1958a. Simulation of genetic systems by automatic digital computers. III. Selection between alleles at an autosomal locus. Aust. J. Biol. Sci. 11: 603–612.

Barker, J.S.F. 1958b. Simulation of genetic systems by automatic digital computers. IV. Selection between alleles at a sex-linked locus. Aust. J. Biol. Sci. 11: 613–625.

Barto, A.G. 1992. Reinforcement learning and adaptive critic methods. pp. 469–491. In: White, D.A. and Sofge, D.A. (eds.). Handbook of Intelligent Control, New York: Van Nostrand Reinhold.

Baschuk, J.J. and Li, X.G. 2000. Modelling of polymer electrolyte membrane fuel cells with variable degrees of water flooding. Journal of Power Sources 86(1): 181–196.

Bellman, R. 1957. Dynamic Programming. Princeton University Press, Princeton, New Jersey.

Bellman, R.E., Kalaba, R.E. and Zadeh, L.A. 1964. Abstraction and Pattern Classification. Memorandum. RM-4307-PR. Santa Monica, CA: RAND Corporation. http//:www.rand.org/pubs/research_memoranda/RM4307.

Bellman, R.E. and Zadeh, L.A. 1970. Decision-making in a fuzzy environment. Management Science 17: 141–164.

Bellman, R.E. and Zadeh, L.A. 1977. Local and fuzzy logics. pp. 103–165. In: Dunn, J.M. and Epstein, G. (eds.). Modern Uses of Multiple-Valued Logic Episteme. Springer, Netherlands.

Benioff, P. 1980. Quantum mechanical Hamiltoniam models of turing machines. Journal of Statistical Physics 29: 515–54.

Benioff, P. 1980. The computer as a physical system: a microscopic quantum mechanical Hamiltonian model of computers as represented by Turing machines. Journal of Statistical Physics 22: 563–591.

Benioff, P. 1982. Quantum mechanical Hamiltoniam models of turing machines that dissipate no energy. Physical Review Letters 48: 1581–1585.

Bennett, C.H. and Brassard, G. 1984. Quantum cryptography: Public key distribution and coin tossing. pp. 175–179. *In*: Proceedings of IEEE International Conference on Computer Systems and Signal Processing. Bangalore, India, IEEE Press, New York.

Berenji, H.R. and Khedkar, P. 1992. Learning and tuning fuzzy logic controllers through reinforcements. IEEE Trans. on Neural Networks 3(5): 724–740.

Berling-Wolff, S. and Wu, J.-G. 2004. Modeling urban landscape dynamics: A Review Ecological Research 19(1): 119–129.

Bernardi, D.M. and Verbrugge, M.W. 1992. A mathematical model of the solid-polymer-electrolyte fuel cell. J. Electrochem. Soc. 139(9): 2477–2491.

Bernstein, E. and Vazirani, U. 1993. Quantum complexity theory. pp. 11–20. *In*: Proceedings of the 25th ACM Symposium on the Theory of Computation (STOC'93). ACM Press.

Bernstein, E. and Vazirani, U. 1997. Quantum complexity theory. SIAM Journal on Computing 26: 1411–1437.

Bersini, H. and Varela, F. 1991. Hints for adaptive problem solving gleaned from immune network. pp. 343–354. *In*: Schwefel, H.P. and Mühlenbein, H. (eds.). Parallel Problem Solving from Nature. Springer-Verlag.

Bevers, D., Wohr, M., Yasuda, K. and Oguro, K. 1997. Simulation of a polymer electrolyte fuel cell electrode. Journal of Applied Electrochemistry 27(11): 1254–1264.

Beyer, H.-G. and Schwefel, H.-P. 2002. Evolution strategies—A comprehensive introduction. Natural Computing 1: 3–52.

Bezdek, J. 2013. The history, philosophy and development of computational intelligence how a simple tune became a monster hit. *In*: Ishibuchi, H. (ed.). Computational Intelligence, Chap. 1, Encyclopedia of Life Support Sciences. Eolss Publishers, Oxford, UK.

Bezdek, J.C. 1981a. Pattern Recognition with Fuzzy Objective Function Algorithms. Kluwer Academic Publishers, pp. 256.

Bezdek, J.C. 1981b. Objective Function Clustering, in Pattern Recognition with Fuzzy Objective Function Algorithms. Springer US: Boston, MA. pp. 43–93.

Bezdek, J.C. 1992. On the relationship between neural networks, pattern recognition, and intelligence. The International Journal of Approximate Reasoning 6: 85–107.

Bian, Y.W., He, P. and Xu, H. 2013. Estimation of potential energy saving and carbon dioxide emission reduction in China based on an extended non-radial DEA approach. Energy Policy 63: 962–971.

Bianco, V., Manca, O. and Nardini, S. 2009. Electricity consumption forecasting in Italy using linear regression models. Energy 34: 1413–1421.

Bickel, B. and Alexa, M. 2013. Computational aspects of fabrication: Modeling, design, and 3D printing. IEEE Computer Graphics and Applications 33(6): 24–25.

Bierman, H.S. et al. 1998. Game Theory with Economic Applications, New York: Addison-Wesley, Inc.

Biswal, L., Chakraborty, S. and Som, S.K. 2009. Design and optimization of single phase liquid cooled micro-channel heat sink. IEEE Transactions on Components and Packaging Technologies 32(4): 876–886.

Bonabeau, E., Dorigo, M. and Theraulaz, G. 2000. Inspiration for optimization from social insect behavior. Nature 406: 39–42.

Boneh, D., Dunworth, C. and Lipton, R. 1996. Breaking DES using a molecular computer. pp. 37–65. 1st DIMACS Workshop on DNA Based Computers, Princeton, 1995. DIMACS Series 27. AMS Press.

Bonizzoni, P., Ferretti, C., Mauri, G. and Zizza, R. 2001. Separating some splicing models. Information Processing Letters 79: 255–259.

Boschma, R.A. and Frenken, K. 2003. Evolutionary economics and industry location. Review of Regional Research 23: 183–200.

Boschma, R.A. 2004. Competitiveness of regions from an evolutionary perspective. Regional Studies 38: 1001–14.

Boschma, R.A. and Frenken, K. 2006. Why is economic geography not an evolutionary science? Towards an evolutionary economic geography. Journal of Economic Geography 6(3): 273–302.

Boverie, S., Cerf, P. and Le Quellec, J. 1994. Fuzzy sliding mode control—application to idle speed control. Proc. 3rd IEEE Int. Conf. on Fuzzy Systems 2: 974–977.

Brehm, T. and Rattan, K.S. 1994. Hybrid fuzzy logic PID controller. Proc. 3rd IEEE Int. Conf. on Fuzzy Systems, Orlando, FL 3: 1682–1687.

Bremermann, H. 1958. The evolution of intelligence: the nervous system as a model of its environment. Technical report, no. 1, contract no. 477(17). Dept. of Mathematics, University of Washington, Seattle, Washington.

Brian Saunders. 2001. NETA: A dramatic change. ELECTRA.

Brubaker, D.I. 1992. Fuzzy-logic basics: intuitive rules replace complex math.

Bullinger, H.-J. and Fischer, D. 1998. New concepts for the development of innovative products: Digital prototyping. Proc. Int. Conf. Industrial Engineering Theories, Applications and Practice, Hong Kong.

Bunn, D. 2001. The agent-based simulation—An application to the new electricity trading arrangements of england and wales. IEEE Trans. Evol. Comp. 5(5).

Burgess, W.L. 2009. Valve regulated lead acid battery float service life estimation using a Kalman filter. Journal of Power Sources 191: 16–21.

Burnett, A., Fan, W.H., Upadhya, P., Cunningham, J., Edwards, H., Munshi, T., Hargreaves, M., Linfield, E. and Davies, G. 2006. Complementary spectroscopic studies of materials of security interest. Proceedings of the SPIE 6402, Pages 64020B.

Calderbank, A.R. and Shor, P.W. 1996. Good quantum error correcting codes exist. Physical Review A 54: 1098–1105.

Caponetto, R., Fortuna, L., Muscato, G. and Xibilia, M.G. 1994. Genetic algorithms for controller order reduction. Proc. 1st IEEE Conf. Evolutionary Computation, Orlando 2: 724–729.

Caracoglia, L. and Jones, N. 2007. Passive hybrid technique for the vibration mitigation of systems of interconnected stays. Journal of Sound and Vibration 307(3-5): 849–864.

Carter, J.H. 2000. The immune system as a model for pattern recognition and classification. Journal of the American Medical Informatics Assocation 7: 28–41.

Cartmell, M.P. 1998. Generating Velocity Increments by Means of a Spinning Motorised Tether 34th AIAA/ASME/SAE/ASEE Joint Propulsion Conference and Exhibit, Cleveland Conference Centre, Cleveland, Ohio, USA, AIAA-98-3739.

Cartmell, M.P. and Ziegler, S.W. 2001. Experimental Scale Model Testing of a Motorised Momentum Exchange Propulsion Tether 37th AIAA/ASME/SEA/ASEE Joint Propulsion Conference and, Exhibit, July 8–11, Salt Lake City, Utah, USA, AIAA 2001-3914.

Cartmell, M.P., Ziegler, S.W. and Neill, D.S. 2003. On the Performance Prediction and Scale Modelling of A Motorised Momentum Exchange Propulsion Tether 20th Symposium Space nuclear, power and propulsion; Space Technology and Applications International Forum 2003, 2–5 February, University of New Mexico, Albuquerque, New Mexico, USA.

Cartmell, M.P. and McKenzie, D.J. 2007. A review of space tether research. Progress in Aerospace Sciences 44: 1–21.

Chang, C.C. 2010. A multivariate causality test of carbon dioxide emissions, energy consumption and economic growth in China. Applied Energy 87: 3533–3537.

Chang, S.L. and Zadeh, L.A. 1972. On fuzzy mapping and control. IEEE Transactions on Systems, Man and Cybernetics 2: 30–34.

Chang, S.S.L. and Zadeh, L.A. 1972. On fuzzy mapping and control. IEEE Trans. Sys., Man and Cyber. 2(1): 30–34.

Charles, L. Cole. 1983. Microeconomics, New York: Harcourt Jovanovich, Inc.

Charles, L. Cole. 1995. Microeconomics, New York: Harcourt Jovanovich, Inc., 2nd Ed.

Chen, G. and Tong, W. 1994. A robust fuzzy PI controller for a flexible-joint robot arm with uncertainties. Proc. 3rd. IEEE Int. Conf. on Fuzzy Systems, Orlando, FL 3: 1554–1559.

Chen, S., Billings, S.A. and Grant, P.M. 1990. Non-linear system identification using neural networks. Int. J. Contr. 51(6): 1191–214.

Chen, S.T., Kuo, H.I. and Chen, C.C. 2007. The relationship between GDP and electricity consumption in 10 Asian Countries. Energy Policy 35: 2611–2621.

Chen, W. 2008. Simulation of Microwave Resonance Field in an Ice Cylinder for a Potential Microwave Ignition System. University of Glasgow, M.Sc. Thesis.

Chen, X., Patterson, B.D. and Settersten, T.B. 2004. Time-domain investigation of OH ground-state energy transfer using picosecond two-color polarization. Spectroscopy Chemical Physics Letters 388(4-6): 358–362.

Chen, Y. 2004. Study on SC6350C Driveline Torsional Vibration, Master Thesis, State Key Laboratory of Mechanical Transmission, Chongqing University.

Chen, Y., Fang, Z., Luo, H. and Deng, Z. 2004. Simulation research on real-time polynomial function supervising PID control based on genetic algorithms. Journal of System Simulation 16(6): 1171–1174.

Chen, Y. and Cartmell, M.P. 2007. Dynamical Modelling of the Motorised Momentum Exchange Tether Incorporating Axial Elastic Effects Advanced Problems in Mechanics, 20–28 June, Russian Academy of Sciences, St Petersburg, Russia.

Chen, Y. and Cartmell, M.P. 2007. Multi-objective optimisation on motorised momentum exchange tether for payload orbital transfer. pp. 987–993. *In*: Proceedings of 2007 IEEE Congress on Evolutionary Computation (CEC), 25–28 September, 2007, Singapore. IEEE Computer Society, Piscataway, USA.

Chen, Y. 2008. SMATLINK—Let Matlab Dance with Mathematica, https://uk.mathworks.com/matlabcentral/fileexchange/20573-smatlink-let-matlab-dancewith-mathematica. Last accessed on Dec. 20, 2017.

Chen, Y. 2009. SGALAB—Simple Genetic Algorithm Laboratory Toolbox, http://www.mathworks.co.uk/matlabcentral/fileexchange/5882. Last accessed on Dec. 20, 2017.

Chen, Y. 2009. Skyhook surface sliding mode control on semi-active vehicle suspension systems for ride comfort enhancement engineering. Scientific Research Publishing 1(1): 23–32.

Chen, Y. and Cartmell, M.P. 2009. Hybrid Fuzzy and Sliding-Mode Control for Motorised Tether Spin-Up When Coupled with Axial Vibration 7th International Conference on Modern Practice in Stress and Vibration Analysis, 8–10 September, New Hall, Cambridge, UK.

Chen, Y. and Cartmell, M.P. 2009. Hybrid Fuzzy Skyhook Surface Sliding Mode Control for Motorised Space Tether Spin-up Coupled with Axial Oscillation Advanced Problems in Mechanics, 30 June–5 July, Russian Academy of Sciences, St. Petersburg, Russia.

Chen, Y. 2010a. Dynamical modelling of a flexible motorised momentum exchange tether and hybrid fuzzy sliding mode control for spin-up. Ph.D. Thesis, Mechanical Engineering Department, University of Glasgow, Glasgow.

Chen, Y. 2010b. Dynamical modelling of space tether: Flexible motorised momentum exchange tether and hybrid fuzzy sliding mode control for spin-up. LAP Lambert Academic Publishing AG & Co., Saarbrucken, Germany.

Chen, Y. and Cartmell, M.P. 2010. Hybrid sliding mode control for motorised space tether spin-up when coupled with axial and torsional oscillation astrophysics and space science. Springer 326(1): 105–118.

Chen, Y. 2011. Fuzzy skyhook surface control using micro-genetic algorithm for vehicle suspension ride comfort. *In*: Mario Koeppen, Gerald Schaefer and Ajith Abraham (eds.). Intelligent Computational Optimization in Engineering: Techniques and Applications, Studies in Computational Intelligence 366: 357–394. Springer: Berlin Heidelberg, 2011.

Chen, Y. 2011. SwarmFish—The Artificial Fish Swarm Algorithm, http://www.mathworks.com/matlabcentral/fileexchange/3202. Last accessed on Dec. 20, 2017.

Chen, Y., Ma, Y., Lu, Z., Peng, B. and Chen, Q. 2011. Quantitative analysis of terahertz spectra for illicit drugs using adaptive-range micro-genetic algorithm. Journal of Applied Physics 110: 044902-10.

Chen, Y., Ma, Y., Lu, Z., Qiu, L.-X. and He, J. 2011. Terahertz spectroscopic uncertainty analysis for explosive mixture components determination using multi-objective micro genetic algorithm. Advances in Engineering Software 42(9): 649–659.

Chen, Y., Ma, Y., Lu, Z., Xia, Z.-N. and Cheng, H. 2011. Chemical components determination via terahertz spectroscopic statistical analysis using micro genetic algorithm. Optical Engineering 50(3): 034401-12.

Chen, Y., Wang, Z.L., Qiu, J., Zheng, B. and Huang, H.Z. 2011. Adaptive bathtub hazard rate curve modelling via transformed radial basis functions. pp. 110–114. *In*: Proceedings of International Conference on Quality, Reliability, Risk, Maintenance, and Safety Engineering (ICQR2MSE 2011), Xian, China.

Chen, Y. 2012. SECFLAB—Simple Econometrics and Computational Finance Laboratory Toolbox, http://www.mathworks.com/matlabcentral/fileexchange/38120. Last accessed on Dec. 20, 2017.

Chen, Y. 2012. SwarmBat-the artificial bat algorithm, http://www.mathworks.de/matlabcentral/fileexchange/39116-swarmbat-the-artificial-batalgorithm-aba. Last accessed on Dec. 20, 2017.

Chen, Y. and Song, Z.J. 2012. Spatial analysis for functional region of suburban-rural area using micro genetic algorithm with variable population size. Expert Systems with Applications 39: 6469–6475.

Chen, Y., Wang, X.Y., Sha, Z.J. and Wu, S.M. 2012. Uncertainty analysis for multi-state weighted behaviours of rural area with carbon dioxide emission estimation. Applied Soft Computing 12: 2631–2637.

Chen, Y., Wang, Z.L., Liu, Y., Zuo, M.J. and Huang, H.Z. 2012. Parameters Determination for Adaptive Bathtub-shaped Curve Using Artificial Fish Swarm Algorithm. The 58th Annual Reliability and Maintainability Symposium, 23–26 Jan, John Ascuaga s Nugget Hotel and Resort, Reno, Nevada, USA.

Chen, Y., Wang, Z.L., Qiu, J. and Huang, H.Z. 2012. Hybrid fuzzy skyhook surface control using multi-objective micro-genetic algorithm for semi-active vehicle suspension system ride comfort stability analysis. Journal of Dynamic Systems, Measurement and Control 134: 041003-14.

Chen, Y. and Zhang, G.F. 2013. Exchange rates determination based on genetic algorithms using Mendel's Principles: Investigation and estimation under uncertainty. Information Fusion 14: 327–333.

Chen, Y., Miao, Q., Zheng, B., Wu, S.M. and Pecht, M. 2013. Quantitative analysis of lithium-ion battery capacity prediction via adaptive bathtub-shaped function. Energies 6: 3082–3096.

Chen, Y., Zhang, G.F., Li, Y.Y., Ding, Y., Zheng, B. and Miao, Q. 2013. Quantitative analysis of dynamic behaviours of rural areas at provincial level using public data of gross domestic product. Entropy 15: 10–31.

Chen, Y., Zhang, G.F., Zheng, B. and Zelinka, I. 2013. Nonlinear spatial analysis of dynamic behavior of rural regions. *In*: Zelinka, I., Chen, G.R., Rossler, O.E., Snasel, V. and Abraham, A. (eds.). Nostradamus 2013: Prediction, Modeling and Analysis of Complex Systems; 3rd–5th June, 2013, Ostrava, Czech Republic. Springer: Advances in Intelligent Systems and Computing 210: 401–412.

Chen, Y. 2014. SwarmDolphin—The Swarm Dolphin Algorithm (SDA) http://www.mathworks. com/matlabcentral/fileexchange/45965.

Chen, Y., Huang, R., He, L.P., Ren, X.L. and Zheng, B. 2014. Dynamical modelling and control of space tethers—A review of space tether research. Nonlinear Dynamics. DOI: 10.1007/s11071-014-1390-5.

Chen, Y., Zhang, G.F., Jin, T.D., Wu, S.M. and Peng, B. 2014. Quantitative modeling of electricity consumption using computational intelligence aided design. Journal of Cleaner Production, http://dx.doi.org/10.1016/j.jclepro.2014.01.058.

Chen, Y. 2017. SwarmFireFly—The Firefly Swarm Algorithm (FFSA). http://www.mathworks.co.uk/ matlabcentral/fileexchange/38931-swarmfirefly-the-fireflyswarm-algorithm-ffsa. Last accessed on Dec. 20, 2017.

Chipperfield, A.J., Fonseca, C.M. and Fleming, P.J. 1992. Development of genetic optimisation tools for multi-objective optimisation problems in CACSD. IEEE Colloq. on Genetic Algorithms for Cont. Sys. Engineering, pp. 3/1–3/6.

Chiu, S., Chand, S., Moore, D. and Chaudhary, A. 1991. Fuzzy logic for control of roll and moment foe a fexible wing aircraft. IEEE Control Systems Mag. 42–48.

Choi, S.K., Grandhi, R.V. and Canfield, R.A. 2007. Reliability-based Structural Design Springer-Verlag London Limited.

Chuang, I.L., Vandersypen, L.M.K., Zhou, X., Leung, D.W. and Lloyd, S. 1998. Experimental realization of a quantum algorithm. Nature 393: 143–146.

Clerc, M. and Kennedy, J. 2002. The particle swarm: Explosion, stability, and convergence in a multi-dimensional complex space. IEEE Transactions on Evolutionary Computation 6: 58–73.

Coello Coello, C.A. 2000. Treating constraints as objectives for single objective evolutionary optimization. Engineering Optimization 32: 275–308.

Coello Coello, C.A. and Pulido, G.T. 2001. A micro-genetic algorithm for multiobjective optimization proceedings of the genetic and evolutionary. Computation Conference, Springer-Verlag, pp. 126–140.

Coello Coello, C.A. and Toscano Pulido, G. 2001a. A microgenetic algorithm for multiobjective optimization. pp. 126–140. *In*: Zitzler, E., Deb, K., Thiele, L., Coello Coello, C.A. and Corne, D. (eds.). First International Conference on Evolutionary Multi-Criterion Optimization. Springer-Verlag. Lecture Notes in Computer Science No. 1993.

Coello Coello, C.A. and Toscano Pulido, G. 2001b. Multiobjective optimization using a micro-genetic algorithm. pp. 274–282. *In*: Spector, L., Goodman, E.D., Wu, A., Langdon, W.B., Voigt, H.-M., Gen, M., Sen, S., Dorigo, M., Pezeshk, S., Garzon, M.H. and Burke, E. (eds.). Proceedings of the Genetic and Evolutionary Computation Conference (GECCO 2001). San Francisco, California. Morgan Kaufmann Publishers.

Coello Coello, C.A., Lamont, G.B. and Van Veldhuizen, D.A. 2007. Evolutionary Algorithms for Solving Multi-Objective Problems Second Edition, Springer.

Coello Coello, C.A. 2015. The EMOO web site. http://delta.cs.cinvestav.mx/~ccoello/EMOO/.

Colorni, A., Dorigo, M. and Maniezzo, V. 1992. An investigation of some properties of an ant algorithm. pp. 509–520. *In*: Proceedings of the Parallel Problem Solving from Nature Conference (ppsn92), Brussels, Belgium, 1992, Elsevier Publishing.

Colorni, A., Dorigo, M. and Maniezzo, V. 1992. Distributed optimization by ant colonies. pp. 134–142. *In*: Varela, F. and Bourgine, P. (eds.). Proceedings of the First European Conference on Artifical Life. MIT Press, Cambridge.

Conn, A., Scheinberg, R.K. and Vicente, L.N. 2009. Introduction to Derivative-Free Optimization, SIAM.

Cook, D.J., Decker, B.K. and Allen, M.G. 2005. Quantitative THz Spectroscopy of explosive materials in optical, terahertz science and technology. Optical Society of America, Orlando, Florida March 14, Paper MA6.

Cooper, M. and Vidal, J. 1994. Genetic design of fuzzy controllers: The cart and jointed-pole problem. Proc. 3rd IEEE Int. Conf. on Fuzzy Systems 2: 1332–1337.

Copeland, D. 2000. Optimization of parallel plate heat sinks for forced convection. pp. 266–272. *In*: Proceedings of 16 Annual IEEE Semiconductor Thermal Measurement and Management Symposium.

Copiello, D. and Fabbri, G. 2009. Multi-objective genetic optimization of the heat transfer from longitudinal wavy fins. International Journal of Heat and Mass Transfer 52(5): 1167–1176.

Coricelli, F. and Dosi, G. 1998. Coordination and order in economic change and the interpretive power of economic theory. pp. 124–47. *In*: Dosi, G., Freeman, C., Nelson, R., Silverberg, G. and Soete, L. (eds.). Technical Change and Economic Theory, New York: Pinter.

Corne, D.W., Knowles, J.D. and Oates, M.J. 2000. The Pareto envelope-based selection algorithm for multiobjective optimization. pp. 839–848. *In*: Schoenauer, M., Deb, K., Rudolph, G., Yao, X., Lutton, E., Merelo, J.J. and Schwefel, H.P. (eds.). Proceedings of the Parallel Problem Solving from Nature VI Conference, Paris, France. Springer.

Corne, D.W., Jerram, N.R., Knowles, J.D. and Oates, M.J. 2001. PESA-II: Region based selection in evolutionary multiobjective optimization. pp. 283–290. *In*: Spector, L., Goodman, E.D., Wu, A., Langdon, W.B., Voigt, H.-M., Gen, M., Sen, S., Dorigo, M., Pezeshk, S., Garzon, M.H. and Burke, E. (eds.). Proceedings of the Genetic and Evolutionary Computation Conference (GECCO-2001), San Francisco, California. Morgan Kaufmann Publishers.

Crosby, J.L. 1960. The use of electronic computation in the study of random fluctuations in rapidly evolving populations. Phil. Trans. Roy. Soc. B 242: 415–417.

Crosby, J.L. 1963. Evolution by computer. New Scientist 17: 415–417.

Crosby, J.L. 1973. Computer Simulation in Genetics. New York: Wiley.

Crouzet, Francois. 1996. France. In Teich, Mikul?; Porter, Roy. The industrial revolution in national context: Europe and the USA. Cambridge University Press. p. 45. ISBN 978-0-521-40940-7. LCCN 95025377.

Crowe, D. and Feinberg, A. 2000. Design for Reliability. CRC Press LLC.

Curtis, H. 2004. Orbital Mechanics for Engineering Students, Elsevier Science Aerospace Engineering Series. Butterworth-Heinemann.

Daley, M.J. and Kari, L. 2002. DNA computing: Models and implementations. Comments on Theoretical Biology 7: 177–198.

Daley, S. and Gill, K.F. 1989. Comparison of a fuzzy logic controller with a P+D control law. Trans. of the ASME, Journal of Dynamic System, Measurement and Control 111(2): 128–137.

Das, I. and Dennis, J. 1997. A closer look at drawbacks of minimizing weighted sums of objectives for pareto set generation in multicriteria optimization problems. Structural Optimization 14: 63–69.

Dasgupta, D. (ed.). 1999. Artificial Immune Systems and their Applications. Springer, Berlin.

Dasgupta, D. and Nino, F. 2008. Immunological Computation: Theory and Applications. Auerbach Publications.

Davidyuk, O., Selek, I., Ceberio, J. and Riekki, J. 2007. Application of micro-genetic algorithm for task based computing. pp. 140–145. *In*: Proceeding of International Conference on Intelligent Pervasive Computing (IPC-07), October, Jeju Island, Korea.

Davies, A.G., Burnett, A.D., Fan, W., Linfield, E.H. and Cunningham, J.E. 2008. Terahertz spectroscopy of explosives and drugs. Materials Today 11(3): 18–26.

de Castro, L.N. and Von Zuben, F.J. 1999. Artificial immune systems: Part I-basic theory and applications. Tech Rep DCA-RT 01/99, School of Computing and Electrical Engineering, State University of Campinas, Brazil.

de Castro, L.N. and Timmis, J. 2002. Artificial Immune Systems: A New Computational Intelligence Approach. Springer.

de Castro, L.N. and Timmis, J. 2003. Artificial immune systems as a novel soft computing paradigm. Soft Computing Journal 7: 526–544.

de Castro, L.N., Von Zubena, F.J. and de Deus, Jr., G.A. 2003. The construction of a Boolean competitive neural network using ideas from immunology. Neurocomputing 50: 51–85.

De Jong, K.A. 1975. An analysis of the behavior of a class of genetic adaptive systems. Doctoral Dissertation. University of Michigan, Michigan, USA.

De Jong, K.A. 1980. Adaptive system design: A genetic approach. IEEE Trans. Sys. Man and Cyber. 10(9): 566–574.

De Jong, K.A., David, D., Fogel, B. and Schwefel, H.-P. 1997. A history of evolutionary computation. pp. A2.3:1–12. In: Bäck, T., Fogel, D.B. and Michalewicz, Z. (eds.). Handbook of Evolutionary Computation. Oxford University Press, New York, and Institute of Physics Publishing, Bristol https://www.researchgate.net/publication/216300863_A_history_of_evolutionary_computation.

De Neyer, M., Gorez, R. and Barreto, J. 1991. Fuzzy controllers using internal models. IMAC Symp. MCTS LILLE, pp. 726–731.

de Rocquigny, E., Devictor, N. and Tarantola, S. 2008. Uncertainty in Industrial Practice John, Wiley & Sons.

de Rooij, M. 2008. The analysis of change, Newton's law of gravity and association models. Journal of the Royal Statistical Society: Series A (Statistics in Society) 171(1): 137–157.

Deb, K. and Kumar, A. 1995. Real-coded genetic algorithms with simulated binary crossover: Studies on multimodal and multiobjective problems. Complex Systems 9: 431–454.

Deb, K. 1999. Solving goal programming problems using multi-objective genetic algorithms. pp. 77–84. In: Proceedings of the 1999 Congress on Evolutionary Computation, Washington, DC. IEEE.

Deb, K. 2001. Multi-Objective Optimization using Evolutionary Algorithms. John Wiley and Sons, Chichester, UK.

Deb, K., Pratap, A., Agarwal, S. and Meyarivan, T. 2002. A fast and elitist multiobjective genetic algorithm: NSGAII. IEEE Trans. on Evolutionary Comput. 6: 182–197.

Deb, K. 2004. Multi-Objective Optimization using Evolutionary Algorithms. John Wiley and Sons, Kanpur, India.

del Valle, Y., Venayagamoorthy, G.K., Mohaghenghi, S., Hernandez, J.C. and Harley, R.G. 2008. Particle swarm optimization: Basic concepts, variants and applications in power systems. IEEE Transactions on Evolutionary Computation 12: 171–195.

Deutsch, D. 1985. Quantum theory, the Church-Turing principle and the universal quantum computer. Proceedings of the Royal Society of London A 400: 97–117.

Deutsch, D. and Jozsa, R. 1992. Rapid solution of problems by quantum computation. Proceedings of the Royal Society of London A 439: 553–558.

Dhillon, B.S. 1979. A hazard rate model. IEEE Trans. Reliability 29(150).

Dhillon, B.S. 2005. Reliability, Quality, and Safety for Engineers, CRC Press LLC.

Ding, S., Li, H., Su, C., Yu, J. and Jin, F. 2013. Evolutionary artificial neural networks: A review. Artificial Intelligence Review 39: 251–260.

Ding, Y.J. and Shi, W. 2006. Progress on widely-tunable monochromatic Thz sources and room-temperature detections of Thz waves. Journal of Nonlinear Optical Physics and Materials 15(1): 89–111.

Dorigo, M., Maniezzo, V. and Colorni, A. 1996. The ant system: Optimization by a colony of cooperating agents. IEEE Transactions on Systems, Man, and Cybernetics 26: 29–41.

Dorigo, M., Di Caro, G. and Gambardella, L.M. 1999. Ant algorithms for discrete optimization. Artifical Life 5: 137–72.

Dorling, C.M. 1985. The Design of Variable Structure Control Systems: Manual for the VASSYD CAD Package. SERC Report (Project GR/C/03119, Grantholders: Zinober, A.S.I. and Billings,

S.A.), Department of Applied and Computational Mathematics, University of Sheffield, Sheffield S10 2TN.

Drubaker, D.I. 1992. Fuzzy-logic basics: Intuitive rules replace complex math. EDNASIA Magazine, pp. 59–63.

Drubaker, D.I. and Sheerer, C. 1992. Fuzzy-logic system solves control problems. EDNASIA Magazine, pp. 57–666.

Ecker, M., Gerschler, J.B., Vogel, J., Käbitz, S., Hust, F., Dechent, P. and Sauer, D.U. 2012. Development of a lifetime prediction model for lithium-ion batteries based on extended accelerated aging test data. Journal of Power Sources 215: 248–257.

Edelman, G.M. and Gally, J.A. 2001. Degeneracy and complexity in biological systems. Proceedings of the National Academy of Sciences 98: 13763–13768.

Edgeworth, F.Y. 1881. Mathematical Physics, P. Keagan, London, England.

Eiden, M.J. and Cartmell, M.P. 2003. Overcoming the challenges: tether systems roadmap for space transportation applications. AIAA/ICAS International Air and Space Symposium and Exposition, Dayton Convention Center, Dayton, Ohio, 14–17 July.

Elbeheiry, E.M. and Karnopp, D.C. 1996. Optimal control of vehicle random vibration with constrained suspension deflection. Journal of Sound and Vibration 189(5-8): 547–564.

Emelyanov, S.V. 1967. Variable Structure Control Systems (in Russian) Moscow: Nauka.

Eric Bond, Sheena Gingerich, Oliver Archer-Antonsen, Liam Purcell and Elizabeth Macklem. 2011. Causes of the Industrial Revolution. The Industrial Revolution February 2003 http://www. ecommerce-digest.com/modeling-trends.html (Last accessed 17 February 2003).

Escher, W., Michel, B. and Poulikakos, D. 2010. A novel high performance, ultra thin heat sink for electronics. Int. J. Heat Fluid Flow 31(4): 586–598.

Evans, M. and Lyons, R. 2002. Order flow and exchange rate dynamics. Journal of Political Economy 110: 170–180.

Fantini, P. and Figini, S. 2009. Applied Data Mining for Business and Industry 2nd Edition, John, Wiley & Sons Ltd.

Farley, B.G. and Clark, W.A. 1954. Simulation of self-organizing systems by digital computer. IRE Transactions on Information Theory 4: 76–84.

Farmer, J.D., Packard, N.H. and Perelson, A.S. 1986. The immune system, adaptation, and machine learning. Physica D: Nonlinear Phenomena 22: 187–204.

Fatai, K., Oxley, L. and Scrimgeour, F.G. 2004. Modelling the causal relationship between energy consumption and GDP in New Zealand, Australia, India, Indonesia, The Philippines and Thailand. Mathematics and Computers in Simulation 64: 431–445.

Feng, W. and Li, Y. 1999. Performance indices in evolutionary CACSD automation with application to batch PID generation. pp. 486–491. In: Proceeding of the 10th IEEE International Symposium on Computer-Aided Control System Design, Hawaii, USA.

Ferguson, B. and Zhang, X. 2002. Materials for terahertz science and technology. Nature Materials 1: 26–33.

Feynman, R.P. 1982. Simulating physics with computers. International Journal of Theoretical Physics 21: 467–488.

Feynman, R.P. 1985. Quantum mechanical computers. Optics News. 11–21.

Feynman, R.P. 1986. Quantum mechanical computers. Foundations of Physics 16: 507–531.

Filho, L.R., Treleaven, P.C. and Alippi, C. 1994. Genetic-algorithm programming environments. IEEE Computer 27: 28–43.

Fisher, R. 1936. The use of multiple measurements in taxonomic problems. Annual Eugenics 7: 179–188.

Fleming, P.J. and Fonseca, C.M. 1993. Genetic algorithms in control systems engineering. Proc. 12th IFAC World Congress 2: 383–390.

Fleming, P.J. and Hunt, K.J. 1995. IEE Colloq. on Genetic Algorithms for Control Systems Engineering Digest, Professional Groups: Computer aided control engineering, Control and Systems theory, and Planning and Control, Savoy Place, London WC2R 0BL, vol. 106, EDS.

Fleming, P.J., Purshouse, R.C. and Lygoe, R.J. 2005. Many-objective optimization: An engineering design perspective. Evolutionary Multi-Criterion Optimization, pp. 14–32.

Flexible Intelligence Group, L.L.C., FlexTool(GA) (MATLAB Toolbox) User's Manual, Version M2.1, Tuscaloosa, AL, June, 1995.

FlexTool(GA) http://www.aip.de/~ast/EvolCompFAQ/Q20_FLEXTOOL.htm.

FlexTool(GA)TM M1.1. 1995. User's Manual, Flexible Intelligence Group, LLC.

Flores Saldivar, A.A. et al. 2016a. Self-organizing tool for smart design with predictive customer needs and wants to realize Industry 4.0. In: World Congress on Computational Intelligence. Vancouver, Canada: IEEE.

Flores Saldivar, A.A. et al. 2016b. Identifying Smart Design Attributes for Industry 4.0 Customization Using a Clustering Genetic Algorithm. In International Conference on Automation & Computing, University of Essex, Colchester city, UK: IEEE.

Flores, A. et al. 2015. Industry 4.0 with cyber-physical integration: A design and manufacture perspective. Proc. Int. Conf. Automation and Computing.

Fogel, D.B. (ed.). 1998. Evolutionary Computation: The Fossil Record. Piscataway, NJ: IEEE Press.

Fogel, D.B. 1991. System Identification through Simulated Evolution: A Machine Learning Approach to Modeling. Ginn.

Fogel, D.B. 1995. Evolutionary Computations: Toward a New Philosophy of Machine Intelligence. Wiley-IEEE Press.

Fogel, D.B. 1995. Phenotypes, genotypes, and operators in evolutionary computation. IEEE International Conference on Evolutionary Computation. Perth, WA, Australia.

Fogel, D.B. and Ghozeil, A. 1997. Schema processing under proportional selection in the presence of random effects. IEEE Transactions on Evolutationary Computation 1: 290–293.

Fogel, D.B. and Fraser, A.S. 2000. Running races with Fraser s recombination. pp. 1217–1222. In: Proc. 2000 Congress on Evolutionary Computation. Piscataway, NJ: IEEE Press, vol. 2.

Fogel, David. 2002. In Memoriam—Alex S. Fraser, IEEE Transactions on Evolutionary Computation 6(5): October 2002.

Fogel, L.J., Owens, A.J. and Walsh, M.J. 1966. Artificial Intelligence through Simulated Evolution. Wiley.

Fogel, L.J. and Atmar, J.W. 1991. Meta-evolutionary programming. pp. 540–545. In: Proceddings of the 25th Asilomar Conference on Signals, Systems and Computers, Pacific Grove, CA.

Fogel, L.J. 1999. Intelligence through Simulated Evolution: Forty Years of Evolutionary Programming. Wiley.

Fonseca, C.M. and Fleming, P.J. 1993. Genetic algorithm for multi-objective optimisation: formulation, discussion and generalization. In: Forrest, S. (ed.). Genetic Algorithm: Proceedings of the Fifth International Conference, Morgan Kaufmann, San Mateo, CA 1: 141–153.

Fonseca, C.M. and Fleming, P.J. 1994. Multiobjective optimal controller design with genetic algorithms. Int. Conf. on Control 1: 745–749.

Forrest, S., Perelson, A., Allen, L. and Cherukuri, R. 1994. Self-nonself discrimination in a computer. pp. 202–212. In: Proceedings of 1994 IEEE Computer Society Symposium on Research in Security and Privacy.

Fraser, A.S. 1957. Simulation of genetic systems by automatic digital computers. I. Introduction, Aust. J. Biol. Sci. 10: 484–491.

Fraser, A.S. 1957. Simulation of genetic systems by automatic digital computers. II. Effects of linkage or rates of advance under selection. J. Biol. Sci. 10: 492–499.

Fraser, A.S. 1958. Monte Carlo analyses of genetic models. Nature 181: 208–209.

Fraser, A.S. 1960a. Simulation of genetic systems by automatic digital computers. 5-linkage, dominance and epistasis. pp. 70–83. *In*: Kempthome, O. (ed.). Biometrical Genetics. New York: Pergamon.

Fraser, A.S. 1960b. Simulation of genetic systems by automatic digital computers. VI. Epistasis, Aust. J. Biol. Sci. 13(2): 150–162.

Fraser, A.S. 1960c. Simulation of genetic systems by automatic digital computers. VII. Effects of reproduction rate, and intensity of selection, on genetic structure. Aust. J. Biol. Sci. 13: 344–350.

Fraser, A.S. 1962. Simulation of genetic systems. J. Theoret. Biol. 2: 329–346.

Fraser, A.S. and Hansche, P.E. 1965. Simulation of genetic systems. pp. 507–516. *In*: Geerts, S.J. (ed.). Major and Minor Loci, Proc. llth Int. Congress on Genetics, Oxford, U.K. Pergamon, vol. 3.

Fraser, A.S., Burnell, D. and Miller, D. 1966. Simulation of genetic systems. X. Inversion polymorphism. J. Theoret. Biol. 13: 1–14.

Fraser, A.S. and Burnell, D. 1967. Simulation of genetic systems. XI. Inversion polymorphism, Amer. J. Human Genet. 19(3): 270–287.

Fraser, A.S. and Burnell, D. 1967. Simulation of genetic systems. XII. Models of inversion polymorphism. Genetics 57: 267–282.

Fraser, A.S. 1968. The evolution of purposive behavior. pp. 15–23. *In*: von Foerster, H., White, J.D., Peterson, L.J. and Russell, J.K. (eds.). Purposive Systems, Washington, DC: Spartan.

Fraser, A.S. 1970. Computer Models in Genetics. New York: McGraw-Hill.

Freeman, C. and Perez, C. 1988. Structural crisis of adjustment, business cycles and investment behavior. pp. 38–66. *In*: Dosi, G. et al. (eds.). Technical Change and Economic Theory, London: Francis Pinter http://www.carlotaperez.org/pubs?s=tf&l=en&a=structuralcrisesofadjustment.

Friedberg, R., Dunham, B. and North, J. 1959. A learning machine: Part II. IBM Journal of Research and Development 3(3): 282–287.

Friedberg, R.M. 1958. A learning machine: part I, IBM. Journal of Research and Development 2(1): 2–13.

Fujiko, C. and Dickinson, J. 1987. Using the genetic algorithm to generate lisp source code to solve the prisoner's dilemma. pp. 236–240. *In*: Proceedings of the Second International Conference on Genetic Algorithms on Genetic algorithms and their application. Lawrence Erlbaum Associates, Inc., Mahwah.

Furia, C.A. 2006. Quantum Informatics: A Survey. Technical Report No. 2006.16. Computer Science at Politecnico di Milano.

Garzon, M.H., Blain, D.R. and Neel, A.J. 2004. Soft molecular computing for biomolecule-based computing. Natural Computing 3: 461–477.

Garzon, M.H., Deaton, R., Rose, J. and Franceschetti, D. 1999. Soft molecular computing. pp. 89–98. *In*: Proceedings of the 5th workshop, MIT, Vol. 54, DIMACS Series. American Mathematical Society.

Ghosh, S. 2002. Electricity consumption and economic growth in India. Energy Policy 30: 125–129.

Gill, J.L. 1963. Simulation of genetic systems. Biometrics 19: 654.

Gill, J.L. 1965. Effects of finite size on selection advance in simulated genetic populations. Aust. J. Biol. Sci. 18: 599–617.

Gill, J.L. 1965. Selection and linkage in simulated genetic populations. Aust. J. Biol. Sci. 18: 1171–1187.

Giuliani, E. 2006. The selective nature of knowledge networks in clusters: evidence from the wine industry. Journal of Economic Geography.

Goebel, K., Saha, B., Saxena, A., Celaya, J.R. and Christophersen, J. 2008. Prognostics in battery health management. IEEE Instrumentation & Measurement Magazine 11: 33–40.

Goldberg, D.E. 1989. Genetic Algorithms in Search, Optimization and Machine Learning. Addison-Wesley Publishing Company, Boston, MA, USA.

Goldberg, D.E. 1989. Genetic Algorithms in Searching, Optimisation and Machine Learning. Addison-Wesley, Reading, MA.

Goldberg, D.E. 1989. Sizing Populations for Serial and Parallel Genetic Algorithms. Proceedings of the Third International Conference on Genetic Algorithms, San Mateo, California 1196: 70–79.

Gottesman, D. 1996. A class of quantum error-correcting codes saturating the quantum Hamming bound. Physical Review A 54: 1862–1868.

GPorg. 2007. http://www.genetic-programming.org/ last accessed on Dec. 20, 2017.

Grefenstette, J.J. 1986. Optimization of control parameters for genetic algorithm. IEEE Trans. on Systems, Man and Cybernetics 16(1): 122–128.

Grefenstette, J.J. 1993. Deception considered harmful. pp. 75–91. In: Whitley, L.D. (ed.). Foundations of Genetic Algorithms 2. Morgan Kaufmann.

Gregor Mendel. 1865. Experiments in Plant Hybridization www.mendelweb.org/Mendel.html (1865) last accessed on Dec. 20, 2017.

Grover, L.K. 1996. A fast quantum mechanical algorithm for database search. pp. 212–219. In: Proceedings of the Annual ACM Symposium on the Theory of Computation (STOC??96). ACM Press.

Grover, L.K. 1997. Quantum mechanics helps in searching for a needle in a haystack. Physical Review Letters 79: 325–328.

Grujicic, M. and Chittajallu, K.M. 2004. Design and optimization of polymer electrolyte membrane (PEM) fuel cells. Applied Surface Science 227(1-4): 56–72.

Guckenheimer, J. 1994. A robust hybrid stabilization strategy for Equilibria. IEEE Trans. on Automatic Control 40(2): 321–325.

Gujarati, D.N. 2004. Basic Econometrics 4th Edition, McGraw-Hill Companies.

Guo, C.X. 2010. An analysis of the increase of CO_2 emission in china-based on SDA technique. China Industrial Economics 12: 47–56.

Hajela, P. and Lin, C.Y. 1992. Genetic search strategies in multicriterion optimal design. Structural Optimization 4: 99–107.

Han, P.Y., Tani, M., Pan, F. and Zhang, X.-C. 2000. Use of the organic crystal dast for terahertz beam applications. Optics Letters 25(9): 675–677.

Han, P.Y., Tani, M., Usami, M., Kono, S., Kersting, R. and Zhang, X.-C. 2001. A direct comparison between terahertz time-domain spectroscopy and far-infrared fourier transform spectroscopy. Journal of Applied Physics 89(4): 2357–2359.

Hangyo, M., Tani, M. and Nagashima, T. 2005. Terahertz time-domain spectroscopy of solids: A review. International Journal of Infrared and Millimeter Waves 26: 1661–1690.

Hann, T.H. and Steurer, E. 1996. Much ado about nothing? exchange rate forecasting: neural networks vs. linear models using monthly and weekly data. Neurocomputing 10: 323–339.

Hansen, M.P. and Jaszkiewicz, A. 1998. Evaluating the quality of approximations to the non-dominated set. Technical Report IMM-REP-1998-7, Technical University of Denmark, March 1998.

Hansen, N. and Ostermeier, A. 1996. Adapting arbitrary normal mutation distributions in evolution strategies: the covariance matrix adaptation. pp. 312–317. In: Proceedings of IEEE International Conference on Evolutionary Computation.

Hanson, S. and Pratt, G. 1988. Reconceptualizing the links between home and work in urban geography. Economic Geography 64: 299–321.

Hanson, S. and Pratt, G. 1995. Gender, Work, and Space London: Routledge.

Haque, M., Rahmani, A. and Egerstedt, M. 2009. A Hybrid, Multi-Agent Model of Foraging
   Bottlenose Dolphins, 3rd IFAC Conference on Analysis and Design of Hybrid Systems,
   September 2009.
Hargreaves, S. and Lewis, R.A. 2007. Terahertz imaging: Materials and methods. Journal of
   Materials, Science: Materials in Electronics 18(Supplement 1/October): 299–303.
Harp, S.A. and Samad, T. 1992. Optimizing neural networks with genetic algorithms. Proc.
   American Power Conf., Chicago IL 54(263): 1138–1143.
Hart, E. and Ross, P. 2001. Clustering moving data with a modified immune algorithm.
   pp. 394–403. In: Boers, E.J.W. (ed.). Applications of Evolutionary Computing Volume 2037 of
   the series Lecture Notes in Computer Science. Springer Berlin Heidelberg.
Haupt, R.L. and Haupt, S.E. 2004. Practical Genetic Algorithms (Second Edition). John Wiley &
   Sons, Inc., Hoboken, New Jersey, USA.
Haupt, R.L. and Haupt, S.E. 2004. Practical Genetic Algorithms: John Wiley & Sons.
Haykin, S. 1994. Neural Network: A Comprehensive Foundation, New York: Macmillan College
   Pub. Co.
He, W., Williard, N., Osterman, M. and Pecht, M. 2011. Prognostics of lithium-ion batteries based
   on Dempster-Shafer theory and the Bayesian Monte Carlo method. Journal of Power Sources
   196: 10314–10321.
Healey, A.J. and Marco, D.B. 1992. Slow speed flight control of autonomous underwater vehicles:
   Experimental results with NPS AUV II. Proc. 2nd International Offshore and Polar Engg.
   Conf., pp. 523–532.
Healey, A.J. and Lienard, D. 1993. Multivariable sliding mode control for autonomous diving and
   steering of unmanned underwater vehicles. IEEE Oceanic Engineering 18(3): 327–338.
Hebb, D. 1949. The Organization of Behavior. Wiley, New York, USA.
Heckerman, D. and Wellman, M.P. 1995. Bayesian networks. Commun. ACM 38(13): 27–30.
Hessburg, T., Lee, M., Takagi, H. and Tomizuka, M. 1993. Automatic design of fuzzy systems using
   genetic algorithms and its application to lateral vehicle guidance. Proc. Int. Soc. for Optical
   Engineering, SPIE 2061: 452–463.
Hessburg, T. and Tomizuka, M. 1994. Fuzzy logic control for lateral vehicle guidance. IEEE Control
   Systems Mag. 14(4): 55–63.
Hilbert, R., Janiga, G., Baron, G. and Thevenin, D. 2006. Multi-objective shape optimization of
   a heat exchanger using parallel genetic algorithms. International Journal of Heat and Mass
   Transfer 49(15): 2567–2577.
Hjorth, U. 1980. A Reliability distribution with increasing, decreasing, constant and bathtub-shaped
   failure rates. Technometrics 22(1): 99–107.
Holl, A.Sz. 2009. Aerodynamic optimization via multi-objective micro-genetic algorithm with
   range adaptation, knowledge-based reinitialization, crowding and ε-dominance. Advances
   in Engineering Software 40(6): 419–430 https://www.sciencedirect.com/science/article/pii/
   S0965997808001282.
Holland, J.H. 1968. Hierarchical descriptions of universal spaces and adaptive systems. Technical
   Report ORA Projects 01252 and 08226. Deparment of Computer and Communication
   Sciences, University of Michigan, Ann Arbor.
Holland, J.H. 1975. Adaptation in Natural and Artificial Systems. University of Michigan Press,
   Ann Arbor, MI, USA.
Holland, J.H. 1992. Adaptation in Natural and Artificial Systems (Second Ed.), Cambridge, MA:
   MIT Press (First Ed., Ann Arbor: The University of Michigan Press, 1975).
Holland, J.H. 1992. Adaptation in Natural and Artificial Systems: An Introduction Analysis with
   Applications to Biology, Control, and Artificial Intelligence, (Cambridge, MA: The MIT
   Press).
Holland, J.J. 1992. Genetic algorithm. Scientific American Magazine, pp. 44–50.

Homaifar, A. and McCormick, E. 1992. Full design of fuzzy controllers using genetic algorithms. Proc. SPIE Conf. on Neural and Stochastic Methods in Image and Signal Processing 1766: 393–404.

Hopfield, J.J. 1982. Neural networks and physical systems with emergent collective computational abilities. Proceedings of the National Academy of Sciences of the USA 79(8): 2554–2558, April 1982.

Horn, J. and Nafpliotis, N. 1993. Multiobjective optimization using the niched pareto genetic algorithm. IlliGAL Report 93005. Illinois Genetic Algorithms Lab. University of Illiniois, Urbana-Champagn.

Horn, J., Nafpliotis, N. and Goldberg, D.E. 1994. A niched pareto genetic algorithm for multiobjective optimization. In: Proceedings of the First IEEE Conference on Evolutionary Computation, IEEE World Congress on Computational Intelligence 1: 82–87.

Horn, J.N. and Goldberg, D.E.N. 1994. A niched Pareto genetic algorithm for multi-objective optimisation. Proceedings of the First IEEE Conference on Evolutionary Computation, IEEE World Congress on Computational Intelligence. Orlando, FL, USA 1: 82–87.

Hornik, K., Stinchcombe, M. and White, H. 1989. Multilayer feedforward networks are universal approximators. Neural Networks 2: 359–366.

Hua, C. and Porell, F. 1979. A critical review of the development of the gravity model. International Regional Science Review 4(2): 97–126.

Huang, H.P. and Lin, W.M. 1993. Development of a simplified fuzzy controller and its relation to PID controller. Journal of the Chinese Institutes of Engineers 16: 59–72.

Hughes, E.J. 2005. Evolutionary many-objective optimisation: many once or one many? In Congress on Evolutionary Computation (CEC), pages 222–227, Edinburgh, UK, 2005. IEEE Computer Society Press.

Hung, J.Y., Gao, W. and Hung, J.C. 1993. Variable structure control: A survey. IEEE Transactions on Industrial Electronics 40(1): 2–22.

Hung, T.C., Yan, W.M., Wang, X.D. and Huang, Y.X. 2012. Optimal design of geometric parameters of double-layered micro-channel heat sinks. International Journal of Heat and Mass Transfer 55(11): 3262–3272.

Hunt, J. and Cooke, D. 1996. Learning using an artificial immune system. Journal of Network and Computer Applications 19: 189–212.

Hunt, K.J. 1992a. Optimal controller synthesis: A genetic algorithm solution. IEEE Colloq. on Genetic Algorithms for Cont. Sys. Engineering, pp. 1/1–1/6.

Hunt, K.J. 1992b. Polynomial LQG and H controller synthesis: A genetic algorithm solution. Proc. 31st IEEE Conf. on Decision and Contr., Tucson, AZ 4: 3604–3609.

Hunt, K.J. 1993. Systems Identification with Genetic Algorithm Solution. Daimler-Benz AG, ALt-Moabit 91b. D-1000, berlin 21, Germany, Internal Report.

Husain, A. and Kim, K.Y. 2008. Multiobjective optimization of a micro-channel heat sink using evolutionary algorithm. Journal of Heat Transfer 130(11): 114505(3 pages).

Husain, A. and Kim, K.Y. 2010. Enhanced multi-objective optimization of a microchannel heat sink through evolutionary algorithm coupled with multiple surrogate models. Applied Thermal Engineering 30(13): 1683–1691.

Hutchison, R. 2009. Encyclopedia of urban studies SAGE Publications, Inc.

Hwang, H.S., Joo, Y.H., Kim, H.K. and Woo, K.B. 1992. Identification of fuzzy control rules utilizing genetic algorithms and its application to mobile robot. IFAC Workshop on Algorithms and Architectures for Real-Time Control 58: 249–254.

Hwang, W. and Thompson, W. 1994. Design of intelligent fuzzy logic controllers using genetic algorithms. Proc. 3rd IEEE Int. Conf. on Fuzzy Systems 2: 1383–1388.

Hwang, Y.R. and Tomizuka, M. 1994. Fuzzy smoothing algorithms for variable structure systems. IEEE Trans. on Fuzzy Systems 2(4): 277–284.

Ichikawa, Y. and Sawa, T. 1992. Neural network application for direct feedback controllers. IEEE Trans. ANN 3: 224–231.

International Organization for Standardization. 1982. ISO/TC108/SC2/WG4 N57 Reporting Vehicle Road Surface Irregularities.

International Organization for Standardization. 1995. ISO 8608:1995 Mechanical Vibration-Road Surface Profiles-Reporting of Measured Data.

International Organization for Standardization. 1997. Mechanical Vibration and Shock—Evaluation of Human Exposure to Whole-Body Vibration-Part 1: General Requirements ISO 2631-1: 1997.

International Organisation for Standardisation. 2008. (1995), ISO/IEC Guide 98-3:2008 the Guide to the Expression of Uncertainty in Measurement (GUM).

IPCC. 1977. Revised 1996 IPCC Guidelines for National Greenhouse Gas Inventories:Reference Manual (Volume 3) NGGIP Publications.

Isaacson, W. 2012. The real leadership lessons of Steve Jobs. Harvard Business Review 4: 92–102.

Isaka, S. and Sebald, A.V. 1992. An optimization approach for fuzzy controller design. IEEE Trans. on System, Man and Cybernetics 22(6): 1469–1473.

Ishibuchi, H., Fujioka, R. and Tanaka, H. 1993. Neural networks that learn from fuzzy If-then rules. IEEE Trans. Fuzzy Systems 1(2): 85–97.

Ishibuchi, H., Murata, T. and Türkşenb, I.B. 1997. Single-objective and two objective genetic algorithms for selecting linguistic rules for pattern classification problems. Fuzzy Sets and Systems 89: 135–150.

Ishibuchi, H., Yoshida, T. and Murata, T. 2002. Balance between genetic search and local search in memetic algorithms for multiobjective permutation flowshop scheduling. IEEE Transactions on Evolutionary Computation 7: 204–223.

Ishigame, A., Furukawa, T., Kawamoto, S. and Taniguchi, T. 1991. Sliding mode controller design based on fuzzy inference for non-linear systems. Proc. Int. Conf. on Industrial Electronics Control and Instrumentation (IECON 91) 3(416): 2096–2101.

Ishigame, A., Furukawa, T.,Kawamoto, S. and Taniguchi, T. 1991. Sliding mode controller design based on fuzzy inference for non-linear system. International Conference on Industrial Electronics, Control and Instrumentation, Kobe, Japan, 28 Oct.–1 Nov. 3: 2096–2101.

Ishigame, A., Furukawa, T.,Kawamoto, S. and Taniguchi, T. 1993. Sliding mode controller design based on fuzzy inference for nonlinear systems. IEEE Trans. Industrial Electronics 40(1): 64–70.

Itkis, U. 1976. Control Systems of Variable Structure. (New York: Wiley).

Itkis, Y. 1976. Control Systems of Variable Structure. New York: Wiley.

Jalili, N. 2002. A comparative study and analysis of semi-active vibration-control systems. Journal of Vibration and Acoustics 124(4): 593–605.

Jamil, M. and Yang, X.-S. 2013. A literature survey of benchmark functions for global optimisation problems. International Journal of Mathematical Modelling and Numerical Optimisation 4: 150–194.

Jang, J.-S.R. 1992. Self-learning fuzzy controllers based on temporal back propagation. IEEE Trans. Neural Networks 3(5): 714–723.

Jiang, R. and Murthy, D.N.P. 1998. Mixture of weibull distributions parametric characterization of failure rate function. Appl. Stochastic Models Data Anal. 14: 47–65.

Ji-Hyeong, H. and Su-Young, C. 2016. Consideration of manufacturing data to apply machine learning methods for predictive manufacturing. In: 2016 Eighth International Conference on Ubiquitous and Future Networks (ICUFN).

John, M. Broder. 2002. California Power Failures Linked to Energy Companies, The New York Times.

Jorgen Moltofta. 1983. Behind the 'bathtub-curve, a new model and its consequences. Microelectronics Reliability 23(3): 489–500.

Justice, K.E. and Gervinski, J.M. 1968. Electronic simulation of the dynamics of evolving biological systems. pp. 205–228. In: Oestreicher, H.L. and Moore, D.R. (eds.). Cybernetics Problems in Bionics, New York: Gordon and Breach.

Kadrovach, B.A., Zydallis, J.B. and Lamont, G.B. 2002. Use of mendelian pressure in a multi-objective genetic algorithm. Proceedings of the 2002 Congress on Evolutionary Computation, Piscataway, New Jersey, 12–17 May 1: 962–967.

Kanamorit, T., Tsujikawa, K., Iwata, Y.T., Inoue, H., Ohtsuru, O., Kishi, T., Hoshina, H., Otani, C. and Kawase, K. 2005. Application of terahertz spectroscopy to abused drug analysis. The Joint 30th, International Conference on Infrared and Millimeter Waves and 13th International Conference on Terahertz Electronics, IRMMW-THz 1: 180–181.

Kang, H.S. et al. 2016. Smart manufacturing: Past research, present findings, and future directions. International Journal of Precision Engineering and Manufacturing-Green Technology 3(1): 111–128.

Kankal, M., Akpinar, A., Kömürcü, M.L. and Özsahin, T.S. 2011. Modeling and forecasting of Turkey s energy consumption using socio-economic and demographic. Applied Energy 88: 1927–1939.

Kari, L. 1997. DNA computing: arrival of biological mathematics. Mathematical Intelligencer 19: 9–22.

Kari, L., Gloor, G. and Yu, S. 2000. Using DNA to solve the bounded post correspondence problem. Theoretical Computer Science 231: 193–203.

Kari, L., Seki, S. and Sośik, P. 2010. DNA computing-foundations and implications. pp. 1073–1127. In: Rozenberg, G., Bäck, T. and Kok, J.N. (eds.). Handbook of Natural Computing. Springer.

Karnopp, D.C., Crosby, M.J. and Harwood, R.A. 1974. Vibration control using semi-active force generators. Journals of Engineering for Industry Transactions of the ASME 94: 619–626.

Karr, C.L., Freeman, L.M. and Meredith, D.L. 1989. Improved fuzzy process control of spacecraft autonomous rendezvous using a genetic algorithm. SPIE Workshop on Intelligent Control and Adaptive Systems 62(1196): 274–288.

Karr, C.L., Freeman, L.M. and Meredith, D.L. 1990. Genetic algorithm based fuzzy control of spacecraft autonomous rendezvous. Proc. 5th Conf. on Artificial Intelligence for Space Applications 62(3073): 43–51.

Karr, C.L., Meredith, D.L. and Stanley, D.A. 1990. Fuzzy process control with a genetic algorithm. Control 90 Symp. at the Annual Meeting of the Soc. for Mining Metallurgy, and Petroleum: Mineral and Metallurgical Processing 32: 53–60.

Karr, C.L. 1991. Design of an adaptive fuzzy logic controller using a genetic algorithm. Proc. 4th Int. Conf. on Genetic Algorithms, pp. 450–457.

Karr, C.L. and E.J. Gentry. 1993. Fuzzy control of pH using genetic algorithms. IEEE Trans. Fuzzy Systems 1(1): 46–53.

Karr, C.L., Sharma, S.K., Hatcher, W. and Harper, T.R. 1993. Fuzzy logic and genetic algorithms for the control of an exothermic chemical reaction. Int. Symp. on Modelling, Simul. and Cont. of Hydrometallurgical Processes 21: 227–236.

Katzenellenbogen, N. and Grischkowsky, D. 1991. Efficient generation of 380 Fs pulses of Thz radiation by ultrafast laser pulse excitation of a biased metal-semiconductor interface. Applied Physics Letters 58(3): 222–224.

Keane, A.J. 1995. Genetic algorithm optimization of multi-peak problems: Studies in convergence and robustness. Artificial Intelligence in Engineering 9: 75–83.

Kennedy, J. and Eberhart, R. 1995. Particle swarm optimization. pp. 1942–1948. In: Proceedings of IEEE International Conference on Neural Networks, 1995. Perth, WA. IEEE.

Kennedy, J. and Eberhart, R. 2001. Swarm Intelligence. Morgan Kaufmann Publisher. San Francisco.

Kephart, J.O. 1994. A biologically inspired immune system for computers. pp. 130–139. *In*: Brooks, R.A. and Maes, P. (eds.). Proceedings of Artificial Life IV: The Fourth International Workshop on the Synthesis and Simulation of Living Systems. MIT Press, Cambridge, MA.

Killeen, W.P., Lyons, R.K. and Moore, M.J. 2001. Fixed versus flexible: lessons from ems order flow. Journal of International Money and Finance 25: 551–579.

Kim, I.Y. and de Weck, O.L. 2005. Adaptive weighted-sum method for bi-objective optimization: Pareto front generation. Structural and Multidisciplinary Optimization 29: 149–158.

Kim, J.H., Kim, K.C. and Chong, E.K.P. 1994. Fuzzy precompensated PID controllers. IEEE Trans. on Control Systems Technology 2(4): 406–411.

Kim, K.D. and Kumar, P.R. 2012. Cyber-physical systems: A perspective at the centennial. Proc. IEEE 100: 1287–1308.

King, P.J. and Mamdani, E.H. 1977. The application of fuzzy control systems to industrial processes. Automatica 13: 235–242.

Kinzel, J., Klawonn, F. and Kruse, R. 1994. Modifications of genetic algorithms for designing and optimizing fuzzy controllers. Proc. 1st IEEE Conf. Evolutionary Computation, Orlando 1: 28–33.

Kiran, M.S., Ozceylan, E., Gunduz, M. and Paksoy, T. 2012. Swarm intelligence approaches to estimate electricity energy demand in Turkey. Knowledge-Based Systems 36: 93–103.

Klepper, S. 2001. Employee startups in high-tech industries. Industrial and Corporate Change 10(3): 639–74.

Klepper, S. 2002. The Evolution of the U.S. Automobile Industry and Detroit as its Capital 9th Congress of the International Joseph A. Schumpeter Society, Gainesville, FL, March.

Klutke, J.A., Kiessler, P.C. and Wortman, M.A. 2003. A critical look at the bathtub curve. IEEE Transactions on Reliability 52(1): 125–129.

Knight, R.W., Hall, D.J., Goodling, J.S. and Jaeger, R.C. 1992. Heat sink optimization with application to microchannels. IEEE Transactions on Components, Hybrids, and Manufacturing Technology 15(5): 832–42.

Knight, T. and Timmis, J. 2002. A multi-layered immune inspired approach to data mining. pp. 182–196. *In*: Lotfi, A., Garibaldi, J. and John, R. (eds.). Proceedings of the 4th International Conference on Recent Advances in Soft Computing. Nottingham, UK.

Knowles, J. and Corne, D. 1999. The Pareto archived evolution strategy: A new baseline algorithm for Pareto multiobjective optimization. Proceedings of the 1999 Congress on Evolutionary Computation 1: 98–105.

Knowles, J.D. and Corne, D.W. 2000. Approximating the nondominated front using the Pareto archived evolution strategy. Evolutionary Computation 8: 149–172.

Konak, A., Coit, D.W. and Smith, A.E. 2006. Multi-objective optimization using genetic algorithms: A tutorial. Reliability Engineering and System Safety 91: 992–1007.

Kotani, M., Ochi, M., Ozawa, S. and Akazawa, K. 2001. Evolutionary discriminant functions using genetic algorithms with variable-length chromosome. pp. 761–766. *In*: Proceedings of 2001 International Joint Conference on Neural Networks.

Koza, J.R. 1992. Genetic Programming. MIT Press.

Koza, J.R. 1994. Genetic Programming II. MIT Press.

Koza, J.R., Bennett, F., Andre, D. and Keane, M.A. 1999. Genetic Programming III: Darwinian Invention and Problem Solving. Morgan Kaufmann Publishers.

Koza, J.R., Keane, M.A., Streeter, M.J., Mydlowec, W., Yu, J. and Lanza, G. 2003. Genetic Programming IV: Routine Human-Competitive Machine Intelligence. Kluwer Academic Publishers.

Kozlowski, J.D., Watson, M.J., Byington, C.S., Garga, A.K. and Hay, T.A. 2001. Electrochemical cell diagnostics using online impedance measurement, state estimation and data fusion techniques. In emph 36th Intersociety Energy Conversion Engineering Conference, July 29–August 2, Savannah, Georgia.

Kozlowski, J.D. 2003. Electrochemical cell prognostics using online impedance measurements and model-based data fusion techniques. In emph 2003 IEEE Aerospace Conference Proceedings, Big Sky, Montana.

Krishnakumar, K.S. 1989. Micro-genetic algorithms for stationary and non-stationary function optimisation. (SPIE) Intelligent Control and Adaptive Systems 1196(26): 289–296.

Kristinsson, K. and Dumont, G.A. 1988. Genetic algorithms in system identification. Third IEEE Int. Symp. Intelligent Conr., Arlington, VA, pp. 597–602.

Kristinsson, K. and Dumont, G.A. 1992. System identification and control using genetic algorithms. IEEE Trans. Sys., Man and Cyber. 22(5): 1033–1046.

Kulikovsky, A.A., Divisek, J. and Kornyshev, A.A. 1999. Modeling the cathode compartment of polymer electrolyte fuel cells: Dead and active reaction zones. J. Electrochem. Soc. 146(11): 3981–3991.

Kull, H. 2015a. Intelligent Manufacturing Technologies in Mass Customization: Opportunities, Methods, and Challenges for Manufacturers, A press: Berkeley, CA, pp. 9–20.

Kull, H. 2015b. Introduction, in Mass Customization: Opportunities, Methods, and Challenges for Manufacturers, A press: Berkeley, CA, pp. 1–6.

Kumar, K.D. 2006. Review of dynamics and control of nonelectrodynamic tethered satellite systems. Journal of Spacecraft and Rockets 43: 705–720.

Kung, C.C. and Lin, S.C. 1992. A fuzzy-sliding mode controller design. Proc. IEEE Int. Conf. on Systems Engineering 153: 608–611.

Kwok, P.D., Tam, P., Sun, Z.Q. and Wang, P. 1991a. Design of optimal linear regulators with steady-state trajectory insensitivity. Proc. IECON 91(3): 2183–2187.

Kwok, P.D., Tam, P., Li, K.C. and Wang, P. 1991b. Analysis and design of fuzzy PID control-systems. Proc. Int. Conf. on Contr. 2: 955–960.

Kwok, P.D. and Sheng, F. 1994. Genetic algorithm and simulated annealing for optimal robot arm PID control. Proc. 1st IEEE Conf. Evolutionary Computation, Orlando 2: 708–713.

Langdon, W.B. and Poli, R. 1998. Fitness Causes Bloat: Mutation. EuroGP 98. Springer-Verlag.

Langdon, W.B., Soule, T., Poli, R. and Foster, J.A. 1999. The evolution of size and shape. pp. 16–190. In: Spector, L., Langdon, W.B., Reilly, U.-M.O. and Angeline, P.J. (eds.). Advances in Genetic Programming 3. MIT Press.

Langdon, W.B. 2017. The Genetic Programming Bibliography. http://www.cs.bham.ac.uk/~wbl/biblio/.

Langdon, W.B. and Gustafson, S.M. 2010. Genetic programming and evolvable machines: ten years of reviews. Genetic Programming and Evolvable Machines 11: 321–338.

LaValle, S. et al. 2011. Big data, analytics and the path from insights to value. In: MIT Sloan Management. MIT Sloan Management Review: North Hollywood, CA, p. 15.

LeCun, Y. 1985. Une procdure d'apprentissage pour rseau a seuil asymmetrique (a learning scheme for asymmetric threshold networks). In Proceedings of Cognitiva 85, pages 599C604, Paris, France.

Lee, C.C. 1990. Fuzzy logic in control systems: fuzzy logic controller part I and II. Syst. Man Cybern. IEEE Trans. 20: 404–435.

Lee, C.C. 1990. Fuzzy logic in control systems: Fuzzy logic controller—Part I. IEEE Trans. Systems, Man and Cybernetics 20(2): 404–418.

Lee, C.C. 1990. Fuzzy logic in control systems: Fuzzy logic controller—Part II. IEEE Trans. Systems, Man and Cybernetics 20(2): 419–435.

Lee, C.C. 2005. Energy consumption and GDP in developing countries: A cointegrated panel analysis. Energy Economics 27: 415–427.

Lee, E.T. and Zadeh, L.A. 1969. Note on fuzzy languages. Information Sciences 1: 421–434.

Lee, J. et al. 2013. Recent advances and trends in predictive manufacturing systems in big data environment. Manufacturing Letters 1(45): 38–41.

Lee, J., Lapira, E., Bagheri, B. and Kao, H.A. 2013. Recent advances and trends in predictive manufacturing systems in big data environment. Manufacturing Letters 1(1): 38–41.

Lee, J., Bagheri, B. and Kao, H.-A. 2015. A cyber-physical systems architecture for Industry 4.0-based manufacturing systems. Manufacturing Letters 3: 18–23.

Lee, J.H. 1993. On methods for improving performance of PI-type fuzzy logic controllers. IEEE Trans. on Fuzzy Systems 1(4): 298–301.

Leete, T., Schwartz, M., Williams, R., Wood, D., Salem, J. and Rubin, H. 1999. Massively parallel DNA computation: expansion of symbolic determinants. pp. 45–58. 2nd DIMACS Workshop on DNA Based Computers, Princeton, 1996. DIMACS Series 44. AMS Press.

Lenstra, A.K. and Lenstra, Jr. H.W. 1994. The Development of the Number Field Sieve. Lecture Notes in Computer Science Volume 1554. Springer.

Levine, D. 1994. A Parallel Genetic Algorithm for the Set Partitioning Problem, Ph.D. Thesis, Argonne National Lab. (Maths. and Comp. Science Division).

Li, K.C., Leung, T.P. and Hu, Y.M. 1994. Sliding mode control of distributed parameter. Automatica 30(12): 1961–1966.

Li, X.K., Hao, X.H., Chen, Y., Zhang, M.H. and Peng, B. 2013. Multi-objective optimizations of structural parameter determination for serpentine channel heat sink. pp. 449–458. In: Proceedings of 16th European Conference, Evo Applications 2013, Vienna, Austria.

Li, X.L., Shao, Z.J. and Qian, J.X. 2002. An optimizing method based on autonomous animate: Fish swarm algorithm. System Engineering Theory and Practice 22(11): 32–38.

Li, X.L. 2003. A New Intelligent Optimization-artificial Fish Swarm Algorithm. Ph.D. thesis, Zhejiang University.

Li, Y. 1995. Modern Information Technology for Control Systems Design and Implementation. Proc. 2nd. Asia-Pacific Conf. Control and Measurement, ChongQing, China.

Li, Y. and Ng, K.C. 1995. Genetic algorithm based techniques for design automation of three-term fuzzy systems, Proc. 6th Int. Conf. Fuzzy Sys. Asso. World Cong., Sao Paulo, Brazil 1: 261–264. (Available on Internet with URL: http://www.mech.gla.ac.uk/ Control/reports. html).

Li, Y. and Rogers, E. 1995. Graph reversal and the design of parallel control and signal processing architectures. Int. J. Contr. 62(2): 271–287.

Li, Y. and Tan, K.C. 1995. Physical Parametric Modelling of Nonlinear Systems by Evolution. Technical Report CSC-95010, Centre for Systems and Control, University of Glasgow. (Available on Internet with URL: http://www.mech.gla.ac.uk/Control/reports.html).

Li, Y., Ng, K.C., Häußler, A., Chow, V.C.W. and Muscatelli, V.A. 1995a. Macroeconomics modelling on UK GDP growth by neural computing. Pre-Prints of IFAC/IFIP/ IFORS/SEDC Symp. Modelling and Contr. of Nat. and Regional Economies, Gold Coast, Australia.

Li, Y., Ng, K.C., Murray-Smith, D.J., Gray, G.J. and Sharman, K.C. 1995b. Genetic algorithm automated approach to design of sliding mode control systems. Int. J. Contr. (in press).

Li, Y., Tan, K.C., Ng, K.C. and Murray-Smith, D.J. 1995c. Performance based linear control system design by genetic evolution with simulated annealing. Proc. 34th IEEE Conf. Decision & Contr., New Orleans, LA (in press).

Li, Y. and A. Haubler. 1996. Artificial evolution of neural networks and its application to feedback control. Artificial Intelligence in Engineering 10(2): 143–152.

Li, Y., Ng, K.C., Murray-Smith, D.J., Gray, G.J. and Sharman, K.C. 1996. Genetic algorithm automated approach to design of sliding mode control systems. Int. J. of Control 63(4): 721–739.

Li, Y. and Ng, K.C. 1996a. A uniform approach to model-based fuzzy control system design and structural optimization. pp. 129–151. In: Herrera, F. and Verdegay, J.L. (eds.). Genetic

Algorithms and Soft Computing. Physica-Verlag Series on Studies in Fuzziness (Vol. 8, ISBN 3-7908-0956-X).

Li, Y. and Ng, K.C. 1996b. A uniform approach to model-based fuzzy control system design and structural optimization. pp. 129–151. *In*: Herrera, F. and Verdegay, J.L. (eds.). Genetic Algorithms and Soft Computing. Physica-Verlag Series on Studies in Fuzziness (Vol. 8, ISBN 3-7908-0956-X).

Li, Y., Tan, K.C. and Gong, M. 1997. Model reduction in control systems by means of global structure evolution and local parameter learning. *In*: Dasgupta, D. and Michalewicz, Z. (eds.). Evolutionary Algorithms in Engineering Applications, Springer Verlag.

Li, Y., Ang, K.H., Chong, G.C.Y., Feng, W.Y., Tan, K.C. and Kashiwagi, H. 2004. CAutoCSD-Evolutionary search and optimisation enabled computer automated control system design. International Journal of Automation and Computing 1(1): 76–88.

Li, Y., Chen, F. and Ouyang, Q. 2004. Genetic algorithm in DNA computing: A solution to the maximal clique problem. Chinese Science Bulletin 49: 967–971.

Li, Y., Chen, W. and Sun, F. 2013. Ignition method and system for internal combustion engine CN 102410126 B https://www.google.com/patents/CN102410126B?cl=en.

Li, Y.F. and Lau, C.C. 1989. Development of fuzzy algorithms for servo systems. IEEE Control Systems Mag., pp. 65–71.

Liang, K.-H., Yao, X. and Newton, C.S. 2001. Adapting self-adaptive parameters in evolutionary algorithms. Applied Intelligence 15: 171–180.

Lim, C.P. and Jain, L.C. 2009. Advances in swarm intelligence. pp. 1–7. *In*: Lim, C.P., Jain, L.C. and Dehuri, S. (eds.). Innovations in Swarm Intelligence, Studies in Computational Intelligence 248. Springer.

Lin, C.T. and George Lee, C.S. 1991. Neural-network-based fuzzy logic control and decision system. IEEE Trans. on Computers 40(12): 1320–1336.

Linder, E.G. 1952. Internal-combustion Engine Ignition. United States 2617841, Patent.

Linkens, D.A. and Abbod, M.F. 1992. Self-organising fuzzy logic control and the selection of its scaling factors. Trans. Inst. Mech. Contr. 14(3): 114–125.

Linkens, D.A. and Nyongesa, H.O. 1992. A real-time genetic algorithm for fuzzy control. IEEE Collq. on Genetic Algorithms for Contr. Sys. Engineering, pp. 9/1–9/4.

Linkens, D.A. and Nyongesa, H.O. 1993. Real-time acquisition of fuzzy rules using genetic algorithms. Annual Review in Automatic Programming 82(17): 335–339.

Linkens, D.A. and Nyongesa, H.O. 1995. Genetic algorithms for fuzzy control Part 1: Offline system developement and application. IEE Proc. Control Theory Appl. 142(3): 161–176.

Linkens, D.A. and Nyongesa, H.O. 1995. Genetic algorithms for fuzzy control Part 2: Online system developement and application. IEE Proc. Control Theory Appl. 142(3): 177–192.

Lipton, R. 1995. DNA solution of hard computation problems. Science 268: 542–545.

Liska, J. and Melsheimer, S. 1994. Complete design of fuzzy logic systems using genetic algorithms. Proc. 3rd IEEE Int. Conf. on Fuzzy Systems 2: 1377–1382.

Liu, G.F., Zhang, Z.Y., Ma, S.H., Zhao, H.W., Ma, X.J. and Wang, W.F. 2009. Quantitative measurement of mixtures by terahertz time-domain spectroscopy. Journal of Chemical Sciences 121(4): 515–520.

Liu, H.B. and Zhang, X.C. 2007. Terahertz Spectroscopy for Explosive, Pharmaceutical, and Biological Sensing Applications NATO Security through Science, Series, Terahertz Frequency Detection and Identification of Materials and Objects, pp. 251–323.

Liu, J. and Diamond, J. 2005. China's environment in a globalizing, World. Nature 435: 1179–1186.

Liu, X.H., Bai, Y.E., Chen, Y., Yang, X.W., Xu, Y.J. and Qiu, L.X. 2012. Non-dominated sorting genetic algorithm for multi-objective optimisation of aquous extraction conditions of floes trollii. Chinese Journal of Health Statistics 29: 846–848.

Liu, Y., Matsuhisa, H. and Utsuno, H. 2008. Semi-active vibration isolation system with variable stiffness and damping control. Journal of Sound and Vibration 313(1-2, 3 June): 16–28.

Lo, K.L. and Khan, L. 2004. Hierarchical micro-genetic algorithm paradigm for automatic optimal weight selection in hinfty loop-shaping robust flexible AC. Transmission System Damping Control Design 151(1): 109–118.

Logsdon, T. 1997. Orbital Mechanics: Theory and Applications. John Wiley and Sons, Inc.

Lu, H. and Yen, G.G. 2002. Rank-density based multiobjective genetic algorithm. Proceedings of the 2002 Congress on Evolutionary Computation 1: 944–949.

Lu, Y.S. and Chen, J.S. 1994. A self-organizing fuzzy sliding-mode controller design for a class of nonlinear servo systems. IEEE Trans. on Industrial Electronics 41(5): 492–502.

Lua, H.K. 1993. Sliding Mode Control for DC Motors. Project Report, Department of Electronics and Electrical Engineering, University of Glasgow, Glasgow, UK.

Lubell, J., Chen, K.W., Horst, J. Frechette, S. and Huang, P. 2012. Model Based Enterprise/Technical Data Package Summit Report, NIST Technical Note 1753 (http://dx.doi.org/10.6028/NIST. TN.1753), National Institute of Standards and Technology, U.S. Department of Commerce, August.

Ludwig, S.A. 2015. Map Reduce-based fuzzy c-means clustering algorithm: implementation and scalability. International Journal of Machine Learning and Cybernetics 6(6): 923–934.

Lyman Ott, R. and Longnecker, M. 2001. An Introduction to Statistical Methods and Data Analysis Fifth Edition, Duxbury.

Lyons, R. 2001. The Microstructure Approach to Exchange Rates, London, England, MIT Press.

Ma, X.J., Zhao, H.W., Liu, G.F., Ji, T., Zhang, Z.Y. and Dai, B. 2009. Qualitative and quantitative analysis of some Saccharides by THZ-TDS spectroscopy and spectral analysis 29(11): 2885–2888.

Maarouf, M., Sosa, A., Galván, B., Greiner, D., Winter, G., Mendez, M. and Aguasca, R. 2015. The role of artificial neural networks in evolutionary optimisation: A review. pp. 59–76. In: Greiner, D., Galván, B., Periaux, J. and Gauger, N. (eds.). Advances in Evolutionary and Deterministic Methods for Design, Optimization and Control in Engineering and Sciences. Computational Methods in Applied Sciences 36. Springer, Switzerland.

MacNish, C. and Yao, X. 2008. Direction matters in high-dimensional optimisation. pp. 2372–2379. In: Proceddings of IEEE Congress on Evolutionary Computation, Hong Kong.

Maharudrayya, S., Jayanti, S. and Deshpande, A.P. 2004. Pressure losses in laminar flow through serpentine channels in fuel cell stacks. Journal of Power Sources 138: 1–13.

Mamdani, E.H. 1974. Application of fuzzy algorithm for control of simple dynamic plant. Proc. IEEE 121(12): 1585–1588.

Mamdani, E.H. and Assilian, S. 1975. An experiment in linguistic synthesis with a fuzzy logic controller. International Journal of Man-Machine Studies 7(1): 1–13.

Mamdani, E.H. 1976. Advances in the linguistic synthesis of fuzzy controllers. International Journal of Man-Machine Studies 8(1): 669–678.

Mamdani, E.H. 1977. Applications of fuzzy logic to approximate teasoning using linguistic synthesis. IEEE Transactions on Computers 26(12): 1182–1191.

Mamdani, E.H., Ostergaard, J.J and Lembessis, E. 1984. Use of fuzzy logic for implementing rule-based control of industrial processes. TIMS/Studies in Management Sciences 20: 429–445.

Mamdani, Ebrahim H. 1974. Application of fuzzy algorithms for control of simple dynamic plant. Proceedings of the Institution of Electrical Engineers 121(12): 1585–1588.

Manderick, B. and Spiessens, P. 1994. How to Select genetic algorithm operators for combinatorial optimization problems by analyzing their fitness landscape. Computational Intelligence Imitating Life, (IEEE Press), pp. 170–181.

Mangeney, J. and Crozat, P. 2008. Ion-irradiated In0:53Ga0:47As photoconductive antennas for Thz generation and detection at 1.55 mm wavelength. Comptes Rendus Physique 9(2): 142–152.

Manin, Y.I. 1980. Computable and uncomputable. Sovetskoye Radio, Moscow. In Russian. Partial English translation in [Man99].

Manin, Y.I. 1999. Classical computing, quantum computing, and Shor s factoring algorithm. Séminaire Bourbaki 41: 375–404.

Mann, J., Connor, R.C., Barre, L.M. and Heithaus, M.R. 2000. Female reproductive success in bottlenose dolphins (Tursiops sp.): life history, habitat, provisioning, and group-size effects. Behavioral Ecology 11(2): 210–219.

Manning, C.D., Raghavan, P. and Schütze, H. 2008. Introduction to Information Retrieval. Cambridge University Press, Cambridge.

Marks, J. 2008. The construction of Mendel's laws. Evolutionary Anthropology 17: 250–253.

Martha McNeil Hamilton and Greg Schneider. 2001. Price Caps Have Questionable Record, Washington Post.

Martin, E.G. and Cockerham, C.C. 1960. High speed selection studies. pp. 35–45. In: Kempthorne, O. (ed.). Biometrical Genetics, London, U.K.: Pergamon.

Martiskainen, M. 2007. Affecting consumer behaviour on energy demand. Final report to EdF Energy.

Masih, A.M.M. and Masih, R. 1996. Energy consumption, real income and temporal causality: results from a multi-country study based on cointegration and error-correction modelling techniques. Energy Economics 18: 165–83.

McClay, C. 2002. The impact of NETA imbalance prices on contracting strategies for generation and demand, Power System Management and Control, Fifth International Conference on (Conf. Publ. No. 488).

Mccllochw, S. and Li, H. 1943. A logical calculus of the ideas immanent in nervous activity. Bull Math Biophys. 10: 115–133.

McGookin, E.W. 1993. Sliding Mode Control of a Submarine. MEng. Thesis, E.E. & E. Dept., University of Glasgow.

Melanie, M. 1998. An Introduction to Genetic Algorithms. MIT Press. Cambridge, Massachusetts, USA.

Mendao, M., Timmis, J., Andrews, P.S. and Davies, M. 2007. The immune system in pieces: Computational lessons from degeneracy in the immune system. pp. 394–400. IEEE Symposium on Foundations of Computational Intelligence, 2007. Honolulu, HI.

Mendes, R., Kennedy, J. and Neves, J. 2004. The fully informed particle swarm: Simpler, may be better. IEEE Transactions on Evolutionary Computation 8: 204–210.

Meredith, D.L., Karr, C.L. and Kumar, K.K. 1993. The use of genetic algorithms in the design of fuzzy logic controllers. Proc. of the Soc. of Photo-optical Instrumentation Engineer (SPIE) 1721(81): 549–555.

Meyer, J.E., Burke, S.E. and Hubbard, T.E. 1993. Fuzzy sliding mode control for vibration damping of flexible structures. Proc. of the Society of Photo-optical Instrumentation Engineering, SPIE 1919(34): 182–183.

Michalewicz, Z. 1996. Genetic Algorithm + Data Structures = Evolution Programs (3rd ed.). Springer-Verlag, New York, USA.

Minsky, M. 1954. Theory of Neural-analog Reinforcement Systems and Its Application to the Brain-model Problem. Ph.D. Thesis. Princeton University, Princeton, USA.

Minsky, M. 1961. Steps toward artificial intelligence. Proc. IRE 49: 8–30.

Minsky, M. and Papert, S. 1969. An Introduction to Computational Geometry. MIT Press.

Misra, A.K. and Modi, V.J. 1982. Dynamics and control of tether connected two-body systems—a brief review. The 33rd International Astronautical Federation, International Astronautical Congress, Paris, France, 27 September–2 October, pp. 219–236.

Misra, A.K. and Modi, V.J. 1986a. A survey on the dynamics and control of tethered satellite systems. NASA/AIAA/PSN International Conference on Tethers, Arlington, VA, 17–19 September.

Mohamed, Z. and Bodger, P. 2005. Forecasting electricity consumption in New Zealand using economic and demographic variables. Energy 30: 1833–1843.

Molaeimanesh, G.R. and Akbari, M.H. 2014. A three-dimensional porescale model of the cathode electrode in polymer-electrolyte membrane fuel cell by lattice Boltzmann method. Journal of Power Sources 258: 89–97.

Möller, D.P.F. 2016. Digital Manufacturing/Industry 4.0, in Guide to Computing Fundamentals in Cyber-Physical Systems: Concepts, Design Methods and Applications. Springer International Publishing, pp. 307–375.

Montaz Ali, M., Khompatraporn, C. and Zabinsky, Z.B. 2005. A numerical evaluation of several stochastic algorithms on selected continuous global optimization test problems. Journal of Global Optimization 31: 635–672.

Montgomery, D.C. and Runger, G.C. 2003. Applied Statistics and Probability for Engineers 3rd, Edition, John Wiley & Sons.

Mühlhäuser, M. 2008. Smart Products: An Introduction in Constructing Ambient Intelligence: AmI 2007 Workshops Darmstadt, Germany, November 7–10, 2007. Revised Papers, M. Mühlhäuser, A. Ferscha, and E. Aitenbichler (eds.). Springer Berlin Heidelberg: Berlin, Heidelberg, pp. 158–164.

Murata, T. and Ishibuchi, H. 1995. MOGA: multi-objective genetic algorithms. In: Proceedings of the 1995 IEEE international conference on evolutionary computation. Perth, WA, Australia. IEEE.

Murdock, T.M., Schmitendorf, W.E. and Forest, S. 1991. Use of genetic algorithm to analyze robust stability problems. Proc. American Control Conf., Evanston, IL 1: 886–889.

Nahata, A., Weling, A.S. and Heinz, T.F. 1996. A wideband coherent terahertz spectroscopy system using optical rectification and electro-optic sampling. Applied Physics Letters 69(16): 2321–2323.

Nandam, P.K. and Sen, P.C. 1992. Control laws for sliding mode speed control of variable speed drives. International Journal of Control 56(5): 1167–1186.

Narayan, P.K. and Smyth, R. 2005. Electricity consumption, employment and real income in Australia evidence from multivariate Granger causality tests. Energy Policy 33: 1109–1116.

National Bureau of Statistics of China, Beijing Statistical Yearbook. 2009. China Statistics Press.

National Bureau of Statistics of China, China Statistical Yearbook. 2009. China Statistics Press.

National Bureau of Statistics of China, Chinese Energy Statistical Yearbook. 2009. China Statistics Press.

National Bureau of Statistics of China, Chongqing Statistical Yearbook. 2009. China Statistics Press.

National Research Council. 2000. Chinese Academy of Sciences, Chinese Academy of Engineering, National Academy of Sciences, Cooperation in the Energy Futures of China and the United States National Academy Press.

Nauck, D. et al. 2008. Predictive customer analytics and real-time business intelligence. pp. 205–214. In: Voudouris, C., Lesaint, D. and Owusu, G. (eds.). Service Chain Management. Springer Berlin Heidelberg.

Nejadkoorki, F., Nicholson, K., Lake, I. and Davies, T. 2008. An approach for modeling CO_2 emissions from road traffic in urban areas. Science of the Total Environment 406(1-2): 269–278.

Nelder, J.A. and Mead, R. 1965. A simplex method for function minimization. The Computer Journal 7(4): 308–313.

Nenortaite, J. and Simutis, R. 2004. Stocks trading system based on the particle swarm optimization, algorithm, workshop on computational methods in finance and insurance, Springer. Lecture Notes in Computer Science 3039/2004: 843–850.

Ng, K.C. 1992. Mapping of Neural Networks onto Parsytec Transputers. Project Report, University of Glasgow.

Ng, K.C. and Li, Y. 1994. Design of sophisticated fuzzy logic controllers using genetic algorithms. Proc. 3rd IEEE Int. Conf. Fuzzy Syst., IEEE World Cong. Compu. Intell., Orlando, FL 3: 1708–1712.

Ng, K.C. and Li, Y. 1994. Design of sophisticated fuzzy logic controllers using genetic algorithms. Proc. 3rd IEEE Int. Conf. on Fuzzy Systems, Orlando, FL 3: 1708–1712. (Available on Internet with URL: http://www.mech.gla.ac.uk/Control/reports.html).

Ng, K.C. 1995. Switching Control Systems and Their Design Automation via Genetic Algorithms (including fuzzy logic control, sliding-mode control, and fuzzy sliding-mode control). Ph.D. Thesis, Department of Electronics and Electrical Engineering, University of Glasgow, Glasgow G12 8LT, U.K.

Ng, K.C., Li, Y., Murray-Smith, D.J. and Sharman, K.C. 1995. Genetic algorithms applied to fuzzy sliding mode controller design. 1st International Conference on Genetic Algorithms in Engineering Systems: Innovations and Applications (GALESIA), pp. 220–225.

NGC Incentives Under NETA. OFGEM, U.K. http://www.ofgem.gov.uk/docs/ngcfeb.pdf, 4, 2000.

Nielsen, M.A. and Chuang, I.L. 2010. Quantum Computation and Quantum Information. 10th Anniversary Edition. Cambridge University Press. Cambridge, UK.

Nix, E.A. and Vose, M.D. 1992. Modelling genetic algorithms with Markov Chains. Annals of Mathematics and Artificial Intelligence 5: 79–88.

Noorkami, M., Robinson, J.B., Meyer, Q., Obeisun, O.A., Fraga, E.S., Reisch, T., Shearing, P.R. and Brett, D.J.L. 2014. Effect of temperature uncertainty on polymer electrolyte fuel cell performance. International Journal of Hydrogen Energy 39(3): 1439–1448.

Norris, S.R. and Crossley, W.A. 1998. Pareto-optimal controller gains generated by a genetic algorithm. AIAA Paper 98-0010. *In*: AIAA 36th Aerospace Sciences Meeting and Exhibit, Reno, Nevada.

Nurcan, S. and Schmidt, R. 2009. Introduction to the First International Workshop on Business Process Management and Social Software (BPMS2 2008), in Business Process Management Workshops: BPM 2008 International Workshops, Milano, Italy, September 1–4, 2008. Revised Papers, D. Ardagna, M. Mecella, and J. Yang (eds.). Springer Berlin Heidelberg: Berlin, Heidelberg, pp. 647–648.

O'Neil, D. 2009. Mendel's Genetics, http://anthro.palomar.edu/mendel/mendel$\textunderscore$1.htm.

Obstfeld, M. and Rogoff, K. 2000. The six major puzzles in international macroeconomics: is there a common cause? NBER Working Papers 7777.

ODell, B. 1977. Fuzzy Sliding Mode Control: A Critical Review Oklahoma State University. Advanced Control Laboratory, Technical Report ACL-97-001.

Ogata, K. 1996. Modern Control Engineering Prentice Hall.

Oh, W. and Lee, K. 2004. Energy consumption and economic growth in Korea: testing the causality relation. Journal of Policy Modeling 26: 973–981.

Oliveira, F., Barat, R., Schulkin, B., Huang, F., Federici, J., Gary, D. and Zimdars, D. 2004. Analysis of THz spectral images of explosives and bio-agents using trained neural networks. pp. 45–50. *In*: Hwu, R.J. and Woolard, D.L. (eds.). Terahertz for Military and Security Applications II, Proc. SPIE 5411.

Oliver, J. 1999. Computation with DNA: matrix multiplication. pp. 113–122. 2nd DIMACS Workshop on DNA Based Computers, Princeton, 1996. DIMACS Series 44. AMS Press.

Palm, R. 1992. Sliding mode fuzzy control. Proc. IEEE Int. Conf. Fuzzy Sys. 167: 519–526.

Palm, R. 1994. Robust control by fuzzy sliding mode. Automatica 30(9): 1429–1437.

Pareto, V. 1896. Cours D Economie Politique, Rouge and Cic, Vol. I and II, Lausanne, 1986.

Park, C.S., Lee, H., Bang, H.C. and Tahk, M.J. 2001. Modified mendel operation for multimodal function optimization. Proceedings of the 2001 Congress on Evolutionary Computation, Seoul, South Korea 2: 1388–1392.

Park, D., Kandel, A. and Langholz, G. 1994. Genetic-based new fuzzy reasoning models with application to fuzzy control. IEEE Trans. Sys., Man and Cyber. 24(1): 39–47.

Parker, D.B. 1985. Learning-logic. Technical Report TR-47, Center for Comp. Research in Economics and Management Sci., MIT.

Passino, K.M. 1993. Bridging the gap between conventional and intelligent control. IEEE Control System Magazine, pp. 12–18.

Passino, K.M. 2000. Distributed optimization and control using only a germ of intelligence. pp. 5–13. *In*: Proceedings of the 2000 IEEE International Symposium on Intelligent Control. IEEE.

Passino, K.M. 2002. Biomimicry of bacteria foraging for distributed optimization and control. IEEE Control Systems Magazine 22: 52–67.

Pathak, R. and Basu, S. 2013. Mathematical modeling and experimental verification of direct glucose anion exchange membrane fuel cell. Electrochimica Acta 113: 42–53.

Pathapati, P.R., Xue, X. and Tang, J. 2005. A new dynamic model for predicting transient phenomena in a PEM fuel cell system. Renewable Energy 30(1): 1–22.

Paule Stephenson. 2001. Electricity market trading, Power Engineering Journal.

Payne, J.E. 2010. A survey of the electricity consumption-growth literature. Applied Energy 87: 723–731.

Payne, R. 2003. Informed trade in spot foreign exchange markets: an empirical investigation. Journal of International Economics 61: 307–329.

Pelletier, F.J. 2000. Review of metamathematics of fuzzy logics. The Bulletin of Symbolic Logic 6: 342–346.

Perret, C., Boussey, J., Schaeffer, C. and Coyaud, M. 2000. Analytic modeling, optimization, and realization of cooling devices in silicon technology. IEEE Transactions on Components and Packaging Technologies 23(4): 665–672.

Peters, G., Weber, C. and Liu, J.-R. 2006. Construction of Chinese Energy and Emissions Inventory Report No.4/2006, Norwegian University of, Science and Technology (NTNU) Industrial Ecology Programme.

Pham, D.T. and Liu, X. 1993. Identification of linear and nonlinear dynamic systems using recurrent neural networks. Artif. Intell. in Eng. 8: 67–75.

Pharoah, J.G. 2005. On the permeability of gas diffusion media used in PEM fuel cells. Journal of Power Sources 144: 77–82.

Pharoah, J.G. 2006. An efficient method for estimating flow in the serpentine channels and electrodes of PEM fuel cells. pp. 547–54. *In*: Proceedings of ASME Conference.

Pisanti, N. 1997. DNA computing: a survey. Technical Report: TR-97-07. Department of Computer Science. University of Pisa. Corso Italia, 40. 56125 Pisa, Italy.

Poli, R. 2000. Why the schema theorem is correct also in the presence of stochastic effects. pp. 487–492. *In*: Proceedings of the 2000 Congress on Evolutionary Computation.

Poli, R., Kennedy, J. and Blackwell, T. 2007. Particle swarm optimization-an overview. Swarm Intelligence 1: 33–57.

Poli, R., Langdon, W.B., McPhee, N.F. and Koza, J.R. 2008. A Field Guide to Genetic Programming. http://lulu.com and freely available at http://www.gp-fieldguide.org.uk. (With contributions by J. R. Koza).

Poole, D.L. and Mackworth, A.K. 2010. Artificial Intelligence: Foundations of Computational Agents. Cambridge University Press.

Porter, B. and Jones, A.H. 1992. Genetic tuning of digital PID controllers. Electronics Letters 28(9): 843–844.

Potter, M.A. and De Jong, K.A. 1994. A cooperative coevolutionary approach to function optimization. pp. 249–257. *In*: Davidor, Y., Schwefel, H.-P. and Männer, R. (eds.). Parallel Problem Solving from Nature-PPSN III Volume 866 of the series Lecture Notes in Computer Science. Springer Berlin Heidelberg.

Preskill, J. 2001. Fault-tolerant quantum computation. pp. 213–269. *In*: Lo, H.-K., Popescu, S. and Spiller, T. (eds.). Introduction to Quantum Computation and Information. World Scientfic.

Pryor, K. 1998. Dolphin Societies—Discoveries and Puzzles. University of California Press.

Psaltis, D., Sideris, A. and Yamamura, A. 1988. A multilayered neural network controller. IEEE Contr. Syst. Mag. 8: 17–21.

Qi, M. and Wu, Y. 2003. Nonlinear prediction of exchange rates with monetary fundamentals. Journal of Empirical Finance 10: 623–640.

Qu, J.S., Wang, Q., Chen, F.H., Zeng, J.J., Zhang, Z.Q. and Li, Y. 2010. Provincial analysis of carbon dioxide emission in China. Quaternary Sciences 30(3): 466–472.

Rabiner, L. and Juang, B.H. 1986. An introduction to hidden Markov models. ASSP Magazine, IEEE 3(12): 4–16.

Radcliffe, N.J. 1997. Schema processing. pp. B.2.5–1.10. *In*: Bäck, T., Fogel, D.B. and Michalewicz, Z. (eds.). Handbook of Evolutionary Computation. Oxford University Press, New York, and Institute of Physics Publishing, Bristol.

Ramadass, P., Haran, B., White, R. and Popov, B.N. 2003. Mathematical modeling of the capacity fade of Li-ion cells. Journal of Power Sources 123: 230–240.

Ramanathan, R. 2005. An analysis of energy consumption and carbon dioxide emissions in countries of the Middle East and North Africa. Energy 30: 2831–2842.

Ramanathan, R. 2006. A multi-factor efficiency perspective to the relationships among world GDP, energy consumption and carbon dioxide emissions. Technological Forecasting and Social Change 73: 483–494.

Ramji, K., Gupta, A., Saran, V.H., Goel, V.K. and Kumar, V. 2004. Road roughness measurements using PSD approach. Journal of the Institution of Engineers 85: 193–201.

Rao, S.S. 1987. Game theory approach for multiobjective structural optimization. Computers and Structures 25: 119–127.

Raymond, C., Boverie, S. and Titli, A. 1994. First evaluation of fuzzy MIMO control laws. Proc. 3rd IEEE Int. Conf. on Fuzzy Systems 1: 545–548.

Rechenberg, I. 1965. Cybernetic Solution Path of an Experimental Problem. Ministry of Aviation, Royal Aircraft Establishment.

Rechenberg, I. 1973. Evolution strategie: Optimierung technischer Systeme nach Prinzipien der biologischen Evolution. Frommann-Holzboog.

Reitz, S., Schmidt, M.A. and Taylor, M.P. 2007. End-user order flow and exchange rate dynamics. The European Journal of Finance 17: 153–168.

Renfer, A., Tiwari, M.K., Tiwari, R., Alfieri, F., Brunschwiler, T., Michel, B. and Poulikakos, D. 2013. Microvortex-enhanced heat transfer in 3D-integrated liquid cooling of electronic chip stacks. International Journal of Heat and Mass Transfer 65: 33–43.

Research Team of China Climate Change Country Study. 2000. China Climate Change Country Study, Tsinghua University Press.

Richard Perez-pena. 2003. Blackout Report Blames Ohio Utility, The New York Times.

Richard, S. and Sutton, Andrew G. 1998. Barto Reinforcement Learning I: Introduction.

Rifkin, J. 2015. The Third Industrial Revolution: How the Internet, Green Electricity, and 3-D Printing are Ushering in a Sustainable Era of Distributed Capitalism, http://www.worldfinancialreview. com/?p=2271.

Rimcharoen, S., Sutivong, D. and Chongstitvatana, P. 2005. Prediction of the stock exchange of thailand using adaptive evolution strategies, Proceedings of the 17th IEEE International Conference on Tools with Artificial Intelligence, IEEE Computer Society, Washington, DC, USA.

Rime, D., Sarno, L. and Sojli, E. 2008. Exchange rate forecasting, order flow, and macroeconomic information. CEPR Discussion Paper No. DP7225. Available at SSRN:http://ssrn.com/abstract=1372545, 2009.

Rochester, N., Holland, J.H., Habit, L.H. and Duda, W.L. 1956. Tests on a cell assembly theory of the action of the brain, using a large digital computer. IRE Transactions on Information Theory 2: 80–93.

Rogers, A. and Prügel-Bennett, A. 1999. Modelling the dynamics of a steady state genetic algorithm. pp. 57–68. In: Banzhaf, W. and Reeves, C. (eds.). Foundations of Genetic Algorithms 5. Springer.

Rogers, E. and Li, Y. (eds.). 1993. Parallel Processing in a Control Systems Environment. Prentice-Hall International, London. Part II: Architectures for Intelligent Control 107–206.

Rogers, E. and Li, Y. (eds.). 1993. Parallel Processing in a Control Systems Environment, London: Prentice Hall International. Special Issue on Intelligent Control. 1993. IEEE Control Systems 13(3).

Rogers, J.L. 2000. A parallel approach to optimum actuator selection with a genetic algorithm. AIAA Paper No. 2000-4484. In: AIAA Guidance, Navigation, and Control Conference, Denver, CO.

Rogoff, K. 1996. The purchasing power parity puzzle. Journal of Economic Literature 34: 647–668.

Rosenblatt, F. 1958. The perceptron: A probabilistic model for information storage and organization in the brain. Psychological Review 65: 386–408.

Rosenblatt, F. 1962. Principles of Neurodynamics. Washington, DC: Spartan Books.

Rowe, A. and Li, X.G. 2001. Mathematical modeling of proton exchange membrane fuel cells. Journal of Power Sources 102(1): 82–96.

Roy, A.E. and Clarke, D. 2003. Astronomy: Principles and Practice, Fourth Edition. Taylor and Francis.

Rudolph, G. 1996. Convergence of evolutionary algorithms in general search spaces. pp. 50–54. In: Proceedings of IEEE International Conference on Evolutionary Computation.

Rudolph, G. 2001. Self-adative mutations may lead to premature convergence. IEEE Transactions on Evolutionary Computation 5: 410–414.

Rumelhart, D.E. and McClelland, J.L. 1986. Parallel Distributed Processing: Explorations in the Microstructure of Cognition, Vol. 1 Foundations. MIT Press Cambridge, MA, USA.

Rumelhart, D.E., Hinton, G.E. and Williams, R.J. 1986. Learning internal representations by error propagation. pp. 318–362. In: Parallel distributed processing: explorations in the microstructure of cognition, vol. 1, MIT Press Cambridge, MA, USA.

Russell, I. 2012. Neural networks module. The UMAP Journal 14:(1).

Saab, S., Badr, E. and Nasr, G. 2001. Univariate modeling and forecasting of energy consumption: the case of electricity in Lebanon. Energy 26: 1–14.

Saha, B., Goebel, K. and Christophersen, J. 2009. Comparison of prognostic algorithms for estimating remaining useful life of batteries. Transactions of the Institute of Measurement and Control 31: 293–308.

Samuels, D., Gershgoren, E., Scully, M., Murnane, M. and Kapteyn, H. 2004. Ultrafast UV Spectroscopy of Dipicolinic Acid The 35th Meeting of the Division of Atomic, Molecular and Optical Physics, May 25–29, Tuscon, AZ.

Sanker, A. et al. 1993. Growing and pruning neural tree networks. IEEE Trans. Computer 42: 291–299.

Saravanan, N. and Fogel, D.B. 1994. Evolving neurocontrollers using evolutionary programming. Proc. 1st IEEE Conf. Evolutionary Computation, Orlando 1: 217–222.

Sari, R. and Soytas, U. 2007. The growth of income and energy consumption in six developing countries. Energy Policy 35: 889–898.

Saviotti, P.P. and Pyka, A. 2004. Economic development by the creation of new sectors. Journal of Evolutionary Economics 14(1): 1–35.

Sayadas, A. and Krishnakumar, K. 1994. GA optimized fuzzy controller for spacecraft attitude control. Proc. 3rd IEEE Int. Conf. on Fuzzy Systems 3: 1708–1712.

Sayers, M.W. 1996. Interpretation of Road Roughness Profile Data Final Report, UMTRI-96-19.

Sayyaadi, H. and Esmaeilzadeh, H. 2013. Determination of optimal operating conditions for a polymer electrolyte membrane fuel cell stack: optimal operating condition based on multiple criteria. International Journal of Energy Research 37(14): 1872–1888.

Schaffer, J.D. 1984. Multiple Objective Optimization with Vector Evaluated Genetic Algorithms. Ph.D. Thesis. Vanderbilt University, Nashville, TN.

Schaffer, J.D. 1985. Multiple-objective optimization using genetic algorithm. pp. 93–100. In: Proceedings of the First International Conference on Genetic Algorithms.

Schlimmer, J.C. 1985. Automonile Data Set, W.S.A. Yearbook (ed.). UCI Machine Learning Repository: United States of America.

Schmidt, R. 2013. Industrie 4.0—revolution oder evolution. Wirtsch. Ostwürtt, pp. 4–7.

Schmidt, R. et al. 2013. Strategic Alignment of Cloud-Based Architectures for Big Data. In: 2013 17th IEEE International Enterprise Distributed Object Computing Conference Workshops.

Schmidt, R. et al. 2015. Industry 4.0—Potentials for Creating Smart Products: Empirical Research Results, in Business Information Systems, 18th International Conference, BIS 2015, Poznań, Poland, June 24–26, 2015, Proceedings, W. Abramowicz (ed.). Springer International Publishing, pp. 16–27.

Schmuttenmaer, C.A. 2004. Exploring dynamics in the far-infrared with terahertz spectroscopy. Chemical Reviews 104: 1759–1779.

Schneider, A., Stillhart, M. and Gunter, P. 2006. High efficiency generation and detection of terahertz pulses using laser pulses at telecommunication. Wavelengths Optics Express 14(12): 5376–5384.

Schöning, C. 2014. Virtual Prototyping and Optimisation of Microwave Ignition Devices for the Internal Combustion Engine. University of Glasgow, Ph.D. Thesis.

Schott, J.R. 1995. Fault tolerant design using single and multicriteria genetic algorithm optimization. Master's thesis. Department of Aeronautics and Astronautics, Massachusetts Institute of Technology, Cambridge, MA, May, 1995.

Schott, J.R. 1995. Fault tolerant design using single and multicriteria genetic algorithm optimization. Technical report, DTIC Document.

Schusterman, R., Thomas, J. and Wood, F. 1986. Dolphin Cognition and Behavior: A Comparative Approach. Lawrence Erlbaum associates, Publishers, Hillsdale, NJ.

Schwefel, H.-P. 1975. Evolutionsstrategie und numerische Optimierung. Ph.D. Thesis. Technische Universität Berlin.

Schwefel, H.-P. 1977. Numerische Optimierung von Computer? Modellen mittels der Evolutionsstrategie. Basel: Birkhäuser.

Schwefel, H.-P. 1981. Numerical Optimization of Computer Models. Wiley, Chichester.

Schwefel, H.P. 1994. On the evolution of evolutionary computation. Computational Intelligence Imitating Life, (IEEE Press), pp. 116–124.

Schwefel, H.-P. 1995. Evolution and Optimum Seeking. John Wiley & Sons.

Sedgewick, R. and Wayne, K. 2011. Algorithms, Addison-Wesley Professional; 4th edition (March 19, 2011).

Seeley, T.D. 1996. The Wisdom of the Hive. Harward University Press.

Sette, S. and Boullart, L. 2001. Genetic programming: principles and applications. Engineering Applications of Artificial Intelligence 14: 727–736.

Settlement Administration Agent User Requirements Specification. OFGEM, UK., 6, 2000.

Sha, L. and Gopalakrishnan, S. 2009. Cyber-Physical Systems: A New Frontier, Machine Learning in Cyber Trust, Springer, pp. 3–13.

Sharman, K.C. and McClurkin, G.D. 1989. Genetic algorithms for maximum likelihood parameter estimation. Proc. IEEE Conf. Acoustics, Speech and Sig. Proc., Glasgow.

Sharman, K.C. and Esparcia-Alcázar, A.I. 1993. Genetic evolution of symbolic signal models. Proc. IEE/IEEE Workshop on Natural Algorithms in Signal Processing, University of Essex.

Sharman, K.C. and McClurkin, G.D. 1993. Genetic algorithms for maximum likelihood coefficient estimation. Proc. IEEE Conf. Acoustics, Speech and Signal Processing, Glasgow.

Sharman, K.C., Esparcia-Alcazar, A.E. and Li, Y. 1995. Evolving digital signal processing algorithms by genetic programming, Proc. First IEE/IEEE Int. Conf. on GA in Eng. Syst.: Innovations and Appl., Sheffield 473–480.

Shen, W., Guo, X.P., Wu, C. and Wu, D.S. 2011. Forecasting stock indices using radial basis function neural networks optimized by artificial fish swarm algorithm. Knowledge-Based Systems 24: 378–385.

Shewchuk, J. 2014. Enabling manufacturing transformation in a connected world. In: Microsoft Internet of Things. Microsoft Corporation: United States, p. 25.

Shi, Y. and Eberhart, R.C. 1998. A modified particle swarm optimizer. pp. 69–73. In: The Proceedings of IEEE International Conference on Evolutionary Computation, Anchorage, AK, USA. IEEE.

Shih, M.C. and Lu, C.S. 1993. Pneumatic servomotor drive a ball-screw with fuzzy-sliding mode position control. Proc. Int. Conf. on Sys., Man and Cyber.: Syst. Engineering in the Service of Human 3(135): 50–54.

Shiu, A. and Lam, P.L. 2004. Electricity consumption and economic growth in China. Energy Policy 32: 47–54.

Shor, P.W. 1994. Algorithms for quantum computation: Discrete log and factoring. pp. 124–134. In: Goldwasser, S. (ed.). Proceedings of the 35th Annual Symposium on the Foundations of Computer Science (FOCS??94). IEEE Computer Society.

Shor, P.W. 1995. Scheme for reducing decoherence in quantum memory. Physical Review A 52: 2493–2496.

Shor, P.W. 1996. Fault tolerant quantum computation. pp. 56–65. In: Proceedings of the 37th Symposium on the Foundations of Computer Science (FOCS??96), IEEE Computer Society.

Shor, P.W. 1997. Polynomial-time algorithms for prime factorization and discrete logarithms on a quantum computer. SIAM Journal on Computing 26: 1484–1509.

Siemens. 2015. Digital Factory-Manufacturing: Self-Organizing Factories, http://www.siemens.com/innovation/en/home/pictures-of-the-future/industry-and-automation/digtial-factory-trends-industry-4-0.html, accessed April 2015.

Sira-Ramirez, H. 1989. Nonlinear variable structure systems in sliding mode: The general case. IEEE Trans. Automatic Control 34(11).

Slotine, J.J. and Sastry, S.S. 1983. Tracking control of non-linear systems using sliding surfaces, with application to robot manipulators. Int. J. Control 38(2): 465–492.

Slotine, J.J. 1984. Sliding mode controller design for non-linear systems. Int. J. Control 40(2): 421–434.

Slotine, J.J.E. 1982. Tracking Control of Non-Linear Systems using Sliding Surfaces with Application to Robot Manipulations Ph.D. Dissertation, Laboratory for Information and Decision Systems, Massachusetts Institute of Technology.

Slotine, J.-J.E. and Li, W.P. 1991. Applied Nonlinear Control, Prentice-Hall International.

Smith, J. and Fogarty, T.C. 1996. Self adaptation of mutation rates in a steady state genetic algorithm. pp. 318–323. In: Proceedings of the Third IEEE Conference on Evolutionary Computation. IEEE Press, Piscataway, NJ.

Smyth, B. and Keane, M.T. 1995. Experiments on adaptation-guided retrieval in a case-based design system. pp. 313–324. In: Veloso, M. and Aamodt, A. (eds.). Case-Based Reasoning: Research & Development, Springer, Berlin, 1995.

Son, J.H. 2009. Terahertz electromagnetic interactions with biological matter and their applications. Journal of Applied Physics 105(10): 102033-102033-10.

Song, I.S., Woo, H.W. and Tahk, H.W. 1999. A genetic algorithm with a Mendel operator for global minimization. Proceedings of the 1999 Congress on Evolutionary Computation, Washington, DC, USA 2: 1521–1526.

Song, I.S., Wang, Q.P., Liu, Z.S., Navessin, T., Eikerling, M. and Holdcroft, S. 2004. Numerical optimization study of the catalyst layer of PEM fuel cell cathode. Journal of Power Sources 126(1-2): 104–111.

Song, J.M., Cha, S.Y. and Lee, W.M. 2001. Optimal composition of polymer electrolyte fuel cell electrodes determined by the AC impedance method. Journal of Power Sources 94(1): 78–84.

Soule, T. 1998. Code Growth in Genetic Programming. Ph.D. Dissertation. University of Idaho.

Soytas, U. and Sari, R. 2003. Energy consumption and GDP: causality relationship in G-7 countries and emerging markets. Energy Economics 25: 33–37.

Soytas, U. and Sari, R. 2007. Ewing BT. Energy consumption, income, and carbon emissions in the United States. Ecological Economics 62: 482–489.

Soytas, U. and Sari, R. 2009. Energy consumption, economic growth, and carbon emissions: Challenges faced by an EU candidate member. Ecological Economics 68: 1667–1675.

Spears, W.M. 1991. A study of crossover operators in genetic programming. Lecture Notes in Artificial Intelligence 542: 409–418.

Spencer, Jr. B.F., Dyke, S.J., Sain, M.K. and Carlson, J.D. 1997. Phenomenological model of magnetorheological damper. Journal of Engineering Mechanics 123(3): 230–238.

Springer, T.E., Zawodzinski, T.A. and Gottesfeld, S. 1991. Polymer electrolyte fuel cell model. J. Electrochem. Soc. 138(8): 2334–2342.

Sridhar, J. and Rajendran, C. 1996. Scheduling in flowshop and cellular manufacturing systems with multiple objectives—A genetic algorithmic approach. Production Planning and Control 7: 374–382.

Srinivas, M. and Patnaik, L.M. 1994. Adaptive probabilities of crossover and mutation in genetic algorithms. IEEE Transactions on Systems, Man and Cybernetics 24(4): 656–667.

Srinivas, M. and Patnaik, L.M. 1994. Genetic algorithms: A survey. IEEE Computer 27(6): 17–26.

Srinivas, N. and Deb, K. 1994. Multiobjective optimization using nondominated sorting in genetic algorithms. Evolutionary Computation 2: 221–248.

Stadler, W. 1988. Fundamentals of multicriteria optimization. pp. 1–25. In: Stadler, W. (ed.). Multicriteria Optimization in Engineering and the Sciences, Plenum Press, New York.

Stanway, R. 1966. The Development of Force Actuators using ER and MR Fluid Technology Actuator Technology: Current Practice and New Developments, IEE Colloquium on (Digest No: 1996/110), Pages 6/1-6/5.

Steane, A.M. 1996. Error correcting codes in quantum theory. Physical Review Letters 77: 793–797.

Stern, D. 1993. Energy and economic growth in the USA: A multivariate approach. Energy Economics 15: 137–150.

Steyn, W.H. 1994. Fuzzy control for a nonlinear mimo plant subject to control constraints. IEEE Trans. Sys., Man and Cyber. 24(10): 1565–1571.

Su, C.Y. and Stepanenko, Y. 1994. Adaptive control of a class of nonlinear systems with fuzzy logic. IEEE Trans. on Fuzzy Systems 2(4): 285–294.

Sumathi, S. and Paneerselvam, S. 2010. Computational Intelligence Paradigms: Theory & Applications using MATLAB. CRC Press.

Sun, F. 2010. Simulation Based a-posteriori Search for an Ice Microwave Ignition System. University of Glasgow, Ph.D. Thesis.

Swan, L.G. and Ugursal, V.I. 2009. Modeling of end-use energy consumption in the residential sector: A review of modeling techniques. Renewable and Sustainable Energy Reviews 13: 1819–1835.

Swann, P. and Prevezer, M. 1996. A comparison of the dynamics of industrial clustering in computing and biotechnology. Research Policy 25: 1139–57.

Tam, V.K., Chen, Y.g. and Lui, K.S. 2006. Using micro-genetic algorithms to improve localization in wireless sensor networks. Journal of Communications, the Academy Publisher 1(4): 1–10.

Tan, K.C., Li, Y., Murray-Smith, D.J. and Sharman, K.C. 1995. System identification and linearisation using genetic algorithms with simulated annealing. Proc. First IEE/IEEE Int. Conf. on GA in Eng. Syst.: Innovations and Appl., Sheffield 164–169.

Tan, K.C., Li, Y., Murray-Smith, D.J. and Sharman, K.C. 1995. System identification and linearisation using genetic algorithms with simulated annealing. First Int. Conf. on GA in Eng. Sys.: Innovations and Appl., Univ. of Sheffield (accepted).

Tan, K.C., Lee, T.H. and Khor, E.F. 2002. Evolutionary algorithms for multi-objective optimization: Performance assessments and comparisons. Artificial Intelligence Review 17: 253–290.

Tan, Y. and Zhu, Y. 2010. Fireworks algorithm for optimization. pp. 355–364. In: Tan, Y., Shi, Y.H. and Tan, K.C. (eds.). Advances in Swarm Intelligence, Volume 6145 of Lecture Notes in Computer Science. Springer Berlin Heidelberg.

Tanaka, K. and Sugeno, M. 1992. Stability analysis and design of fuzzy control systems. Fuzzy Sets and Systems 45: 135–156.

Tenti, P. 1996. Forecasting foreign exchange rates using recurrent neural networks. Applied Artificial Intelligence 10: 567–581.

Teodorović, D. and Dell'orco, M. 2005. Bee colony optimization—A cooperative learning approach to complex transportation problems. Advanced OR and AI Methods in Transportation, pp. 51–60.

Teodorović, D., Lučić, P., Marković, G. and Dell Orco, M. 2006. Bee colony optimization: Principles and Applications. pp. 151–156. 8th Seminar on Neural Network Applications in Electrical Engineering, Neurel-2006, Belgrade.

Test Functions Index, 2017. http://infinity77.net/global_optimization/test_functions.html#.

The National Academy of Sciences (NAS). 2004. Teaching about Evolution and the Nature of Science, National Academy of Sciences Press.

The New Electricity Trading Arrangements. 1999a. Ofgem DTI Conclusions, document, UK.

The New Electricity Trading Arrangements. 1999b. OFGEM, London, http://www.ofgem.gov.uk/docs/reta3.pdf, July.

The Online Home of Artificial Immune Systems. 2015. www.artificial-immune-systems.org[BDO Industry 4.0 Report - IMechE (2016)] BDO Industry 4.0 Report - IMechE ,https://www.imeche.org/policy-and-press/reports/detail/industry-4.0-report.

The Review of the First Year of NETA. OFGEM, London, UK., July 2002.

Thorndike, E.L. 1898. Animal intelligence: An experimental study of the associative processes in animals. Psychological Monographs: General and Applied 2(4): 109 https://archive.org/details/animalintelligen00thoruoft.

Thorndike, E.L. 1911. Animal Intelligence: Experimental Studies. New York: Macmillan.

Thrun, S.B. 1992. The role of exploration in learning control with neural networks. In Handbook of Intelligent Control: Neural, Fuzzy and Adaptive Approaches. Van Nostrand Reinhold.

Timmis, J., Neal, M. and Hunt, J. 2000. An artificial immune system for data analysis. Biosystems 55: 143–150.

Timmis, J. and Neal, M. 2001. A resource limited artificial immune system for data analysis. Knowledge Based Systems 14: 121–130.

Timmis, J., Knight, T., de Castro, L.N. and Hart, E. 2004. An overview of artificial immune systems. pp. 51–86. In: Paton, R., Bolouri, H., Holcombe, M., Parish, J.H. and Tateson, R. (eds.). Computation in Cells and Tissues: Perspectives and Tools for Thought. Natural Computation Series. Springer.

Timmis, J., Andrews, P., Owens, N. and Clark, E. 2008. An interdisciplinary perspective on artificial immune systems. Evolutionary Intelligence 1: 5–26.

Ting, C.S., Li, T.H.S. and Kung, F.C. 1994. Fuzzy-sliding mode control of nonlinear system. Proc. 3rd IEEE Int. Conf. on Fuzzy Systems 3: 1620–1625.

Tong, R.M. 1977. A control engineer reviews of fuzzy control. Automatica 13: 559–569.

Tso, G.K.F. and Yau, K.K.W. 2007. Predicting electricity energy consumption: A comparison of regression analysis, decision tree and neural networks. Energy 32: 1761–1768.

Tu, X.C. and Tu, C.Y. 1988. A new approach to fuzzy control. Fuzzy Logic in Knowledge-Based Systems, Decisions and Controls, pp. 307–315.

Tuckerman, D.B. and Pease, R.F.W. 1981. High-performance heat sinking for VLSI. IEEE Electron Device Letters 2(5): 126–129.

Turing, A.M. 1952. The chemical basis of morphogenesis. Phil. Trans. R. Soc. London B 237: 37–72.

Turkakar, G. and Okutucu-Ozyurt, T. 2012. Dimensional optimization of micro-channel heat sinks with multiple heat sources. International Journal of Thermal Sciences 62: 85–92.

Turksen, I.B. and Tian, Y. 1993. Constraints on membership functions of rules in fuzzy expert systems. 2nd. IEEE Int. Conf. on Fuzzy Systems 2: 845–850.

Tzafestas, S.G. 1994. Fuzzy systems and fuzzy expert control: An overview. The Knowledge Engineering Review 9(3): 229–268.

Utkin, V.A. 1978. Sliding Modes and Their Application in Variable Structure Systems Moscow: Nauka (in Russian) (also Moscow: Mir, 1978, in English).

Utkin, V.I. 1977. Variable structure systems with sliding modes. IEEE Trans. Automatic Control, Survey Paper 22(2).

Utkin, V.I. and Yang, K.D. 1978. Methods for constructing discontinuity planes in multi-dimensional variable structure systems. Automatic Remote Control, Guest Editor 39(10): 1466–1470.

Utkin, V.L. 1993. Special issue on sliding mode control. International Journal of Control, Guest Editor 53(5): 1003–1259.

Van Veldhuizen, D.A. and Lamont, G.B. 1998. Multiobjective evolutionary algorithm research: A history and analysis. Technical Report TR-98-03. Department of Electrical and Computer Engineering, Graduate School of Engineering, Air Force Institute of Technology, Wright-Patterson AFB, OH.

Van Veldhuizen, D.A. 1999. Multiobjective Evolutionary Algorithms: Classifications, Analyses, and New Innovations. Ph.D. Thesis, Department of Electrical and Computer Engineering. Graduate School of Engineering. Air Force Institute of Technology, Wright-Patterson AFB, OH, May 1999.

Van Veldhuizen, D.A. and Lamont, G.B. 2000. On measuring multiobjective evolutionary algorithm performance. In Proc. IEEE Cong. Evol. Comput., pp. 204–211.

Varšek, A., Urbancic, T. and Filipic, B. 1993. Genetic algorithms in controller design and tuning. IEEE Trans. on Sys., Man and Cyber. 30(5): 1330–1339.

Venugopal, V. and Narendran, T.T. 1992. A genetic algorithm approach to the machine-component grouping problem with multiple objectives. Computers and Industrial Engineering 22: 469–480.

Victoria Furió, Andrés Moya and Rafael Sanjuán. 2005. The cost of replication fidelity in an RNA virus. PNAS 102: 10233–10237.

Viljamaa, P. and Koivo, H. 1994. Tunning of a multivariable PI-like fuzzy logic controller. Proc. 3rd IEEE Int. Conf. on Fuzzy Systems 1: 388–393.

Virtual Library of Simulation Experiments. 2017. Test Functions and Datasets http://www.sfu.ca/~ssurjano/optimization.html.

Vose, M.D. and Liepins, G.E. 1991. Punctuated equilibria in genetic search. Complex Systems 5: 31–44.

Walther, M., Fischer, B., Schall, M., Helm, H. and Jepsen, P.U. 2000. Farinfrared vibrational spectra of all-Trans, 9-cis and 13-cis retinal measured by Thz time-domain spectroscopy. Chemical Physics Letters 332(3-4): 389–395.

Waltz, M.D. and Fu, K.S. 1965. A heuristic approach to reinforcement learning control systems. IEEE Trans. Automatic Control 10(4): 390–398.

Wang, K.S., Hsu, F.S. and Liu, P.P. 2002. Modeling the bathtub shape hazard rate function in terms of reliability. Reliability Engineering and System Safety 75(33): 397–406.

Wang, L., Husar, A., Zhou, T.H. and Liu, H.T. 2003. A parametric study of PEM fuel cell performances. International Journal of Hydrogen Energy 28(11): 1263–1272.

Wang, L.X. and Mendel, J.M. 1992. Generating fuzzy rules by learning from examples. IEEE Trans. Sys., Man and Cyber. 22(6): 1414–1427.

Wang, L.X. 1993. Stable adaptive fuzzy control of nonlinear systems. IEEE Trans. Fuzzy Systems 1(2): 146–155.

Wang, P. and Kwok, D.P. 1992. Optimal fuzzy PID control based on genetic algorithm. Proc. 1992 Int. Conf. on Industrial Electronics, Control, Instrumentation and Automation 286(3): 977–981.

Wang, P. and Kwok, D.P. 1993. Auto-tunning of classical PID controllers using an advanced genetic algorithm. Proc. Int. Conf. on Industrial Electronics, Control, Instrumentation and Automation 286(3): 1224–1229.

Wang, Y. and Wang, C.Y. 2006. Ultra large-scale simulation of polymer electrolyte fuel cells. Journal of Power Sources 153(1): 130–135.

Wang, Y. and Feng, X.H. 2008. Analysis of reaction rates in the cathode electrode of polymer electrolyte fuel cell i. single-layer electrodes. J. Electrochem. Soc. 155(12): A8–A16.

Wang, Y. and Feng, X.H. 2009. Analysis of reaction rates in the cathode electrode of polymer electrolyte fuel cell ii. dual-layer electrodes. J. Electrochem. Soc. 156(3): B403–B409.

Wang, Y., Chen, K.S., Mishler, J., Cho, S.C. and Adroher, X.C. 2011. A review of polymer electrolyte membrane fuel cells: Technology, applications, and needs on fundamental research. Appl. Energy 88(4): 981–1007.

Watanabe, Y., Kawase, K., Ikari, T., Ito, H., Ishikaw, Y. and Minamide, H. 2004. Component analysis of chemical mixtures using terahertz spectroscopic. Imaging Optics Communications 234(1-6): 125–129.

Weber, A.Z. and Newman, J. 2004. Modeling transport in polymer-electrolyte fuel cell. Chemical Reviews 104: 4679–4726.

Webster, C.S. 2012. Evolutionary Intelligence Alan Turing's unorganized machines and artificial neural networks: his remarkable early work and future possibilities 5: 35–43.

Wei, L., Fwa, T.F., ASCE, M. and Zhe, Z. 2005. Wavelet analysis and interpretation of road roughness. Journal of Transportation Engineering 131(2): 120–130.

Weichert, A., Riehn, C., Barth, H.-D., Lembach, G., Zimmermann, M. and Brutschy, B. 2001. Implementation of a high-resolution two-color spectrometer for rotational coherence spectroscopy in the picosecond time domain. Review of Scientific Instruments 72(6): 2697–2708.

Weiss, M., Neelis, M., Blok, K. and Patel, M. 2009. Non-energy use of fossil fuels and resulting carbon dioxide emissions: Bottom-up estimates for the world as a whole and for major developing countries. Climatic Change 95(3-4): 369–394.

Werbos, P.J. 1974. Beyond Regression: New Tools for Prediction and Analysis in the Behavioral Sciences. Ph.D. Thesis, Harvard University.

Wesseh, P.K. and Zoumara, B. 2012. Causal independence between energy consumption and economic growth in Liberia: Evidence from a non-parametric bootstrapped causality test. Energy Policy 50: 518–527.

Whitacre, J.M. 2010. Degeneracy: a link between evolvability, robustness and complexity in biological systems. Theoretical Biology and Medical Modelling 7: 6.

Whitley, D., Rana, S., Dzubera, J. and Mathias, K.E. 1996. Evaluating evolutionary algorithms. Artificial Intelligence 85: 245–276.

Whitley, D. 2001. An overview of evolutionary algorithms: practical issues and common pitfalls. Information and Software Technology 43: 817–831.

Widrow, B. and Hoff, Jr. M.B. 1960. Adaptive Switching Circuits, 1960 IRE WESCON Convention Record, pp. 96–104.

Wiesner, S. 1983. Conjugate coding. SIGACT News 15: 78–88.

Wikipedia. 2011. Computer-automated design.

Wikipedia. 2015a. Digital Prototyping,http://en.wikipedia.org/wiki/Digital_prototyping, accessed April 2015.

Wikipedia. 2015b.Computer-Automated Design, http://en.wikipedia.org/wiki/CAutoD, accessed April 2015.

Wikipedia. 2018. https://en.wikipedia.org/wiki/Industry_4.0.

Wikipedia. 2018. Industrial Revolution https://en.wikipedia.org/wiki/Industrial_Revolution, accessed January 2018.

Wolpert, D.H. and Macready, W.G. 1997. No free lunch theorems for optimization. IEEE Trans. Evol. Comput. 1: 67–82

Wong, J.Y. 2001. Theory of Ground Vehicles 3rd Edition, John Wiley & Sons.

Worasucheep, C. and Chongstitvatana, P. 2009. A multistrategy differential evolution algorithm for financial. Prediction with Single Multiplicative Neuron, ICONIP 2009, Part II, LNCS 5864: 122–130.

Wu, Q. and Zhang, X.-C. 1997. Free-space electro-optics sampling of mid-infrared pulses. Applied Physics Letters 71(10): 1285–1287.

Xie, G.N., Li, S.A., Sunden, B., Zhang, W.H. and Li, H.B. 2014. A numerical study of thermal performance of microchannel heat sinks, with multiple length bifurcation in laminar liquid flow. Numerical Heat Transfer-Part A 65: 107–126.

Xie, G.N., Zhang, F.L., Zhang, W.H. and Sunden, B. 2014. Constructal design and thermal analysis of microchannel heat sinks with multistage bifurcations in single-phase liquid flow. Applied Thermal Engineering 62: 791–802.

Xu, J. and Tan, G.-J. 2007. A review on DNA computing models. Journal of Computational and Theoretical Nanoscience 4: 1219–1230.

Xu, Y.J., Shi, X.F., Yang, X.W., Liu, X.H., Chen, Y. and Qiu, L.X. 2012. Optimisation of extraction technology of schisandrae chinensis fructus by multi-objective genetic algorithm. Chinese Pharmaceutical Journal 47: 669–673.

Yadav, R.N., Kalra, P.K. and John, J. 2007. Time series prediction with single multiplicative neuron model. Applied Soft Computing 7: 1157–1163.

Yang, B. and Burns, N. 2003. Implications of postponement for the supply chain. International Journal of Production Research 41(9): 2075–2090.

Yang, X.S. 2005. Engineering optimizations via nature-inspired virtual bee algorithms. pp. 317–323. In: Mira, J. and Álvarez, J.R. (eds.). Artificial Intelligence and Knowledge Engineering Applications: A Bioinspired Approach, First International Work-Conference on the Interplay Between Natural and Artificial Computation. Springer, Heidelberg.

Yang, X.S. 2008. Firefly algorithm. Nature-Inspired Metaheuristic Algorithms 20: 79–90.

Yang, X.S. 2008. Nature-Inspired Metaheuristic Algorithms. Luniver Press.

Yang, X.S. 2010. A new metaheuristic bat-inspired algorithm. pp. 65–74. In: Cruz, C., Gonzlez, J., Krasnogor, G.T.N. and Pelta, D.A. (eds.). Nature Inspired Cooperative Strategies for Optimization (NISCO 2010), Studies in Computational Intelligence, vol. 284. Springer Verlag, Berlin.

Yang, X.S. 2010. Engineering Optimization: An Introduction with Metaheuristic Applications. Wiley & Sons, New Jersey.

Yen, G.G. and Lu, H. 2003. Dynamic multiobjective evolutionary algorithm: adaptive cell-based rank and density estimation. IEEE Transactions on Evolutionary Computation 7: 253–274.

Yerramalla, S., Davari, A., Feliachi, A. and Biswas, T. 2003. Modeling and simulation of the dynamic behavior of a polymer electrolyte membrane fuel cell. Journal of Power Sources 124(1): 104–113.

Yi Chen, Rui Huang, Xianlin Ren, Liping He and Ye He. 2013. History of the tether concept and tether missions: A review. ISRN Astronomy and Astrophysics, Volume 2013, Article ID 502973, 7 pp.

Yi, K., Wargelin, M. and Hedrick, K. 1993. Dynamic tire force control by semi-active suspensions. Journal of Dynamic Systems, Measurement, and Control 115(3): 465–474.

Ying, M. 2010. Quantum computation, quantum theory and AI. Artificial Intelligence 174: 162–176.

Yohanis, Y.G., Mondol, J.D., Wright, A. and Norton, B. 2008. Real-life energy use in the UK: How occupancy and dwelling characteristics affect domestic electricity use. Energy and Buildings 40: 1053–1059.

Yoon, B., Holmes, D.J., Langholz, G. and Kandel, A. 1994. Efficient genetic algorithms for training layered feed forward neural networks. Info. Sciences 76: 67–85.

Yun, Li., Gregory, K.H.A., Chong, C.Y., Wenyuan Feng, Kay Chen Tan and Hiroshi Kashiwagi. 2004. CAutoCSD-evolutionary search and optimisation enabled computer automated control system design. International Journal of Automation and Computing 1(17): 76–88.

Zadeh, L.A. 1962. From circuit theory to system theory. Proceedings of the IRE 50: 856–865.

Zadeh, L.A. 1965. Fuzzy Sets. Information and Control 8: 338–353.

Zadeh, L.A. 1965a. Fuzzy sets. Information and Control 8: 338–353.

Zadeh, L.A. 1965b. Fuzzy sets and systems. pp. 29–39. In: Fox, J. (ed.). System Theory. Polytechnic Press, Brooklyn, New York, USA.

Zadeh, L.A. 1966. Shadows of fuzzy sets. Problems of Information Transmission 2: 37–44.

Zadeh, L.A. 1968a. Fuzzy algorithm. Information and Control 12: 94–102.

Zadeh, L.A. 1968b. Probability measures of fuzzy events. Journal of Mathematical Analysis and Applications 23: 421–427.

Zadeh, L.A. 1971a. Toward a theory of fuzzy systems. pp. 209–245. In: Kalman, R.E. and Declaris, N. (eds.). Aspects of Network and System Theory. Holt, Rinehart and Winston, New York, USA.

Zadeh, L.A. 1971b. Similarity relations and fuzzy orderings. Information Sciences 3: 177–200.

Zadeh, L.A. 1971c. Quantitative fuzzy semantics. Information Sciences 3: 159–176.

Zadeh, L.A. 1971d. Toward fuzziness in computer systems: Fuzzy algorithms and languages. pp. 9–18. In: Boulaye, G. (ed.). Architecture and Design of Digital Computers. Dunod, Paris.

Zadeh, L.A. 1972a. On Fuzzy Algorithms. ERL Memorandum M-325, University of California, Berkeley.

Zadeh, L.A. 1972b. A rationale for fuzzy control. Journal of Dynamic Systems, Measurement, and Control 94(1): 3–4.

Zadeh, L.A. 1973. Outline of a new approach to the analysis of complex systems and decision processes. IEEE Trans. Sys., Man and Cyber. 3(1): 28–44.

Zadeh, L.A. 1973a. Outline of a new approach to the analysis of complex systems and decision processes. IEEE Transactions on Systems, Man, and Cybernetics 3(1): 28–44.

Zadeh, L.A. 1973b. A system-theoretic view of behavior modification. pp. 185–194. In: Wechsler, H. (ed.). Beyond the Punitive Society. Wildwood-House, London, UK.

Zadeh, L.A. 1973c. Outline of a new approach to the analysis of complex systems and decision processes. IEEE Transactions on Systems, Man and Cybernetics 3: 28–44.

Zadeh, L.A. 1974a. Fuzzy logic and its application to approximate reasoning. Information Processing (Proceedings of International Federation for Information Processing Congress) 74: 591–594.

Zadeh, L.A. 1974b. On the analysis of large scale systems. pp. 23–37. *In*: Gottinger, H. (ed.). Systems Approaches and Environment Problems. Vandenhoeck and Ruprecht, Gottingen, Germany.

Zadeh, L.A. 1975a. The concept of a linguistic variable and its application to approximate reasoning-I. Information Sciences 8(3): 199–249.

Zadeh, L.A. 1975b. Fuzzy logic and approximate reasoning. Synthese 30: 407–428.

Zadeh, L.A. 1975c. The concept of a linguistic variable and its application to approximate reasoning-II. Information Sciences 8(4): 301–357.

Zadeh, L.A. 1975d. The concept of a linguistic variable and its application to approximate reasoning-III. Information Sciences 9(1): 43–80.

Zadeh, L.A. 1975e. Calculus of fuzzy restrictions. pp. 1–39. *In*: Zadeh, L.A., Fu, K.S., Tanaka, K. and Shimura, M. (eds.). Fuzzy Sets and their Applications to Cognitive and Decision Processes. Academic Press, New York, USA.

Zadeh, L.A. 1976a. The linguistic approach and its application to decision analysis. pp. 339–370. *In*: Ho, Y.C. and Mitter, S.K. (eds.). Directions in Large-Scale Systems. Plenum Press, New York, USA.

Zadeh, L.A. 1976b. Fuzzy sets and their application to pattern classification and clustering analysis. pp. 251–299. *In*: Van Ryzin, J. (ed.). Classification and Clustering: Proceedings of an Advanced Seminar Conducted, the Mathematics Research Center, the University of Wisconsin at Madison, USA.

Zadeh, L.A. 1978. Possibility theory and its application to information analysis. pp. 173–182. *In*: Proceeding International Colloquium on Information Theory. CNRS, Paris.

Zadeh, L.A. 1978a. PRUF—A meaning representation language for natural languages. International Journal of Man-Machine Studies 10: 395–460.

Zadeh, L.A. 1978b. Fuzzy sets as a basis for a theory of possibility. Fuzzy Sets and Systems 1: 3–28.

Zadeh, L.A. 1979a. Fuzzy sets and information granularity. pp. 3–18. *In*: Gupta, M., Ragade, R. and Yager, R. (eds.). Advances in Fuzzy Set Theory and Applications. North-Holland, New York, USA.

Zadeh, L.A. 1979b. A theory of approximate reasoning. pp. 149–194. *In*: Hayes, J., Michie, D. and Mikulich, L.I. (eds.). Machine Intelligence, Vol. 9. Halstead Press, New York, USA.

Zadeh, L.A. 1981. Fuzzy probabilities and their role in decision analysis. *In*: Proceedings of 4th MIT/ONR Workshop on Command, Control, and Communications. MIT, Cambridge, MA, USA.

Zadeh, L.A. 1982. Fuzzy systems theory: A framework for the analysis of humanistic systems. pp. 25–41. *In*: Cavallo, R. (ed.). Systems Methodology in Social Science Research. Frontiers in Systems Research. Springer, Netherlands.

Zadeh, L.A. 1982. Test-score semantics for natural languages and meaning representation via PRUF. pp. 281–349. *In*: Edge. R. (ed.). Empirical Semantics. Brockmeyer, Bochum, Germany.

Zadeh, L.A. 1983. Linguistic variables, approximate reasoning and dispositions. Medical Information 8: 172–186.

Zadeh, L.A. 1983. The role of fuzzy logic in the management of uncertainty in expert systems. Fuzzy Sets and Systems 11: 197–198.

Zadeh, L.A. 1984. A theory of commonsense knowledge. pp. 257–295. *In*: Skala, H.J., Termini, S. and Trillas, E. (eds.). Aspects of Vagueness. Springer, Netherlands.

Zadeh, L.A. 1984. Making computer think like people. IEEE Spectrum 8: 26–32.

Zadeh, L.A. 1984a. Precisiation of meaning via translation into PRUF. pp. 373–401. *In*: Vaina, L. and Hintikka, J. (eds.). Cognitive Constraints on Communication-Representations and Processes. Springer, Netherlands.

Zadeh, L.A. 1984b. Fuzzy probabilities. Information Processing and Management 20: 363–372.

Zadeh, L.A. 1985. Syllogistic reasoning in fuzzy logic and its application to usuality and reasoning with dispositions. IEEE Transactions on Systems, Man and Cybernetics 15: 754–763.

Zadeh, L.A. 1986. Outline of a computational approach to meaning and knowledge representation based on the concept of a generalized assignment statement. pp. 198–211. *In*: Winter, H. (ed.). Artificial Intelligence and Man-Machine Systems. Lecture Notes in Control and Information Sciences. Springer, Berlin Heidelberg.

Zadeh, L.A. 1986. Outline of a theory of usuality based on fuzzy logic. pp. 79–97. *In*: Jones, A., Kaufmann, A. and Zimmermann, H. (eds.). Fuzzy Sets Theory and Applications. NATO ASI Series Volume 177. Springer, Netherlands.

Zadeh, L.A. 1987. Dispositional logic and commonsense reasoning. pp. 375–389. *In*: Proceedings of the Second Annual Artificial Intelligence Forum. NASA Ames Research Center. Moutain View, CA, USA.

Zadeh, L.A. 1988. A computational theory of dispositions. pp. 215–241. *In*: Turksen, I.B., Asai, K., Ulusoy, G. (eds.). Computer Integrated Manufacturing. NATO ASI Series. Springer, Berlin Heidelberg.

Zadeh, L.A. 1988. Dispositional logic. Applied Mathematics Letters 1: 95–99.

Zadeh, L.A. 1988. Fuzzy sets, usuality and commonsense reasoning. pp. 289–309. *In*: Vaina, L.M. (ed.). Matters of Intelligence. Synthese Library Volume 188. Springer, Netherlands.

Zadeh, L.A. 1989. Knowledge representation in fuzzy logic. IEEE Transactions on Knowledge and Data Engineering 1: 89–100.

Zadeh, L.A. 1992. The calculus of fuzzy if-then rules. AI Expert 7: 23–27.

Zadeh, L.A. 1993. Fuzzy logic, neural networks and soft computing. pp. 320–321. *In*: Natke, H.G. (ed.). Safety Evaluation Based on Identification Approaches Related to Time-Variant and Nonlinear Structures. Springer, Fachmedien.

Zadeh, L.A. 1994. Soft computing and fuzzy logic. IEEE Software 11: 48–56.

Zadeh, L.A. 1996. Fuzzy languages and their relation to human and machine intelligence. pp. 148–179. *In*: Klir, Y. (ed.). Fuzzy Sets, Fuzzy Logic and Fuzzy Systems. World Scientific, Singapore.

Zadeh, L.A. 1996. Fuzzy logic = computing with words. IEEE Transactions on Fuzzy Systems 4: 103–111.

Zadeh, L.A., Klir, G.J. and Yuan, B. 1996. Fuzzy Sets, Fuzzy Logic, and Fuzzy Systems: Selected Papers. World Scientific. Singapore.

Zadeh, L.A. 1997. Toward a theory of fuzzy information granulation and its centrality in human reasoning and fuzzy logic. Fuzzy Sets Syst. 90: 111–127.

Zadeh, L.A. 2002. From computing with numbers to computing with words: From manipulation of measurements to manipulation of perceptions. pp. 81–117. *In*: MacCrimmon, M. and Tillers, P. (eds.). The Dynamics of Judicial Proof. Studies in Fuzziness and Soft Computing Volume 94. Physica-Verlag HD.

Zadeh, L.A. 2002. Toward a perception-based theory of probabilistic reasoning with imprecise probabilities. Journal of Statistical Planning and Inference 105: 233–264.

Zadeh, L.A. 2005. Toward a generalized theory of uncertainty (GTU)-an outline. Information Science 172: 1–40.

Zadeh, L.A. 2006. Generalized theory of uncertainty (GTU)-principal concepts and ideas. Computational Statistics and Data Analysis 51: 15–46.

Zadeh, L.A. 2008. Is there a need for fuzzy logic? Information Sciences 178: 2751–2779.

Zadeh, L.A. 2012. Computing with Words-Principal Concepts and Ideas. Studies in Fuzziness and Soft Computing Volume 277. Springer, Berlin Heidelberg.

Zadeh, L.A. 2013. Toward a restriction-centered theory of truth and meaning (RCT). Information Science 248: 1–14.

Zadeh, L.A. 2014. A note on similarity-based definitions of possibility and probability. Information Sciences 267: 334–336.

Zadeh, L.A. 2015. Fuzzy logic-a personal perspective. Fuzzy Sets and Systems. doi:10.1016/j. fss.2015.05.009.

Zadeh, L.A. 2015. The information principle. Information Science 294: 540–549.

Zhang, X.P. and Cheng, X.M. 2009. Energy consumption, carbon emissions, and economic growth in China. Ecological Economics 68: 2706–2712.

Zhang, Z.Y., Ji, T., Yu, X.H., Xiao, T.Q. and Xu, H.J. 2006. A method for quantitative analysis of chemical mixtures with THz time domain spectroscopy Chinese. Physics Letters 23(8): 2239–2242.

Zhao, L. and Yang, Y. 2009. Expert systems with applications: pso-based single multiplicative neuron model for time series prediction. Expert Systems with Applications 36: 2805–2812.

Zhong, H., Redo-Sanchez, A. and Zhang, X.C. 2006. Identification and classification of chemicals using terahertz reflective spectroscopic focal plane. Imaging System Optics, Express 14(20): 9130–9141.

Zhou, A., Qu, B.-Y., Li, H., Zhao, S.-Z., Suganthan, P.N. and Zhang, Q. 2011. Multiobjective evolutionary algorithms: A survey of the state of the art. Swarm and Evolutionary Computation 1: 32–49.

Zhou, F.X. and Fisher, D.G. 1992. Continuous sliding mode control. Int. J. Control 55(2): 313–327.

Zhou, Q., Nielsen, S. and Qu, W. 2008. Semi-active control of shallow cables with magnetorheological dampers under harmonic axial support motion. Journal of Sound and Vibration 311(3-5): 683–706.

Ziegler, S.W. and Cartmell, M.P. 2001. Using motorised tethers for payload orbital transfer. Journal of Spacecraft and Rockets 38(6): 904–913.

Ziegler, S.W. 2003. The Rigid-body Dynamics of Tethers in Space Ph.D. Dissertation. Department of Mechanical Engineering, University of Glasgow.

Zimmermann, K.H. 2002. Efficient DNA sticker algorithms for NP—complete graph problems. Computer Physics Communications. 144: 297–309.

Zitzler, E. and Thiele, L. 1998. Multiobjective optimization using evolutionary algorithms—A comparative study. pp. 292–301. In: Eiben, A.E. (ed.). Parallel Problem Solving from Nature V, Amsterdam, The Netherlands, September 1998. Springer-Verlag. Lecture Notes in Computer Science No. 1498.

Zitzler, E. 1999. Evolutionary Algorithms for Multiobjective Optimization: Methods and Applications, Diss. ETH No. 13398, ETH Zurich, Switzerland.

Zitzler, E. and Zitzler, L.1999. Multiobjective evolutionary algorithms: A comparative case study and the strength pareto approach. IEEE Transactionson Evolutionary Computation 3: 257–271.

Zitzler, E., Deb, K. and Thiele, L. 1999. Comparison of Multiobjective Evolutionary Algorithms: Empirical Results. Technical Report 70, Computer Engineering and Networks Laboratory (TIK), Swiss Federal Institute of Technology (ETH) Zurich, Gloriastrasse 35, CH-8092 Zurich, Switzerland.

Zitzler, E., Deb, K. and Thiele, L. 2000. Comparison of multiobjective evolutionary algorithms: Empirical results. Evolutionary Computation 8(2): 173–195.

Zitzler, E., Laumanns, M. and Thiele, L. 2001. SPEA2: Improving the strength Pareto evolutionary algorithm. Technical Report 103, Computer Engineering and Networks Laboratory (TIK), Swiss Federal Institute of Technology (ETH) Zurich, Gloriastrasse 35, CH-8092 Zurich, Switzerland.

Glossary

Use the template *glossary.tex* together with the Springer document class SVMono (monograph-type books) or SVMult (edited books) to style your glossary in the Springer layout.

glossary term Write here the description of the glossary term. Write here the description of the glossary term. Write here the description of the glossary term.

glossary term Write here the description of the glossary term. Write here the description of the glossary term. Write here the description of the glossary term.

glossary term Write here the description of the glossary term. Write here the description of the glossary term. Write here the description of the glossary term.

glossary term Write here the description of the glossary term. Write here the description of the glossary term. Write here the description of the glossary term.

glossary term Write here the description of the glossary term. Write here the description of the glossary term. Write here the description of the glossary term.

Index